INSTRUMENTS
and
MEASUREMENT

INSTRUMENTS
and
MEASUREMENT

M. R. Bottaccini
The University of Arizona

Charles E. Merrill Publishing Company
A Bell & Howell Company
Columbus, Ohio

Richard J. Grosh, Consulting Editor

Published by
Charles E. Merrill Publishing Co.
A Bell & Howell Company
Columbus, Ohio 43216

International Standard Book Number: 0-675-08889-5

Library of Congress Card Catalog Number: 73-85467

1 2 3 4 5 6 7—80 79 78 77 76 75

Printed in the United States of America

To Frank and Lina and Fannie
Who came first

PREFACE

The design and use of accurate instruments has been the major duty of the experimental physicist since the beginning of the science. The engineer, on the other hand, has always belonged to the "approximate" school of measurement. This does not mean that he does not appreciate precision and accuracy; it simply means that he has no choice. His art has to be applied to the circumvention of the need for accurate measurement, indeed, to the circumvention of the need for any precise knowledge. The engineer is grateful for a good measurement, if he can get it at a reasonable expenditure of time and money, but he is so accustomed to building with tools not completely understood by science that he has developed a neurotic disbelief in accuracy, and automatically treats every measurement as a "rough approximation" or even as a "figure of merit."*

**A figure of merit* is a number assigned to a physical configuration to permit rough comparison to another configuration. Figures of merit are usually intended to fool the front office into accepting a particular design in place of another one. Figures of merit have somewhat the same relevance to design as the planks of a political platform have to post-election performances.

Since the end of the Second World War, however, a subtle change has been noticed in the engineering art. It has become evident that "safety-factor" design is inefficient, costly and no longer necessary. Wherever optimum behavior and automatic operation are fundamental to design, measurement has become the foundation of practice.

It is not surprising, therefore, that instrumentation literature has proliferated to the point of absurdity. The periodical literature of all engineering disciplines is full of measurement technique information, and the specialized journals on instrumentation and control are increasing in number in a most alarming fashion. Textbooks, monographs and collections on instrumentation come off the presses almost daily. The result is a bewildering hodge-podge of limited information and of specific answers to restricted problems.

On the other hand, books designed to teach the principles of measurement and the foundations of instrument design are quite rare. Thus it is apparent that there is need for a new textbook on the design of accurate instruments and on the science of measurement. This work is an attempt to fill the gap and to provide both a design manual for the practicing engineer and a textbook for seniors and first-year graduate students in the sciences and engineering. The text grew from a course taught by the author at the University of Arizona for the past ten years to graduate students and advanced undergraduates. The present text represents the culmination of considerable experimentation in subject matter, although the material has been used without major change for the past three years.

Since this is primarily a textbook, the author has added a number of problems to each chapter. Most of the problems are short aids to the understanding of the theory. A number of longer design problems have been included, since it is the author's belief that only through hard labor is engineering practice communicated.

CONTENTS

1

THE TAXONOMY OF MEASUREMENT

1.1 Introduction

Physical configurations, that is, combinations of mass particles and of quantities of energy at a given time, are investigated in terms of certain arbitrarily selected *describing properties*. These properties appear relevant to the investigator and can be so defined that other investigators are able to identify the property from the description. Whenever the technologist speaks of length, velocity, time, force, temperature, current and neutron flux, he expects to be understood by his colleagues.

Yet many describing properties are not definable in words. For example, the "definition" of *force* can only be given in terms of its effects upon another property called *momentum*, which in turn is "definable" in terms of *mass*, *length* and *time*. But mass is best defined in terms of force. The circle is complete and the definition is wanting.

A solution to the definition dilemma is to operate with effects (or measurements). One can define, for example, a specific volume of a specific material as a unit of mass and then define the unit of force as that "thing" which causes the linear position of the mass to change one unit of distance per unit of time for each unit of time the force is applied to the mass. The investigator actually does not know what force is, yet he can produce a force and he can compare the force acting on a body of mass m_1, accelerating with acceleration a_1, to the force acting on another body of mass m_2 and accelerating at a_2.

Comparison of properties in various configurations is the substance of experimental science and the foundation of engineering. Such comparisons are accomplished by the association of a number (hopefully unique) to the property in question. Devices constructed for the purpose of associating numbers to properties are called *instruments*, and the art of designing and building instruments is called *instrument engineering*.

To most users of instruments the devices are secondary to the results of the measurement sequence. Completely immersed in the study of a phenomenon, a scientist or an engineer will often invent a "one-shot" solution of great ingenuity which he will then report casually as a footnote to his report. So much of this casual invention goes in the world that many people deny the existence of such a field as "instrument" engineering and state that measurement is everybody's business. Actually modern scientific and engineering practice has become so complex that few workers design their own instruments. Most of them buy a finished device from one of the more than 5,500 manufacturers[1] of instruments and often hire a consultant to devise a meaningful test sequence.[2] In modern physics the physicist leaves the design of spectrometers, counters and accelerators to specialists. Chemical balances and electron microscopes are beyond the design ability of most chemists, and even the simple micrometer would tax the shop facilities of many engineering firms.

Instrument engineering is indeed a discipline with clearly defined duties, although it is not necessarily practiced by trained engineers. Any person who designs measurement devices is practicing instrument engineering and should be competent in three major subdivisions of his profession:

(a) Theory of Measurement,
(b) Theory of Instrument Design and
(c) Instrumentation.

1.1.1 The Theory of Measurement is a collection of mathematical techniques, partly deterministic and partly statistical, which permit the inference of a "best" measurement from a set of measurements. The theory of measurement is concerned with random error, with error propagation, with smoothing and presentation of data, and with the design of optimum experimental sequences.

The instrument designer should know what can be measured and how well it can be measured as a foundation for rational design and as an aid in designing test sequences.

1.1.2 The Theory of Instrument Design is the study of the components of an instrument and the relationship into which they are placed. The theory is, therefore, the study of the physical restrictions of measurement and of the intrinsic sources of error. The elements of an instrument form a system which must be analyzed to discover the performance limitations; thus the theory of instrument design is a branch of system analysis[3] and consists of techniques for the discovery of the relationships between input and output.

1.1.3 Instrumentation is the *art* of the application to specific problems of the theories of measurement and of instrument design.[4] To practice instrumentation the engineer combines ingenuity and experience with the theories. Instrumentation is largely a matter of knowledge of the state of the art: a good memory, a well-orgainzed collection of handbooks and an efficient filing system are the basic tools of the instrumentation specialist.

In the United States the word "instrumentation" has come to mean the design and the use of instrument systems rather than the design of individual instruments. The emphasis seems to be on the word "use." In this book instrumentation will be discussed only in illustrative examples as a demonstration of the power of the basic theories.

1.2 Detection of Physical Quantities: Terminology

The process of measurement is philosophically complicated and can give rise to profound epistemological difficulties. It is possible to divide and subdivide the measurement procedure into smaller and smaller routines and finally to descend into hairsplitting legalisms and theological arguments. There are as many ways of defining "measurement" and "instrument" as there are practitioners; each author claims philosophical universality and unbounded intellectual precision. Actually any taxonomic scheme is a temporary agreement between author and reader: the author offers the classification and the definitions and promises uniformity of meaning throughout the book; the reader agrees to use the definitions in his reading. If afterward the reader finds the definitions and classification schemes to his liking, he can adopt them for his own use—if he does not like them, he can make his own.

*Measurement** is the process of associating a number to a physical phenomenon. Given a particular combination of masses and energies at a specific instant of time, it is usually possible for a trained person to describe the *physical configuration* in terms of certain *describing properties* which presumably fix the configuration. Another trained person should then be able to reproduce the physical configuration

*Precise definitions of the italicized words will be found in the definition summary at the end of this chapter.

from the description. If to a describing property of a physical configuration it is possible to connect a unique number y_n, then the configuration is said to have a *measurable property*, or *measurand*.[5] The number y_n is called the *expected value* of the property.

"Expected value" is the modern phrase employed by the statistician for the ideal, "true" value of the measurand. It must be understood that our definition is broader since it does not require the expected value to coincide with the mathematical *expectation* (or theoretical mean) as is usually done in statistical theory.[6] In this book the expected value of the measurand is the ideal, theoretically perfect number assignable to a measurable property.

The number y_n (given in an appropriate system of units) is the only quantitative characterizer of the property and, as will be discussed below, cannot be determined exactly by an instrumental system. y_n is inferred from the output of an instrument by mathematical means. It will be shown in chapter 6 that there is always a chance that a finite number of measurements cannot be related to the expected value.

A *measurent* is a number obtainable by an instrument as an estimate of the *measurand* y_n. Since it is possible to obtain repeated measurements of y_n and since each measurement is burdened with error and may differ from any other measurement, the symbol z_{nk} ($k = 1, 2, 3, \ldots$) will serve to indicate the kth measurement obtained by an instrument as an estimate of y_n. y_n is the input and z_{nk} is the output of the instrument.

An instrument is itself a collection of masses and energies and is, therefore, a physical configuration with its appropriate complement of properties. Measurement is achieved by selecting a property of the instrument as an *indication* of the measurand and the association of a number to the indication by means of a *scale*. Thus an instrument connects an indication ξ_{nk} to a measurand y_n. ξ_{nk} is interpreted as a measurement z_{nk} through the association of a scale.

Examples

(a) Electron drift is the input property to a galvanometer, and angular displacement of an indicator needle is the output indication.

(b) Acceleration is the input property to a strain gage accelerometer, and current is the output indication.

(c) The effect of gravitation on mass is the input property to a spring balance, and the linear displacement of a point of the spring is the output indication.

Association of a number to a property requires the creation of a *unit*. This is done by placing the input system in an agreed reproducible *reference configuration*. The output indication $\xi_n^{(0)}$, which corresponds to the reference input $y_n^{(0)}$, is connected to the scale reference marking $z_n^{(0)} = 1$. If the instrument were error free, then $y_n^{(0)} = 1$, and thus the reference property $y_n^{(0)}$ is called the *unit* of the property y_n. The process of selecting the unit and marking the scale of an indicator is called *calibration*.

In practice the reference configuration is not at the disposal of the engineer. Through years of practice and international negotiations a reference configuration has been selected for most measurable properties. Such a configuration is called a *standard* and is reproducible from a legal-technical description. Thus the standard of length, the meter, has been defined in terms of a given multiple of the wavelength of a certain single frequency of light under certain restrictive but reproducible conditions.[7] All other length units refer to the standard meter.

1.3 Precision and Accuracy

Through much practical experience and analysis it has been shown that it is not reasonable to expect absolutely perfect measurements. This is not to deny the possibility of obtaining the exact expected value: rather it is to affirm the impossibility of ever knowing that the expected value has been found. The best that the designer or the user of instruments can say is that the instrument probably determines the expected value within certain approximate limits.

The central problem of instrument design is to estimate the confidence the user can place on measurement, or to estimate the tendency toward error inherent in the instrument. The designer is required to obtain numbers which he can use to compare the effectiveness of differing designs in meeting error performance specifications.

The three primary indicators of performance are the *accuracy*, the *precision* and the *certainty*, each of which can be associated with errors inherent to the instrument.

1.3.1 Every measurement z_{nk} is burdened with error and is merely an approximation to the ideal, theoretically exact expected value y_n. To every measurement there corresponds a set of errors

$$\epsilon_{nk} = z_{nk} - y_n \qquad k = 1, 2, \ldots \tag{1.3.1}$$

or an equivalent set of dimensionless errors

$$e_{nk} = \frac{z_{nk}}{y_n} - 1 \tag{1.3.2}$$

A measurement is said to be *accurate* or error free if $e_{nk} = 0$. The quantity

$$B_{nk} = 1 - |e_{nk}| \tag{1.3.3}$$

is called the *accuracy* of the measurement. B_{nk} is an indicator of what fraction of y_n is measured by the instrument. Often B_{nk} is presented as a percentile rather than a decimal.

The number B_{nk} is usually greater than zero and less than one. For wildly erratic measurements, however, the accuracy can be a negative number. The most accurate mesurements occur whenever $B_{nk} = 1$. Each measurement will have connected with it a specific B_{nk}, in general different from any other B_{nK} $(k \neq K)$ associated with the same measurement in an "identical" configuration. It is obvious that since $-\infty < B_{nk} \leq 1$, any individual B_{nk} is meaningless as an overall measurement of instrument performance. Some function

$$B_n = B(B_{n1}, B_{n2}, \ldots, B_{np})$$

is needed to measure the performance of the instrument in the repeated measurement of y_n.

Instrument manufacturers usually quote an overall accuracy figure in their advertised specifications. How a characteristic instrumental accuracy is found and under what conditions is almost never stated in advertisements and is subject to a certain amount of commercial interpretation. The usual statement is that an instrument has an "accuracy" of (say) 1.00%. Actually this number is the percentile error or *inaccuracy* to be expected, that is, a statement that any given measurement, under certain unspecified conditions, will not differ from the expected value more than $\pm 0.01y_n$. The most common method of determining accuracy is to treat B_{nk} $(k = 1, 2, \ldots)$ as a sequence of randomly distributed quantities and to define some statistic as the accuracy of the instrument.

1.3.2 If to a given input there corresponds a set of measurements z_{nk} and if it is possible to find a number z_n which is the most likely value of z_{nk}, that is, z_n is a number which by some technique we can select as the scale indication which best corresponds (for the given instrument) to the measurand y_n, then z_n is called the expected value of the output, or simply *expected output*. Each z_{nk} differs from z_n by a deviation η_{nk},

$$\eta_{nk} = z_{nk} - z_n \tag{1.3.4}$$

which indicates how close the individual measurement is to the expected output. In a well-designed system $\eta_{nk} \approx 0$ and the z_{nk} will cluster closely to z_n. An instrument which minimizes η_{nk} is said to be *precise* and the quantity

$$C_{nk} = 1 - |\beta_{nk}| \tag{1.3.5}$$

in which $\beta_{nk} = \eta_{nk}/y_n$ is called the *precision* of the measurement.

It is clear from the definitions that although an accurate mesurement is necessarily precise, a precise measurement need not be accurate. If an instrument is used to obtain a sequence of measurements z_{nk} $(k = 1, 2, \ldots, K)$ of some property in a fixed configuration, there will be "scatter" in the z_{nk}. It is usually possible (see chapter 6) to obtain a number $\bar{z}_n$ from the set of measurements in such a fashion that $\bar{z}_n$ is an estimate of z_n, the expected output. If the instrument is properly built and calibrated $\bar{z}_n \approx z_n \approx y_n$. Unfortunately it often happens that we obtain a consistent z_{nk} with negligible deviation (i.e., the instrument is precise) while at the same time $z_n \neq y_n$ (i.e., the instrument is not accurate). For example, in an otherwise properly built instrument the scale can be untrue or displaced. A measurement obtained with a meter stick can be highly precise ($C = 1.00$) and yet always have $B \leq 0.99$ if the scale consists of a 99-cm stick divided into 100 equal divisions.

1.3.3 As discussed in Section 1.3.2, the expected value and the expected output need not be numerically equal. A displaced scale or a change in an instrument component can cause $z_n \neq y_n$. The error is inherent in the system and is called the *systemic error*, α_n:

$$\alpha_n = z_n - y_n \quad \textbf{(1.3.6)}$$

and its dimensionless equivalent

$$\gamma_n = \frac{z_n}{y_n} - 1 \quad \textbf{(1.3.7)}$$

If γ_n is zero, a measurement is said to be *certain*. With γ_n it is now possible to define the *certainty* of a set of measurement:

$$D_n = 1 - |\gamma_n| \quad \textbf{(1.3.8)}$$

Certainty is quite different from accuracy and precision. The former is the difference between two expected values and, if obtainable, is characteristic of the instrument while the latter two are, at this point, connected with individual measurement. An accurate measurement must be both certain and precise; a certain measurement need not be precise and a precise measurement need not be certain. The 99-cm "meter" stick for which $C = 1.00$ and $B \leq 0.99$ is precise and uncertain. In this instance $D = B$.

The equation

$$e = \beta + \gamma \quad \textbf{(1.3.9)}$$

gives an exact relationship between the dimensionless error, deviation and systemic error of the measurement (Fig. 1.1).

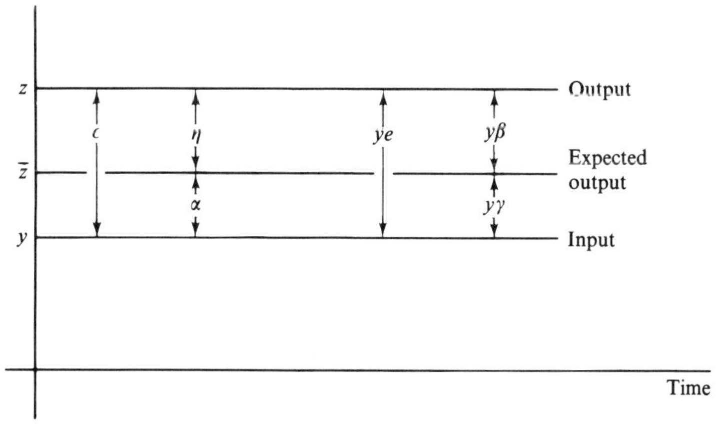

FIGURE 1.1 Error

The quantities B, C, D, however, are not connected to each other uniquely. Their relationship depends on the signs of β and γ. It is easy to show that

$$(1 - B)^2 = (1 - D)^2 + (1 - C)^2 \pm 2(1 - C)(1 - D) \qquad \textbf{(1.3.10)}$$

The positive sign corresponds to $\beta\gamma > 0$ and the negative sign to $\beta\gamma < 0$.

Equation (1.3.9) is actually a relationship between absolute errors, inaccuracy and uncertainty

$$i_B^2 = i_C^2 + i_D^2 \pm 2i_C i_D \qquad \textbf{(1.3.11)}$$

where

$$i_B = |e| = 1 - B \qquad \text{inaccuracy}$$

$$i_C = |\beta| = 1 - C \qquad \text{imprecision}$$

$$i_D = |\gamma| = 1 - D \qquad \text{uncertainty}$$

In the form given, B, C, D, i_B, i_C and i_D are not particularly useful. Thus, in the case of i_B, it is evident that if K measurements were made of i_{Bk} ($k = 1, 2, \ldots, K$) then an individual i_{Bk} would differ from any other inaccuracy in the set of measurements. Any specific i_{Bk} would not measure the overall inaccuracy of an instrument since the error-generating conditions of a particular measurement are not necessarily reproduced in a subsequent measurement. The only certain (though impossible) procedure is to make an infinite set of measurements and then to compute a representative inaccuracy from the aggregate. Almost any given function will do. For example, the average of the infinite set of inaccuracies is a reasonable representative of the whole. For reasons which will be made clear in chapter 6, a more acceptable measure of the inaccuracy is obtained by summing $(i_{Bk})^2$ rather than i_{Bk}. The instrument with the largest value of

$$\lim_{n \to \infty} \frac{1}{n} \sum_{n=1}^{n} (i_{Bk})^2$$

is assumed to be the least accurate. For a definite number of measurements we define the sum

$$\frac{1}{K} \sum_{n=1}^{K} (i_{Bk})^2$$

to be an *estimate* of the infinite sum and is used as an estimate of inaccuracy. If $\bar{p}$ is the expected value of an infinite set of numbers p_k ($k = 1, 2, \ldots$) then the quantity

$$\sigma_p^2 = \lim_{K \to \infty} \frac{1}{K} \sum_{k=1}^{K} (p_k - \bar{p})^2 \qquad \textbf{(1.3.12)}$$

is called the variance of the set p_k. It is possible to define a set of variances

$$\sigma_B^2 = \lim_{m \to \infty} \frac{1}{m} \sum_{k=1}^{m} (i_{Bk})^2$$

$$\sigma_C^2 = \lim_{m \to \infty} \frac{1}{m} \sum_{k=1}^{m} (i_{Ck})^2$$

$$\sigma_D^2 = \lim_{m \to \infty} \frac{1}{m} \sum_{k=1}^{m} (i_{Dk})^2$$

Each variance is an indication of the ability of the instrument to carry out the function of measurement. (See chapter 6 for further discussion.) The accuracy, precision and certainty of the instrument can now be defined.

Definition 1.3.1 THE ACCURACY, PRECISION AND CERTAINTY OF AN INSTRUMENT ARE GIVEN BY

$$B_n = 1 - \sigma_B$$

$$C_n = 1 - \sigma_C$$

$$D_n = 1 - \sigma_D$$

RESPECTIVELY.

A complete definition of B_n is

$$B_n = 1 - \sqrt{\lim_{K \to \infty} \frac{1}{K} \sum_{k=0} \left(\frac{z_{nk}}{y_n} - 1\right)^2}$$

in which the positive value of the square root is to be taken. Similar definitions can be written for C and D. The relationship (1.3.10) can now be written as

$$\frac{1}{m} \sum_{k=1}^{m} i_{Bm}^2 = \frac{1}{m} \sum_{k=1}^{m} i_{Ck}^2 + \frac{1}{m} \sum_{n=1}^{K} i_{Dk}^2 \pm \frac{1}{m} \sum_{k=1}^{K} i_{Ck} k_{Dk}$$

Let the limit sum of the last term be R. Then in the limit

$$\sigma_B^2 = \sigma_C^2 + \sigma_D^2 + R$$

In chapter 6 it will be argued that in some instances $R \longrightarrow 0$ and thus as $m \longrightarrow \infty$, $\sigma_B^2 = \sigma_C^2 + \sigma_D^2$ which implies

$$(1 - B)^2 = (1 - C)^2 + (1 - D)^2 \qquad \textbf{(1.3.13)}$$

1.4 Transduction

An instrument presents a number (or something equivalent to a number) to the user. The output number bears a functional relationship to the expected value of the measurand. It is necessary, therefore, that there be some physical contact between the instrument and the configuration to be measured. If, for example, it were required to measure the dimensions of a machine part by means of a shadowgraph comparator, the configuration would then consist of only the machine part. The instrument would consist of a light source, a set of condensing lenses and a ground glass screen. A fixed relationship between the dimensions of the shadow cast on the screen and the dimensions of the machine part would exist, and it would be possible to estimate the expected values by measurements performed on the screen.

The instrument in this instance is composed of a *source of energy*, the light source, and a *detector*, the glass screen. The contact between the instrument and the configuration is provided by the light beam. At first sight one would think that the instrument does not affect the machine part. Actually the "particles" of light bombard the surface of the solids and cause minute surface changes. Such a set of changes is negligible in this instance, but may not always be so. If the object illuminated (not necessarily by visible light) were of subatomic size, the effects of the illumination would deform the configuration to the point of rendering the measurement meaningless.

Often an instrument will supply no energy to the configuration as in, for example, the detection of fluid velocity by means of a Pitot-static tube.[8] Here a physical object is placed in the configuration and the particles of fluid must move out of their normal trajectory to avoid the obstacle. Such a change of motion will require energy, from which we conclude that the presence of the instrument affects the behavior of the system. Most flow meters withdraw energy from the fluid; in some instances larger pumping systems have to be installed to accommodate the instruments.

Still another example of energy withdrawal is given by the resistance strain gage.[9] The gage, usually a piece of wire, is attached to a structural member such that the gage deforms with the member. A small amount of strain energy must be withdrawn to operate the gage. If mounted on a large beam the gage would have unimportant effects on the property being measured; if mounted on a piece of aluminum foil, however, the gage could affect severely the local elastic behavior of the configuration. From the foregoing arguments it is possible to formulate two important axioms.

Axiom 1.4.1 THE CONTACT BETWEEN AN INSTRUMENT AND THE CONFIGURATION TO BE MEASURED ALWAYS AFFECTS THE CONFIGURATION.[10]

Axiom 1.4.2 DETECTION OF A PROPERTY BY AN INSTRUMENT REQUIRES A SOURCE OF ENERGY.

The energy abstracted from the configuration, or from an external source, is converted by the instrument into an output. The process of abstracting energy from a system and of transforming the energy so as to make it useful to another system is called *transduction*. A *transducer* is a device which abstracts energy from one system and makes it available in usable form to another system. Transducers are the principal parts of all instruments. Quite often the instrument is coterminous with the transducer. Frequently the instrument is formed from an *input transducer*, an *output transducer* and a collection of intermediate elements known as *modifiers* or *conditioners*.

The input transducer, usually called called the *detector*, obtains energy from the measurand and generates a *signal* which is then transmitted through a set of internal modifiers to the output transducer or *display* which makes the signal available to the human operator or a piece of equipment. The operation of transducers and modifiers usually requires external energy sources which are called secondary energy sources. (See Figure 1.2.)

A transducer may actually reduce the total energy available to the configuration or else may obtain energy from an external source which impinges upon the measurand. Basically this implies that either a transducer abstracts *intrinsic* energy from the measurand or else it abstracts energy supplied by a *primary* energy source, the so-called illuminator. *Illumination* in its most general sense is the process of supplying energy to a measurand solely for the purpose of measurement. For example, an X-ray tube provides illumination to a casting, the flaws of which are detected by the fluoroscope, a transducer which transforms energy in the X-ray range into light energy.

The transducer generates a *signal* which bears an "error-burdened functional relationship" to the expected value of the measurand. The signal is transmitted in various forms through the elements of the instrument, or the signal flow path (sometimes called the energy path). From transducer through modifiers to display, the instrument transmits *information* which should always bear a functional relationship to the measurand expected value.

The definitions of transducer, modifier (conditioner), detector and display are quite vague and in modern commercial usage are becoming vaguer. The problem is that it is almost impossible to agree on what is a transducer. As an example, let us examine the pressure-measuring device shown in Figure 1.3. The manufacturer of such a device would probably call it a transducer and state that a given number of pressure units on the diaphram would create a number of current units on the output. The combination of pressure transducer, secondary energy source and oscilloscope would form a complete instrument. On the other hand, the diaphram and post combination could be called a transducer which transforms the intrinsic energy of the configuration (a fluid) into strain energy. The strain gages would then be called a first modifier designed to change strain energy into electrical energy. The amplifier would then be a second modifier designed to increase the amplitude of the signal (change low voltage electrical energy to high voltage electrical energy) and present it to the display. Which portion of an instrument is a detector or a transducer can become a matter of wildly varying opinion.[11]

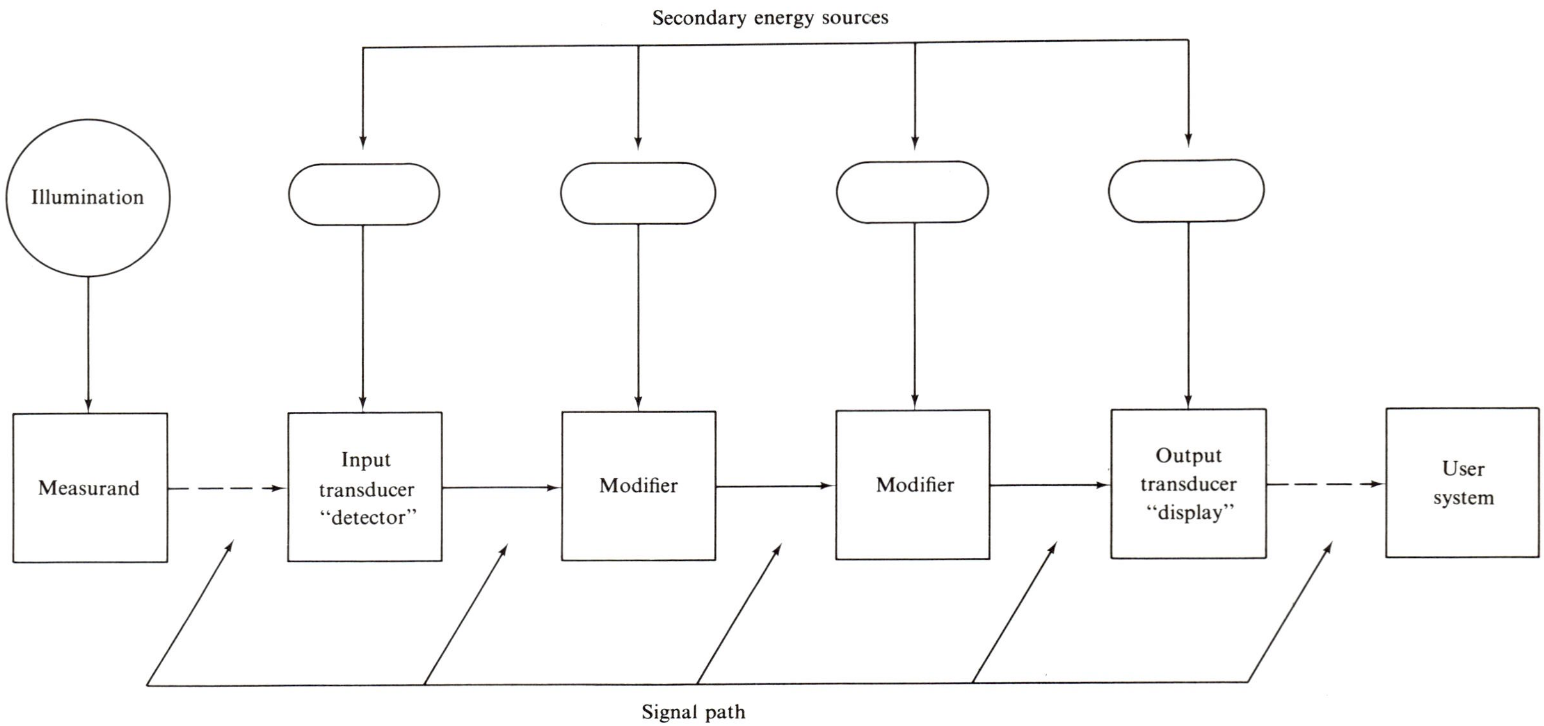

FIGURE 1.2 Energy path or signal flow in a typical open-loop instrument

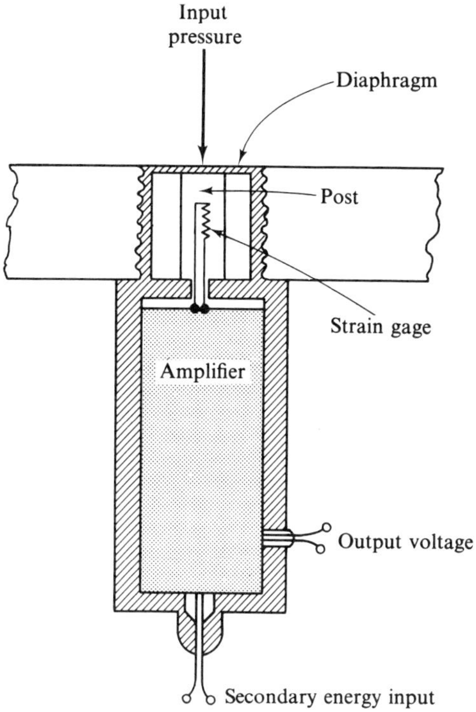

FIGURE 1.3 A self-contained pressure detector

Examples *Detectors*: Bourdon Tubes, Pressure Diaphrams, Galvanometer Coils, Radio Antennas, Pitot Tubes, Thermocouples, Litmus Paper.

Modifiers: Amplifiers, Relays, Hydraulic Valves, Gears and Linkages, Transformers.

Displays: Galvanometers, Cathode Ray Tubes, Counters.

A d'Arsonval movement galvanometer (see chapter 3) is a complete instrument whenever it is used as a current detector in an electrical circuit; it is a modifier whenever it is used as a transducer to change electrical energy into mechanical energy; and it is a display, its most common function, whenever it is used as the output of an instrument. A galvanometer can be detector, modifier or display. It can be treated as a complex instrument or as a portion of an instrument chain; it can be an indicator, and it can become an electro-mechanical amplifier when a position detector is placed on the output needle.

Describing a specific device as belonging intrinsically to a certain functional category can be an exercise in taxonomic futility, and the isolation of a portion of an instrument as a detector, another as a modifier, and still another as a display is often difficult and usually unnecessary.

1.5 Threshold Matching and Modifiers

An instrument must perform the tasks of detection and display; it is not necessary, however, to use separate devices to perform the tasks. A graduated scale placed in contact with a machine part is detector, display and calibrated scale all in one. Detection is accomplished by physically lining up the end of the scale with one edge of the configuration and then by reading a scale graduation lined up with another edge. Such a procedure is burdened with error of observation (see chapter 6): too much reliance is placed on the ability of the human eye to compare geometric figures. An improvement may be accomplished by relieving the eye of the duty of detection through separation of the functions of detection and display. The calipers (Fig. 1.4) provide such a separation. The jaws of the calipers form the detector, and the graduated scale and the hairline form the display.

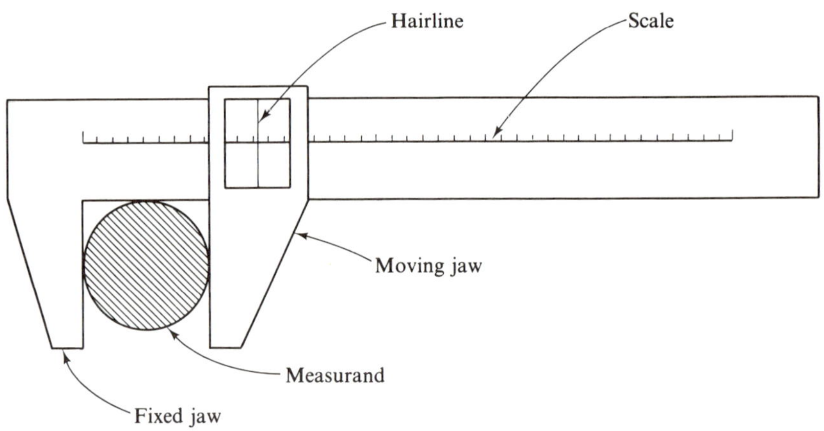

FIGURE 1.4 The simple calipers

Repeated use of the calipers would soon show another limitation: although the eye can detect some small changes in displacement between the smallest scale graduations, the brain cannot connect a unique number to the discernible change. Improvement can be accomplished by the introduction of an intermediate device which helps the eye and the brain connect a number to the displacement threshold. Such a device is the vernier scale attached to the moving jaw. The vernier allows comparison of the smallest discernible movement, the *visual threshold*, to the smallest detectable input, the *input threshold*.

Further improvement can be accomplished by expansion of the scale motion, that is, by generation of an output displacement larger than the input displacement. The designer can *amplify* the motion by gear and linkage mechanisms, or else he could expand the linear motion by first converting it to angular motion as is done in the screw micrometer. Here a small advance of the micrometer spindle is reproduced as a large advance of screw circumference to make a subliminal linear motion discernible to the eye. In this instance two modifiers have been placed between the detector (the jaws) and the display (the output scale), namely, the screw "amplifier" and the vernier.

In the linear caliper, definite but subliminal changes cannot be detected. The screw amplifier then is a device for matching the full capabilities of the detector jaws to the full capabilities of the display scale, that is, matching the resolution of the display to the resolution of the detector.

The words "resolution" and "threshold" are nearly synonymous and are often used interchangeably. A *threshold* is the smallest detectable change of a property starting the test near the zero value of the measurand, while *resolution* is the smallest detectable change of a property starting from a nonzero value of the measurand.[12]

Resolution is not a constant but is dependent on test conditions. A good example is given by the wire resistance potentiometer. This device consists of a wire helix upon which rides a slide contact: the resistance between the sliding contact and one end of the helix is presented to the circuit. As motion is imparted to the slide holder, no electrical change is noticed. The contact is sticking to the potentiometer wire by friction and the motion is taken up by elastic deformation of the slider and the potentiometer wire. Eventually the slide begins to move, but no electrical change is recorded until the contact moves from one turn of the helix to another. Thus the threshold depends on the friction coefficient, the elastic properties of the materials, the rate of force application to the slide and the width and separation of the wire in the helix.

Once motion is established and the friction stabilized only the wire thickness enters into the picture so that the resolution of the system is simply a geometric property. If, on the other hand, the motion is reversed, the slider will come to a halt, the strain on the system will be released and then will be applied in the opposite direction, the static friction will be overcome, and then a change in electrical properties will be noticed as soon as the slide moves from one helix element to another. It is also clear that if the slide is near the end of a helix element the threshold will be lower than if the slide were near the beginning.

Insofar as the designer is concerned, only the worst possible condition is to be considered in instrument design or use. Such a requirement runs counter to the commercial desire to put the best foot forward, so that, in general, there exists considerable diasgreement in the definition of threshold. To bring order into the operation, the Aircraft Industries Association of America published a document which purports to define threshold: "*Threshold is the largest absolute value of the minimum input that produces an output equal to some specified percent (at least 50 percent) of the output expected using the numerical scale factor* " (Ref. 64, note 13). The 50% figure is interpretable in many ways but is usually applied to threshold testing under sinusoidal inputs and represents an attempt to determine the meaning of "minimum detectable output" in a legally repeatable manner. Application of the definition is again open to interpretation and normally leads to unrealistically low (and nearly useless) threshold values.[13] To avoid difficulties we as designers will define the threshold in a less optimistic manner.

Definition 1.5.1 THRESHOLD IS THE LARGEST REPEATABLE ABSOLUTE VALUE OF THE MINIMUM INPUT CHANGE TO WHICH CAN BE ASSIGNED A UNIQUE VALUE OF THE CHANGE OF EXPECTED OUTPUT WHENEVER THE INSTRUMENT IS USED IN A PRESCRIBED, REPEATABLE MEASUREMENT SITUATION.

The threshold is obtained experimentally by starting from a given input and increasing the input until a change is detected which can be used (that is, can be connected uniquely with the known measurand). The test is repeated many times and at different input levels. The largest repeatable absolute value of the input is recorded as the threshold.

Implicit in the definition of threshold is the concept of "discernibility." The user of the instrument must be able to discern, or to assign a unique number to, the input charge. A galvanometer designed to give a 0.1-degree deflection to a one-inch indicator needle as an indication of input threshold may serve as a desk instrument—fifty feet away, the same galvanometer would be completely useless to the observer. If, on the other hand, the observer had an intermediate modifier, say, a telescope, then the visual threshold of the user would be matched to the threshold of the measurement system.

It is for the above reason that the threshold definition contains the phrase "whenever the instrument is used in a prescribed, repeatable measurement situation." Whenever the instrument is used outside the prescribed situation, it may well happen that the threshold of that instrument and the threshold of the user are no longer the same. If the threshold of the instrument is z_T and the threshold of the observer is z_{T0}, then in a properly designed and used instrument it is expected that the *threshold ratio* $T_0 = z_T/z_{T0}$ should be equal to one. If T_0 were greater than one, a threshold input would induce an instrument movement greater than needed. In such a case the instrument is still reasonably well designed even though the user would be inclined to read the output with more significant figures than is warranted by the capabilities of the transducer. If, on the other hand, $T_0 < 1$ the instrument would be either improperly used or improperly designed because the user would not be making use of the full potentialities of the detector.

Proper design requires $T_0 \approx 1$ for economic as well as aesthetic reasons. For $T_0 < 1$, more money has been spent on the transducer than is needed, and for $T_0 > 1$, the display is better than required.

We have demonstrated one of the major uses of intermediate modifiers. If the intrinsic ability of a detector to measure an input change is not matched by the ability of the display to detect the output of the detector, a modifier may be used to make $T_0 \approx 1$.

1.6 Other Reasons for Using Modifiers

Modifiers are also introduced in a system to *isolate* components from each other, to carry out a *computation* on the signal and to *filter* noise.

1.6.1 Devices used in instruments are rarely used at peak efficiency, that is, at the maximum ratio of output power to input power. Usually the output power is very small, and the functional relationship between input signal and output signal is required to be essentially undisturbed by output energy withdrawal. If some device is connected to the output of a transducer in such a fashion that a substantial amount of energy is withdrawn, then the loading device affects the output of the transducer and the signal is *disturbed.*

There are two mechanisms which can cause signal disturbance:

(a) If the functional relationship between input and output is valid (within specified limits) for a given energy withdrawal range and if the output device requires energy outside the range, then the element is said to be disturbed through *impedance mismatch.* If an output device does not withdraw energy improperly from the input device (within acceptable limits), then the output device is said to be impedance matched to the input system.

(b) Sometimes a device gives a proper signal, that is, it maintains a functional relationship between input and output only if the input signal amplitude lies between two specified limits. If the input signal amplitude lies outside the limits, then the output is not in the correct functional relationship to the input and is said to be disturbed through input *saturation.*

Electric amplifiers develop errors through input saturation and through output impedance mismatch. The output signal is proportional to the input signal only as long as the input signal amplitude lies within the linear response region of the vacuum tube curve. The amplifier is usually designed to operate into a specified impedance: if the output load impedance is improperly selected the signal will be disturbed.

If two devices α_1 and α_2 are mismatched, it becomes necessary to introduce a modifier between α_1 and α_2 in such a fashion that α_1 is optimally loaded and α_2 is driven by a signal of the proper form and amplitude. In electrical parlance one says that an *isolator* is used to match impedances. Thus a low impedance load can be matched to an electric circuit by an electronic amplifier. A good example of this last procedure is the electronic voltmeter. The usual d'Arsonval movement voltmeter can affect the behavior of the measured circuit drastically. Introduction of an electronic amplifier, say, a cathode follower with an amplification factor near one, between the measurand and the galvanometer, will accomplish isolation. The amplifier will present a high input impedance (minimum power drain) to the measurand while generating sufficient power to drive the galvanometer.

It is clear that the isolator can perform the functions of threshold matcher (signal amplifier) as well as impedance matcher. In the electronic voltmeter, for example, an amplifier with a gain of one serves simply as an impedance matcher; a similar amplifier with a gain greater than one may also match the thresholds.

Example 1.6.1 In the circuit shown in Figure 1.5, $E = 6$ volts, $R_1 = 120$ ohms and $R_2 = 500$ ohms. To measure the voltage drop across R_1, a voltmeter of internal resistance R_M is connected across R_1. It is clear that the voltmeter draws current from the circuit and hence will give an erroneous value for the voltage drop V. Assuming the voltmeter to be perfect, that is, to have an accuracy of 1, determine the two values of R_M which will give measurement accuracies of 1 and 0.99 respectively.

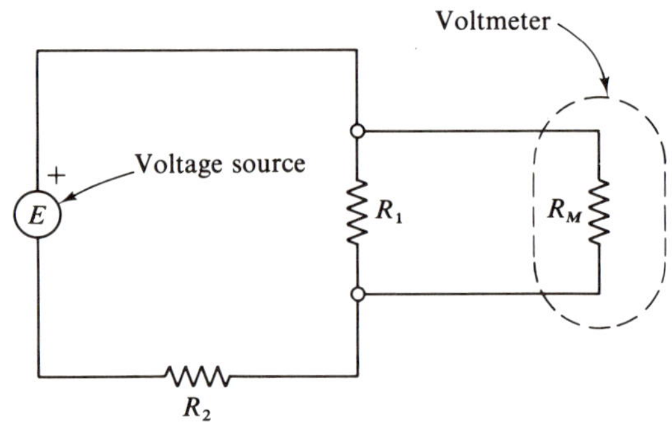

FIGURE 1.5 Circuit for Example 1.6.1

Solution This is a problem of impedance matching. The voltage drop across R_1 without a meter is given by

$$V_0 = \frac{ER_1}{R_1 + R_2}$$

while the meter loads the circuit and gives the voltage drop

$$V = \frac{ER_1}{R_1 + R_2 + (R_1 R_L/R_M)}$$

The accuracy of the instrument is [Equation (1.3.3)]

$$B = 1 - \left|\frac{V}{V_0} - 1\right|$$

which for $1 \geq V/V_0 \geq 0$ becomes $B = V/V_0$.

B can be one only if $R_M \longrightarrow \infty$, that is, the voltmeter presents infinite impedance to the circuit. If, on the other hand, $B \geq 0.99$ were acceptable, $R_M \geq 9{,}600$ ohms would be sufficient.

1.6.2 All sorts of computations can be introduced in the signal chain by intermediate modifiers. Thus an intermediate circuit in an ac voltmeter will compute the rms voltage; a modifier subtracts the unwanted lateral acceleration component from forward acceleration in an accelerometer, and a multiplier is used to compute the cross correlation of two signals in a radar ranger. The most common computation carried out in modern instruments is the linearization of signals.

The mathematical description of a physical system may be linear or nonlinear, but the behavior of the system is always nonlinear. Translated into practical terms, the previous sentence means that analysis and design of instrument systems through the use of ordinary linear differential equations is always an approximation. It does not mean, however, that instruments which exhibit strong nonlinearities are bad, or even that they should be avoided. A nonlinear scale with uneven divisions is not particularly harmful as long as the measurements in the neighborhood of the smallest scale subdivisions meet the specifications for minimum local precision. One must admit, however, that there is a certain aesthetic appeal to even graduations and that there may be an increase of observational accuracy if the operator were not required to think in terms of nonlinear interpolation.

Linearization is often fairly expensive; it is usually better to leave the scale uneven and to design the instrument for $T_0 \approx 1$ in the neighborhood of the smallest subdivisions. Such a procedure is not always possible.

1.6.3 A problem which plagues all measurement is "noise" or unwanted random signals superimposed on the basic measurement (Fig. 1.6). In laboratory practice it is common to record the output and then to

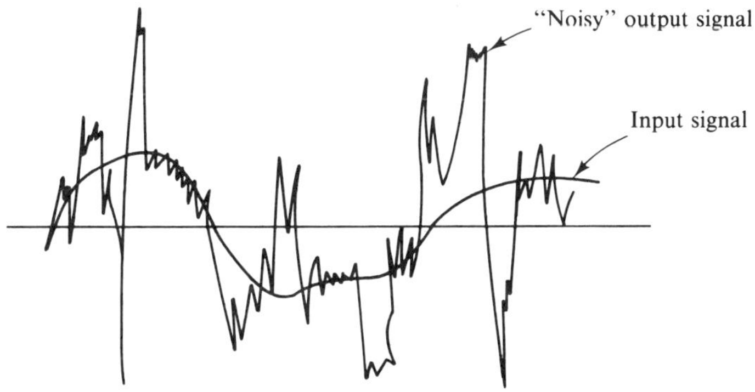

FIGURE 1.6 Noise

recover the signal by statistical analysis. Often statistical analysis on each test is too expensive and too time-consuming. Under such circumstances the signal is recovered by filtering. A filter is a device which reduces the amplitude of signals of a specified frequency band while leaving other signals approximately unchanged. If the frequency spectra of the noise and the signal are quite different (see chapters 5, 6 and 9), it is often possible to filter out the noise by means of an intermediate filter and to record the "clean" input signal.

1.7 Element and Error

Although some inaccuracy is inevitable in measurement, it is possible to avoid gross error by proper design and careful use of the instrument. The word "gross" here implies "normally larger than specified," rather than "occasionally too large." All measurement is occasionally burdened with large error introduced as a random external variation superimposed on the signal (noise and error of observation), and it is always burdened with dynamic error (inability of the instrument to follow a time-varying measurand). Noise can be filtered, usually at the expense of instrument response time and sensitivity, or else it can be computed out of the signal by stochastic (statistical) methods. Dynamic errors are inherent to the instrument and are the result of such useful natural phenomena as inertia, friction, resistance and capacitance. The designer can usually minimize dynamic errors by deterministic (nonstochastic) analysis of various solutions to the measurement problem and by selection of the system which can be shown by analysis to have the fastest response time.

Deterministic analysis, which includes both mathematical modeling and physical experimentation, also points the way to the avoidance of systemic errors, that is, of errors connected with the inability of the instrument to maintain the required, unique, functional relationship between input and output.

An instrument is usually analyzed in terms of subassemblies which may be sources of systemic error if improperly designed. Such error-causing subassemblies shall be called *elements.*[14]

Definition 1.7.1 ANY SUBASSEMBLY OF AN INSTRUMENT WHICH CAN BE ISOLATED AS A SOURCE OF ERROR SHALL BE CALLED AN ELEMENT OF THE INSTRUMENT.

The element is completely arbitrary. A particular collection of components may form an error-generating assembly: the same components may be combined with adjoining components to form another element with respect to another form of error. Thus the question of discovering elements resolves itself into the discovery of simple analyzable parts of the instrument. There may be other more reasonable combinations, but for the designer the key word is "analyzable."

The designer isolates an element intuitively and then follows the error generated by the element along the information chain, under the assumption that all other elements are operating ideally. The system is adjusted to minimize the error from each element in its turn. Most instruments do not have as many independent degrees of freedom as there are elements, and hence it is often impossible to adjust the design for an absolute optimum. Eventually the system interactions must be considered and a design compromise made. Since no designer can consider all the possible interactions and couplings, the final design is not perfect. To remedy the situation the "system" approach has been suggested, that is, the complete system equations are to be written and then the element parameters and error inputs are to be varied in combinations. Such analysis can be carried out with a computer.[15] The complete system technique is only a minor extension of the element by element procedure; regardless of the method the elements have to be identified.

The input and the output of an element are related to each other by a physical relationship which is approximately representable by a mathematical performance relationship. Error is introduced into the system if the physical relationship is wrong (misdesigned) or if the properties of the elements change in such a manner that the performance relationship is not unique.

Definition 1.7.2 THE COMPONENTS[16] OF AN ELEMENT ARE CONNECTED IN PATTERNS IN SUCH A FASHION THAT A PERFORMANCE RELATIONSHIP IS ESTABLISHED BETWEEN THE INPUT TO THE ELEMENT AND THE OUTPUT.

The mathematical representation of the performance relationship is called a *model.*

Definition 1.7.3 COMPONENTS AND PATTERNS SPECIFICALLY DESIGNED TO KEEP A PERFORMANCE RELATIONSHIP INTACT ARE CALLED MAINTAINERS.

Definition 1.7.4 COMPONENTS AND PATTERNS DESIGNED TO RETURN A DISTURBED PERFORMANCE RELATIONSHIP TO ITS ORIGINAL FORM ARE CALLED RESTORERS.

The word "pattern" in Definitions 1.7.2, 1.7.3 and 1.7.4 is quite relevant. In many instruments it is not possible to discover any component or subassembly specifically designed as a maintaining or restoring unit. Every component may be placed in the system only to generate the required performance relationship. The final pattern, however, may be intrinsically self-restoring or self-maintaining with respect to certain disturbances.

Definition 1.7.5 ANY DEVICE WHICH TENDS TO RETARD OR SLOW DOWN A CHANGE IN THE PERFORMANCE RELATIONSHOP IS CALLED A DAMPER.

Dampers, restorers and maintainers are the guarantors of accuracy. Instruments are designed to have a performance relationship. *Accurate* instruments must also have maintainers, restorers and dampers since most errors are the results of improperly maintained performance relationships.

Definition 1.7.6 IF THE PERFORMANCE RELATIONSHIP OF AN ELEMENT IS MOMENTARILY DISTURBED BY AN UNWANTED EXTERNAL INPUT, THE ELEMENT IS SAID TO BE DISTURBED AND IS SAID TO GENERATE A RANDOM ERROR.

Definition 1.7.7 IF AN ELEMENT RESPONDS ACCURATELY TO A CONSTANT INPUT, BUT INTRODUCES ERROR INTO THE INFORMATION CHAIN IN RESPONSE TO A TIME VARIABLE INPUT, THEN THE ELEMENT IS SAID TO BE DELAYED AND TO GENERATE A DYNAMIC ERROR.

Definition 1.7.8 IF THE PERFORMANCE RELATIONSHIP OF AN ELEMENT IS NOT MAINTAINED BECAUSE THE PROPERTIES OF ITS COMPONENTS TEND TO CHANGE CONTINUOUSLY WITH TIME (I.E., THE CHANGE IS A CONTINUOUS FUNCTION OF TIME ALMOST EVERYWHERE), THEN THE ELEMENT IS SAID TO BE INDETERMINATE AND IS SAID TO GENERATE A DRIFT ERROR.

Definition 1.7.9 IF THE PERFORMANCE RELATIONSHIP OF AN ELEMENT IS NOT UNIQUELY DEFINED BUT IS DRAWN RANDOMLY FROM A SPECIFIED SET OF RELATIONSHIPS, THEN THE ELEMENT IS SAID TO BE INCOMPLETE AND IS SAID TO GENERATE AN AMBIGUITY ERROR.

Definition 1.7.10 IF AN ELEMENT IS DESIGNED TO MAINTAIN THE RELATIONSHIP $z = f(y)$ BUT INSTEAD MAINTAINS $z = g(y) \neq f(y)$ THEN THE ELEMENT IS SAID TO BE FALSE AND TO GENERATE A SYSTEMIC ERROR.

Examples

(a) Mismatched impedances cause systemic errors and the mismatched components are connected into a false element. Introducing a coupling modifier without changing the original components will result in a true element.

(b) A wire resistor in an electric circuit will change resistance with temperature, and thus continuously with time. The circuit performance relationship will be affected by drift error.

(c) The friction force between two parts is a function of velocity since the friction coefficient can vary between the static and dynamic values. Consequently, the final position of two sliding parts is somewhat indeterminate and the element is incomplete.

(d) Gears are subjected to backlash and thus to ambiguity.

1.8 Feedback

Most instruments are of the open-loop (or open-chain) type, that is, there exists an energy chain from detector to display, but no deliberate energy return from the display to the transducer or from any element to a preceding element in the path. Ideally a change in the measurand would be seconded by a change in the output signal, while a change of the display configuration would not affect the detector. In a properly designed open-loop system the error induced by the disturbance of an element should be transmitted forward but not backward along the energy path. It is occasionally advantageous for an energy path to be deliberately established which transmits a signal from the output of an element back to the input of a previous element. Such an operation is said to establish *feedback* (Figs. 1.7a and 1.7b). Feedback can be used to correct error [for example, drift (chap. 8)], to minimize nonlinearities (chap. 3) or to stabilize the motion (chap. 4).

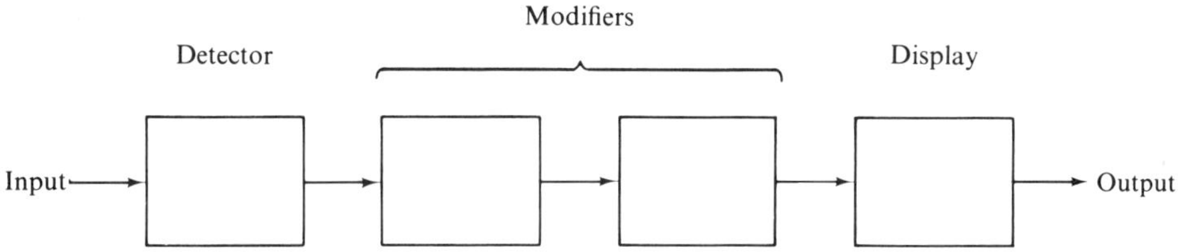

FIGURE 1.7a Open-loop signal chain or path

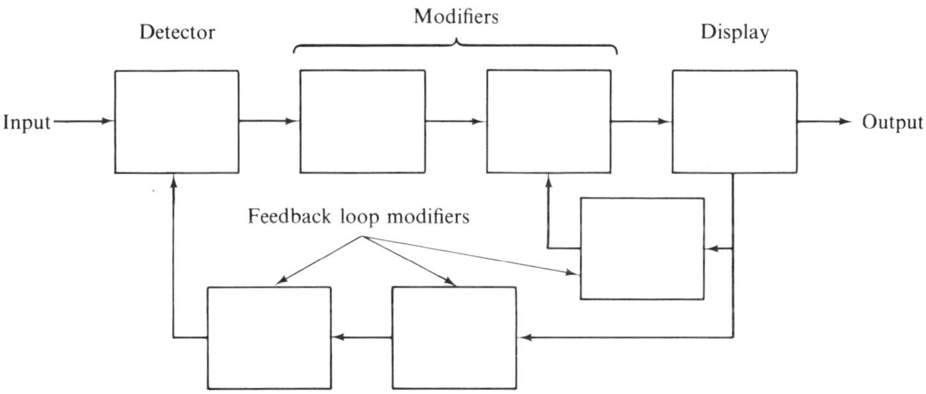

FIGURE 1.7b Closed-loop signal path

The theory and practice of the synthesis and analysis of feedback systems are too extensive to be discussed thoroughly in this book. For the details the reader will usually be referred to the appropriate literature.[17]

1.9 Planning and Designing the Instrument

The usual specifications which face the instrument designer are quite vague. The sponsor is interested in measuring a property under certain circumstances and at a cost not to exceed a certain number of dollars.

The engineer is quite free to design as he pleases as long as the instrument measures the given property in the specified environment. The selection of detector, display and modifiers is left to the judgment of the designer. Roughly speaking, the design procedure consists of selecting a detector which responds to a measurand and selecting a display which meets the conditions of use. Intermediate devices are then introduced to match the input device to the output device. An error analysis is then carried out and the design "optimized." A prototype is built and tested in the laboratory and then an environmental package is designed.

Instrument designers do not usually leave the production design to production engineers. The case, the mounting, the materials to be used and the exterior shape are often as important as the energy chain elements. For example, an improperly suspended gyroscope is nearly useless, and a poorly ventilated vacuum tube voltmeter eventually develops serious systemic errors.

1.9.1 The selection of a detector can make or break an instrument. Improvement of the performance of a commercial instrument is tied to the improvement of detection methods. The beliefs that display methods are much more advanced than detection methods and that innovations in instrument design are fundamentally innovations in transducer design are instrumentation clichés. Quite often the invention of a new form of detector generates a whole family of new instruments, as happened with the invention of the thermistor. Most measurement problems are solved with "state of the art" detectors. It is not necessary, therefore, to teach how one may invent new detectors: serendipity[18] is outside our jurisdiction, inventiveness is God given, and experience helps a little, but genius is better.

Faced with a novel measurement problem, the designer may eventually come up with a completely new idea. Usually he will find a new mixture of tried components. Initially he will search through handbooks, manuals and advertisements for a detector principle, transducer or complete instrument to meet the design specifications.

"The day has gone," said T. N. Whitehead,[19] "when the maker of 'philosophical' instruments, as they were called, was an artist mechanic who achieved, by incredible labor and skill, a unique instrument with its individual merits and defects; not only do we require an accuracy often far beyond his powers, but this must be achieved in the face of a drop in manual skill and at a smaller cost of human effort."

Thirty years later we can say "amen" to Whitehead and can add that unless a new measurand is invented or found, completely unique instruments are no longer possible nor necessary. Most *known* detection principles have been exploited.[20] With the exception of a few special problems the selection of a detector becomes a matter of literature search. This has been recognized for quite some time, and the enormity of the task has been recognized almost as long. All sorts of specialized literature, transducer handbooks and their analogues, have come into existence, as well as some literature on the methodology of transducer literature search.[21] For the ordinary designer the use of handbooks and descriptive handbooks is usually sufficient (References 2, 3, 4, 5, 6, 7, 8, 9, 11 in Appendix A); indeed the author has managed to get along quite nicely only with the aid of references 3, 5, 6 and 8. An invaluable aid is the *Review of Scientific Instruments* and its indices.

For obscure measurands and peculiar environmental conditions the search becomes difficult. A specially set up library and a specialized filing scheme have to be used. Such a process was proposed by Pearlstein[21] and modified by Stein.[22] Stein attached a unique number to certain energy categories and then assigned a unique number to each type of transducer according to input energy, output energy and secondary energy sources. Such a scheme can serve as a method of general library classification to aid in literature search.

Although the energy classification method is very flexible as a general system, a classification according to measurand is easier to use for more modest applications. The fact that a device converts mechanical to electrical energy is not as relevant to the engineer as the fact that the device is a pressure detector. As an aid to design the author of this book has compiled a "dictionary" of measurands (Appendix A). To each measurand is appended a list of references.

To the dictionary there should be added a "state of the art" listing of commercially available transducers. The task was accomplished by the Instrument Society of America (ISA) with the publication of the *ISA Transducer Compendium* which lists 37,700 individual transducers from 556 manufacturers. The *Compendium* is a companion to this book. The *Compendium* is probably the most extensive single source of transducer information currently available to the engineer. The *Compendium* is classified according to measurand in approximately the same manner as the dictionary.

1.9.2 The selection of the display or output transducer is usually, though not always, a simple affair. If the signal is to be used by a human, the output is almost always visual and occasionally acoustic. If the signal is to be used by a nonbiologic system, the output is dictated by the system and is usually electrical, occasionally solid-mechanical or fluid-mechanical. The selection of a display thus reduces to the selection of a transducer which matches the energy output of the signal chain to the energy input of the user.

Four types of users are found at the end of a signal chain:

(a) Human Observer,
(b) Control Mechanism,
(c) Computer and
(d) Recording Device.

The human observer usually wishes to *see* the output display. Occasionally, as in alarm systems and beat frequency oscillators, the signal can be aural. Conceivably the signal could be olfactory or tactile, although no instrument known to the author uses smell or touch in the display.

Control systems can be operated by almost any energy input: visual (by means of photoelectric cells); aural (through the use of microphones); mechanical; electrical hydraulic and pneumatic. The designer is required to match the energy available in the signal chain to the energy required by the control mechanism. This is a job for a transducer and hence solvable by use of the *Compendium*.

Computers see instrument information in the same fashion as control mechanisms. If the measurement is continuous and the computer uses discrete digital information, the linking element will be a digitizer, that is, a device for the transformation of continuous information into digital information.

Recording devices have become the most common scientific tools in the world. In most modern laboratories one usually finds commercial strip chart recorders of some type and often a number of specialized recorders of local manufacture. The input to recording devices is usually electrical and occasionally mechanical.

For recording frequencies less than 100 cycles per second (100 Hz), the trace-on-paper recording galvanometer is the universally used

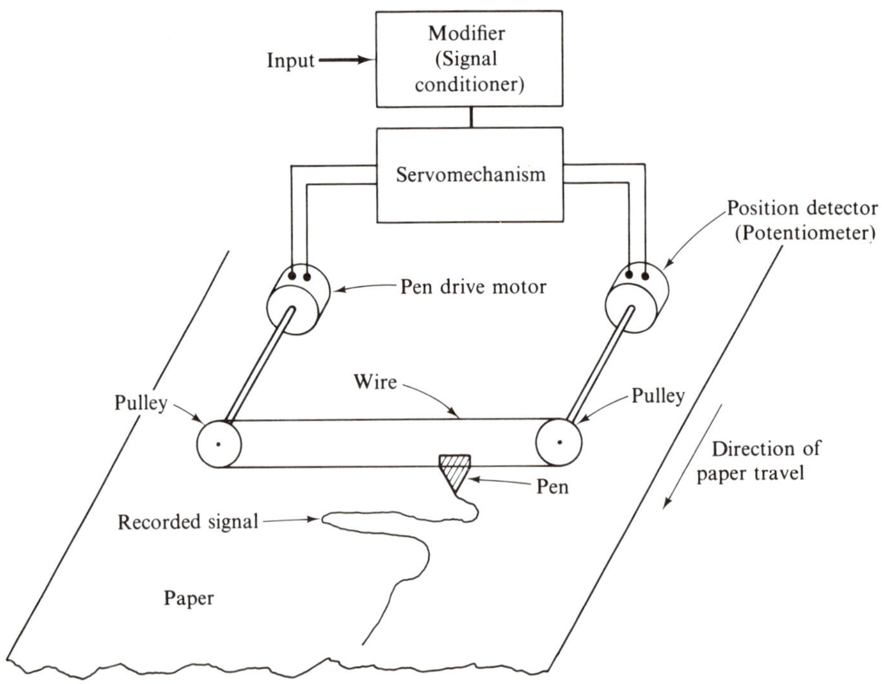

FIGURE 1.8 Servomechanism-driven recorder

recording device. To the indicator needle of a d'Arsonval movement galvanometer (chap. 3) a pen is attached in some fashion such that the excursions of the needle can be recorded as a continuous trace on a strip of paper. A variation of the recording galvanometer is the motor driven, servo-controlled linear galvanometer (Fig. 1.8). A signal applied to the control winding of the motor will drive the pen until the comparator potentiometer signal equals the input signal. Thus the pen will stop at a distance from the motor proportional to the input signal. Two such devices can be combined to form an X-Y recorder, that is, a device which records two signals as functions of each other on a cartesian plane.[23]

For frequencies higher than 100 Hz but less than 5,000 Hz the light-lever galvanometer can be used.[24] This device is a low inertia, low damping d'Arsonval movement with a mirror mounted on the armature. A light beam is reflected from the mirror in such a fashion that a movement of the armature is indicated by a movement of the beam. The beam is then used to trace a line on sensitized paper.

Highest frequency signals (greater than 5,000 Hz) can only be recorded on magnetic tape[25] or by means of a recording oscilloscope.[26] If the designer needs such high performance equipment he should obtain either some of the excellent commercial devices currently available or else the help of a specialist in tape recorders or oscilloscopes.

1.10 Design Procedure

The methodology of engineering design is highly individualistic and often quite disorganized. Each engineer develops his own method of coping with problems, and each engineering group develops its own design philosophy as a synthesis of the idiosyncrasies of the component engineers. It would be presumptuous for anyone to tell a colleague how to create. Experience has shown, however, that the design of instruments is accomplished by answering certain questions about the instrument. Two distinct major steps must be taken: the specifications must be gathered and expressed numerically, and equipment must be found to meet the specifications. Often a firm grasp of the specifications is the only major engineering task. The selection of the equipment takes care of itself (just as often it does not). In the following section there appears a list of questions which is an outline of an orderly design procedure; answering the questions in sequence will lead to a proper instrument.

1.11 Question Sequence for Instrument Design

- 1.0 What Are The Specifications?
  - 1.1 *What is to be measured?*
    - 1.1.1 What is the maximum amplitude of the measurand?

1.1.2 What is the required threshold?
1.1.3 What dynamic behavior is expected of the measurand?

1.2 *Who or what is to use the instrument?*
1.2.1 Is the user human or is it a nonbiologic system?
1.2.2 What form of energy input does the user require?
1.2.3 Is the output to be used immediately or recorded?

1.3 *What is the environment in which the instrument is to be used?*
1.3.1 Is the instrument to be used in the field or in the laboratory?
1.3.2 Is the instrument to be used by skilled or unskilled personnel?
1.3.3 What secondary energy sources will be available in the environment?
1.3.4 Are there natural hazards that face the instrument?
1.3.4.1 Humidity or dryness?
1.3.4.2 High or low temperatures?
1.3.4.3 Lightning?
1.3.4.4 Salt water?
1.3.5 Are there local man-made hazards?
1.3.5.1 Explosive atmospheres?
1.3.5.2 Corrosive atmospheres?
1.3.5.3 Nuclear radiation?
1.3.6 Is the measurement to be read at a remote station?
1.3.6.1 How far?
1.3.6.2 Is a wire link possible or should one use wireless methods?
1.3.7 Shall the instrument operate without maintenance?
1.3.8 Is the instrument to be subjected to shock or vibration?

2.0 How Is The Design To Be Accomplished?

2.1 *Is there a commercially available detector which meets the specifications?*
2.1.1 If 2.1 is answered *no*, is there a commonly known method of measurement which can be utilized?
2.1.2 If 2.1.1 is answered *no*, is there a principle which may, after some research, be used to invent a detector?

2.2 *Repeat sequence 2.1 with the word "detector" replaced by "display."*

2.3 *How is the input device to be matched to the output?*
2.3.1 Is the threshold matched?
2.3.2 What is the required sensitivity? Is amplification needed?
2.3.3 Is an intermediate transducer needed?
2.3.4 What about feedback?

2.4 *Is there a dynamical error?*
2.4.1 What is the response of the system to the three test functions?
2.4.2 Is there a frequency connected sensitivity change in the operation range?

- 2.4.3 Is there a phase shift?
- 2.5 *What are the sources of error?*
  - 2.5.1 Are there any false elements? (If there are, stop and start from the beginning!)
  - 2.5.2 Are there any disturbable elements?
  - 2.5.3 Are there any unnecessary delayed elements?
  - 2.5.4 Are there any indeterminate elements?
  - 2.5.5 Are there any incomplete elements?
- 2.6 *What is the effect of disturbable elements?*
  - 2.6.1 What is the input spectrum?
  - 2.6.2 Does a random input at a disturbable element affect the output signal of the instrument?
  - 2.6.3 If the answer to 2.6.2 is *yes*, can the performance be improved by redesign?
  - 2.6.4 If redesign is not possible can a filter be used?
- 2.7 *Is there any possibility of drift?*
  - 2.7.1 Does the selection of higher quality components help?
  - 2.7.2 Is it possible to compensate for drift?
  - 2.7.3 Will feedback help?
- 2.8 *Will the system show marked nonlinearities?*
  - 2.8.1 Does the nonlinearity generate a problem of scale selection?
  - 2.8.2 Does the nonlinearity introduce zero shift, hysteresis, amplitude dependent phase shift or any other deleterious nonlinear effects?
  - 2.8.3 Can nonlinearity be reduced by intermediate modifiers?
  - 2.8.4 Can nonlinearity be reduced by feedback?

In addition to the questions in the above list the designer will have to answer questions about human factors, such as, "Is the scale unambiguous?" and questions about environment and materials. Such questions are mostly outside the scope of this book and will not be included in our list. The text does contain detailed answers to most of the questions posed in our list or at least a guide to the pertinent literature. Detailed information on components, however, is not given below. For them the reader is referred to Appendix A (the bibliography), to the Compendium and to commercial indices such as "Sweet's Catalogues"[27] and the "Electronics Master."[28]

TABLE 1.1 Nomenclature

B_{nk}	Accuracy of the kth measurement of the nth measurand
B_n	Overall accuracy of the instrument with respect to the nth measurand
C_{nk}	Precision of the kth measurement of the nth measurand
C_n	Overall precision of the instrument with respect to the nth measurand
D_n	Certainty of the instrument with respect to the nth measurand
e_{nk}	Dimensionless error of the kth measurement of the nth measurand

TABLE 1.1 (Continued)

i_B	Inaccuracy
i_C	Imprecision
i_D	Uncertainty
z_n	Expected output
y_n	Expected value of the measurand
$y_n^{(0)}$	Reference input property
$\bar{z}_n$	Estimate of the expected output
$z_n^{(0)}$	Scale unit (scale reference property)
z_{nk}	kth measurement of the nth measurand
α_n	Systemic error corresponding to the nth measurand
β_{nk}	Dimensionless deviation
γ_n	Dimensionless systemic error
ϵ_{nk}	Total error of the kth measurement of the nth measurand
σ	Variance
σ_B	Variance of the accuracy
σ_C	Variance of the precision
σ_D	Variance of the certainty
η_{nk}	Deviation of the kth measurement of the nth measurand
$\xi_n^{(0)}$	Reference indication
ξ_{nk}	Indication corresponding to z_{nk}

1.12 Definition Summary

Accuracy—If the expected value of the nth measurand is y_n and the kth measurement corresponding to y_n is z_{nk}, then the accuracy B_{nk} of the measurement is

$$B_{nk} = 1 - \left| \frac{z_{nk}}{y_n} - 1 \right|$$

The accuracy of the instrument with respect to the nth measurand is defined as a function

$$B_n = B_n(B_{n1}, B_{n2}, \ldots, B_{nk})$$

which converges to some unique number as $k \longrightarrow \infty$.

Ambiguity—If the performance relationship of an element is not uniquely defined but is drawn randomly from a specified set of relationships, then the element is said to be *incomplete* and is said to generate an ambiguity error.

Calibration—The process of associating a unit and a *scale* to an *indicator*.

Certainty—If the expected output is z_n and the expected input y_n, then the certainty D_n is

$$D_n = 1 - \left| \frac{z_n}{y_n} - 1 \right|$$

Component—Components are the subelements of an element. Components are connected in *patterns* in such a fashion that a performance *relationship* is established between the input to the element and the output. The division of a system into elements and components is a matter of convenience and is largely arbitrary.

Configuration—A physical configuration is a specific combination of masses and energies at a given instant of time.

Damper—Any device which tends to retard or slow down a change in a configuration through the mechanism of energy absorption.

Detector—A detector is a *transducer* which is in direct contact with the configuration which generates the measurand. Detectors abstract energy from the configuration and its *illuminator* to generate a *signal.*

Deviation—The difference between the kth measurement of the nth measurand and the expected value of the output.

Display—A display is a transducer which makes the signal understandable to the user.

Distortion—If a device is designed to obey the relationship $z = f(y)$ and if under certain circumstances the functional relationship is not satisfied, then the signal is said to be distorted.

Drift—If the performance relationship of an element is not maintained because the properties of its components tend to change continuously with time, then the element is said to be *indeterminate* and to generate a drift error.

Dynamical Error—If an element responds accurately to a constant input, but introduces error into the information chain in response to a time variable input, then the element is said to be *delayed* and to generate a dynamic error.

Element—Any subassembly of an instrument which can be isolated as a source of error.

Energy Source, Intrinsic—Energy generated by the input configuration and used by the detector to generate a signal.

Energy Source, Primary—An energy source which *illuminates* the configuration and supplies the physical contact between the measurand and the transducer.

Energy Source, Secondary—Energy that is used in the instrumental system to operate the elements but that is not directly concerned with the transmission of information.

Error—Difference between some estimate of the output and the *expected value of the input.*

Expected Value of the Input—An ideal "perfect" or uniquely correct number assignable to the input property.

Expected Value of the Output—A unique number assignable to the output as a response to the input. The expected value of the output need not be equal to the expected value of the input.

False Element—Cf. *systemic error.*

Illumination—The process of making energy available to an input configuration from an external source.

Incomplete—Cf. *ambiguity.*

Indicator—A device which connects an output physical property to an input physical property.

Indeterminate—Cf. *drift.*

Information—Cf. *signal.*

Impedance—The effective load which corresponds to the energy withdrawn by the output device from the input device. If the energy used by the output system is within the capabilities of the input device, the two elements are said to be impedance matched.

Isolator—A device connected between two elements A_1 and A_2 in such a fashion that A_1 will not be overloaded and A_2 will receive the optimum power needed for operation.

Instrument—A device which connects a unique number to a *measurand* or *property* of a physical configuration.

Maintainers—*Components*, *patterns* and *elements* specifically designed to keep a performance relationship intact.

Measurement—A number obtained as an estimate of the input expected value.

Mismatched Energy—Cf. *impedance.*

Modifier—A device introduced in the signal chain to change the local characteristics of the signal.

Pattern—The combination of elements which produces a given performance relationship.

Performance Relationship—In an *element*, the functional relationship between input and output.

Precision—If the *expected output* is z_n and the kth measurement of the nth measurement z_{nk}, then the precision is defined as

$$C_{nk} = 1 - \left| \frac{z_{nk} - z_n}{y_n} \right|$$

in which y_n is the *expected value of the input.* Precision measures the scatter of individual measurements. The precision of the instrument is some number, often a statistic, obtained by some transformation

$$C_n = \lim_{k \to 0} F_n(C_{n1}, C_{n2}, \ldots, C_{nk})$$

Property—One of a set of recognizable elements of a physical configuration which serve to describe the configuration uniquely at a given time.

Random Error—If the performance relationship of an element is momentarily disturbed by an unwanted external input drawn from a distribution, the element is said to be *disturbed* and generates a *random error.*

Reference Configuration—A reference configuration selected arbitrarily for the purpose of assigning to the *unit* $y_n^{(0)} = 1$ the corresponding scale *reference marking* $z_n^{(0)} = 1$.

Restorer—*Components*, *elements* and *patterns* designed to return a disturbed performance relationship to its original form.

Scale—An arbitrary set of numbers connected to the output property. The scale is connected to the input property by *calibration*.

Signal—An energy quantity which bears a functional relationship to the expected value of the measurand. A signal is said to carry *information*.

Systemic Error—If an element which is required to maintain the relationship $z = f(y)$ maintains instead $z = f(y) \neq f(y)$, then the element is said to be *false* and to generate a *systemic error* (cf. *precision*).

Transducer—A device which abstracts energy from a particular system and transforms the energy to make it available to another system.

Threshold—The largest repeatable absolute value of the minimum input change to which can be assigned a unique value of the change of expected output whenever the instrument is used in a prescribed, repeatable measurement situation.

Threshold Ratio—The ratio of the threshold of the instrument in a given situation to the threshold of the observer.

Threshold, Visual—The minimum change in a property discernible by the eye at a given distance and in a fixed, repeatable measurement situation.

Variance—If an infinite set of numbers p_k has an expected value $\bar{p}$, then the quantity

$$\sigma_p^2 = \lim_{K\to\infty} \frac{1}{K} \sum_{k=1}^{K} (p_k - \bar{p})^2$$

is called the variance of the set p_k.

PROBLEMS

1.1 A gage block is 1.000000 $\pm$ 0.000004 inches long, i.e., its length is known to five digits. The block is used to calibrate an electronic length-measuring instrument. The following measurements are made:

Test #	1	2	3	4	5
Output	1.003	1.001	0.997	1.000	0.999

Test #	6	7	8	9	10
Output	1.006	1.007	1.005	1.004	1.005

(a) Estimate the accuracy of the instrument.
(b) Assume that z_n is approximated by the arithmetic mean of z_{nk}. What is the precision of the instrument?
(c) Estimate the systemic error.

1.2 A mechanical system with an output torque of 10 in-oz at 3 rpm is to drive another mechanical system which runs at 3 in-oz and 15 rpm. Are the impedances of these devices matched? What should an intermediate modifier do?

1.3 The first device in problem 1.2 has a breakaway torque, i.e., a minimum *input* torque at which it will move, of 0.001 in-oz. The second device has an input breakaway torque of 0.0037 in-oz. What should the minimum torque amplification be? Can you make any conclusion about the threshold ratio?

1.4 Assume that the breakaway torque of the first device corresponds to 0.115 unit of signal change and that the second device has an output change of 0.01 unit corresponding to its breakaway torque. Assume that the user has a threshold of 0.02 unit. Give the input output characteristics of an intermediate modifier that will best meet *all* the conditions of problems 2.2 and 2.3 as well as threshold matching.

NOTES

References and citations are keyed to the general bibliography in Appendix A.

[1]Ref. 13, p. X.

[2]The "one-shot" devices keep on coming; they are the source of most new developments in measurement technique.

[3]Ref. 19, 20, 21, 22.

[4]Ref. 20, pt. II.

[5]To the thermodynamicist the separation of property from its measurement may seem a little strange, or at least contrary to the practice of identifying a measurement as a property of a closed system. Yet it is clear that the phenomenon called *pressure* is not a measurement; it is a property of the system to which a number (also called pressure) may be assigned by an instrument.

[6]Ref. 18, pp. 170–74; Ref. 17, pp. 196–200.

[7]Ref. 36. 1 meter = 1,553,164.13 wavelengths of red cadmium light. See also Ref. 37.

[8]Ref. 8, p. 53.

[9]Ref. 7, pp. 13–18.

[10]That all measurement affects the measurand is a conclusion drawn from the "principle of uncertainty" of Heisenberg (Ref. 21, p. 278). The Heisenberg principle establishes a "floor" under all measurements.

[11]Compare, for example, Ref. 13 with Ref. 11.

[12]Ref. 64, p. 94.

[13]See Ref. 64.

[14]The concept of element is due to T. M. Whitehead (Ref. 15, pp. 2–7). Whitehead's definitions have been modified to meet modern requirements.
[15]Ref. 22.
[16]Whitehead uses the word "link" to mean both "pattern" and "component" (Ref. 15, pp. 2–3).
[17]Ref. 16.
[18]From an 18th-century tale "Three Princes of Serendip" (no source available).
[19]Ref. 15, pp. IX–X.
[20]There are, it is hoped, many unknown principles still to be discovered.
[21]Ref. 12.
[22]Ref. 11.
[23]Ref. 43.
[24]Ref. 39, 48.
[25]Ref. 40.
[26]Best information on modern cathode ray oscilloscopes is found in manufacturers' promotional literature. No journal reference known to the author has the slightest connection with current equipment.
[27]Ref. 42.
[28]Ref. 41.

2

MATHEMATICAL PRELIMINARIES

2.1 Modeling

The analysis of ideal systems prior to construction is conducted through *mathematical modeling*. A set of integro-differential equations is written in such a fashion that the variables represent the input and output properties of the system and such that the mathematical statements represent as closely as possible the transformation from the expected value of the input properties of the expected value of the output. It is implicit, therefore, that the behavior of the output as a function of the input is predictable and reproducible. The properties of the input and the relationship of the input to the output at any time define the *state* of the complete system. By state we mean a physical description of the condition of every element and component at a given instant of time, or equivalently, a mathematical listing of a unique set of numbers which describes completely the configuration of instrument and measurand. Any set of functions of time which at any time completely define the state of the system are called the *state variables* of the system. If it is possible to write a mathematical model such that given the state variables at an arbitrarily selected time t_0 the state at any subsequent time $t > t_0$ can be computed, then the system is said to be *totally predictable* and the model is said to be *deterministic.* If, on the other hand, a system is subjected to unpredictable changes but a predictable, unique statistic related to the state is computable,[1] then the system is said to be *statistically predictable* and the model is called *stochastic.*[2]

An instrument system must necessarily be a predictable system, either total or statistical, since the output of the instrument must depend uniquely on the properties of the measurand configuration and upon a known unique functional transformation. From the state of some portion of the instrument (say, an output indicator needle) and from the state of the instrument (inferred by modeling), the operator should be able to infer at least a portion of the measured configuration. That portion of the input configuration which can be measured is said to be *observable*[3] with respect to the instrument.

The previous discussion of observability may seem somewhat academic to the inexperienced designer. It is well known, however, that certain mechanical modes in vibrating systems can be suppressed by a proper choice of filter components.[4] The effect of this deliberate isolation or decoupling is to render a vibration mode unobservable at some point in the machine. If it is possible to suppress an unwanted vibration deliberately, it is just as possible to suppress accidentally or to render unobservable a mode which one is required to measure.

2.2 Ordinary Differential Equations and the Concept of State

Creation of a mathematical model requires much practice and knowledge. To be able to deal with complicated systems which contain mechanical, electrical, thermal and hydraulic components requires a fundamental knowledge of physical laws and their mathematical representation. In modern American universities the engineering senior has acquired[5] a sufficient knowledge of elementary thermodynamics, fluid mechanics, heat transfer, electric circuit analysis, dynamics and elasticity to be able to write reasonable relationships between the input and the output of a given element. He is limited, however, by the nature of the assumptions made in his "engineering science" courses. In general, he is not able to deal with systems representable by nonlinear differential equations and has no idea of the effects of parameter distribution in a system. The last limitation is rather unfortunate since configurations are originally definable in terms of local field properties (for example, resistivity, fluid velocity, strain), and in terms of expressions connecting the local properties to the neighboring properties through consideration of local rates of change in space and time (for example, continuity and compatibility equations). The exact representation of a distributed system requires the creation of a set of partial integro-differential equations. Such complete equations are difficult to solve and often give more information than is needed. The designer will usually simplify matters through experiment and through approximate analysis by deciding which of the properties are important and which can be neglected. The selection of relevant properties and the rejection of others place a set of constraints on design. These constraints must be known and are to be kept in mind throughout the design procedure. A simple example of such a constraint is the description of the rotation of a shaft.[6] Although the relevant parameter is rotation, and the shaft is considered rigid (i.e., elastic properties and vibrations are neglected), it is clear that under certain circumstances the neglect of the elastic properties of the shaft can lead to mechanical failure. Even without consideration of failure, a relationship which connects the input and output of the shaft ($z_{in} = z_{out}$) without consideration of elasticity may be adequate when used in the estimation of instrument sensitivity and gross response but will fail completely when used to study the threshold properties of the instrument.[7]

Often the distribution of a property in a component or an element will not be relevant to the overall study of the instrument. A finite set of quantities can be used (often under restrictions) to characterize the total "lumped" effect of the distributed property in connecting the input to the output.[8] A simple example of the simplification achieved is found in the determination of the gross motion of a wheel. An instrument support roller exhibits a local vector velocity $\mathbf{V}(x, y, z, t)$ at every point on the wheel and at every instant of time. $\mathbf{V}$ is a vector combination of the linear velocity $\mathbf{U}$ of the hub, of the gross angular velocity $\boldsymbol{\omega}$ of the wheel taken about the hub, of the vector distance $\boldsymbol{\rho}$ from the hub to some point on the wheel and of the local variations in velocity $\mathbf{v}'$ introduced by elastic distortion of the material of the wheel. Thus we can write

$$\mathbf{V} - \mathbf{U} + \boldsymbol{\omega} \mathbf{x} \boldsymbol{\rho} + \mathbf{v}'$$

The vector $\mathbf{V}$ has a different length and a different direction at every point on the wheel and at every instant of time. Yet if the wheel is considered as the means of support of an instrument carriage, ideal behavior would require $\mathbf{v}'$ to be zero, or equivalently, $\mathbf{v}'$ to be considered an error introduced by the choice of materials. From the point of view of wheel function as a carriage support, the element is described by the lumped hub velocity $\mathbf{U}$, which, in the absence of slipping, is related to the angular velocity by the relationship $\boldsymbol{\omega} \mathbf{x} \mathbf{R} = \mathbf{U}$, $\mathbf{R}$ being the radius of the wheel. As another example we can point out that although a piece of wire has a resistivity per unit length (a distributed property), the electrical designer is usually concerned with the total (or lumped) resistance of a finite length of wire or resistor.

The advantage gained through the use of lumped parameters is that the system performance equations usually reduce to ordinary differential equations. Consequently the instrument is represented by a set of ordinary differential equations with a number of known inputs (forcing functions) and a number of known outputs (system variables).

In a completely defined system there must be as many equations as there are outputs so that it should be possible by successive substitutions to obtain a set of differential equations, each equation containing only one output and driven by a functional combination of the various inputs. If a solution of the individual "uncoupled" equations exists, it should then be possible to obtain a set of nondifferential expressions connecting the several inputs to each of the outputs.

Example 2.2.1 A system obeys the nonlinear ordinary differential equations

$$\frac{d^2z_1}{dt^2} + \frac{dz_2}{dt} + z_1^2 = y_1(t) \tag{2.2.1}$$

$$\frac{d^2z_2}{dt^2} + z_1 + z_2^2 = y_2(t) \tag{2.2.2}$$

Write two equations, each with y_1 and y_2 as inputs and either z_1 or z_2 alone as outputs.

Solution Solve Equation (2.2.2) for z_1 and substitute into Equation (2.2.1).

$$\frac{d^4z_2}{dt^4} + \left(2z_2 + \frac{dz_2}{dt} - 2z_2^2 + y_2\right)\frac{d^2z_2}{dt^2} + \frac{d^2y_2}{dt^2}\left(2\frac{dz_2}{dt} + z_2^2 - y_2\right)\frac{dz_2}{dt} + y_1 - (z_2 - y_2)^2 = 0$$

Although messy, this equation is of the fourth order in z_2 and does not show z_1 explicitly. To obtain the equation for z_1 we can solve Equation (2.2.1) for z_2 to give us

$$z_2 = \int^t \left(\frac{y_1}{z_1} - z_1 - \frac{1}{z_1}\frac{d^2z_1}{dt^2}\right) dt + C$$

in which the integral symbol represents the antiderivative and C is a constant to be evaluated from the initial conditions. Substitution of z_2 into Equation (2.2.2) gives an integro-differential equation in z_1.

The procedure of substitution employed in Example 2.2.1 is often impossible to perform and is usually unnecessary. It does demonstrate, however, that a coupled system in n variables is equivalent to a set of n uncoupled systems and that each variable can be studied independently.[9] Let us assume that the quantity z_1 exists and is analytic, that is, it is continuous and exhibits continuous derivatives of all orders. Then, in the region of analyticity, it is possible to expand z_1 in terms of the Taylor series in the time t referred to on initial time t_0.[10]

$$z_1 = z_1(t_0) + \left.\frac{dz_1}{dt}\right|_{t_0}(t - t_0) + \cdots + \frac{1}{n!}\left.\frac{d^nz_1}{dt^n}\right|_{t_0}(t - t_0)^n + \cdots \tag{2.2.3}$$

It follows from Equation (2.2.3) that the quantity z_1 is known completely for all times $t > t_0$ if the function and all its derivatives are known at time $t = t_0$. (The same conclusion can be made about z_2.) It is not necessary to know all the derivatives to evaluate Equation (2.2.3). If as in Example 2.2.1, z_1, z_2, dz_1/dt and dz_2/dt are known at time $t = t_0$, then all the derivatives of z_1 and z_2 can be computed with the aid of Equations (2.2.1) and (2.2.2).

Example 2.2.2 In the system discussed in Example 2.2.1, the following conditions apply at the time $t = 0$:

$$z_1 = z_2 = \frac{dz_1}{dt} = \frac{dz_2}{dt} = 0$$

The functions y_1 and y_2 have the form

$$y_1 = 4e^{-t} \qquad y_2 = e^t$$

Obtain the first four terms of the Taylor series expansion of z_1 in the neighborhood of $t_0 = 0$.

Solution The first two terms of Equation (2.2.3) are zero by definition. At time $t = 0$ we find immediately from Equation (2.2.1) that

$$\left.\frac{d^2z_1}{dt^2}\right|_0 = y_1(0) = 4$$

If we now perform a differentiation on Equation (2.2.1) and substitute the initial condition we obtain

$$\left.\frac{d^3z_1}{dt^3}\right|_0 = -4$$

Differentiating twice, again we find that

$$\left.\frac{d^4z_1}{dt^4}\right|_0 = +4 \qquad \left.\frac{d^5z_1}{dt^5}\right|_0 = -52$$

To evaluate the fifth derivative we had to introduce the second derivative of z_2 from Equation (2.2.2). Since for small $t > 0$, t^5 is a very small number, it follows that initially z_1 and z_2 are nearly uncoupled (in the neighborhood of the initial conditions and near the initial timc) and that coupling effects are not noticeable until t^5 becomes significant. For example, at $t = 0.5$ second, the fifth-order term provides approximately only 3% of z_1. With the computed derivative we now can write the answer to our problem

$$z_1 = 2t^2 - \frac{2}{3}t^3 + \frac{1}{6}t^4 - \frac{13}{30}t^5 + \cdots$$

That only four initial conditions are needed to fix the state of the system for all time is expectable from the theory of ordinary differential equations. The equation in z_1 is only of the third order but contains an arbitrary constant so that four conditions are needed. The equation in z_2 is of the fourth order and also requires four initial conditions. Thus we have established (intuitively at least) that a system representable by n ordinary differential equations in n variables, if solvable, has solutions of the n variables which are only functions of the initial conditions and of time. This intuitive concept is the result of the Cauchy-Lipschitz theorem which will be discussed in the next section.

Consider now any four initial conditions for the problem discussed in Example 2.2.2, say, $z_1(t_0)$, $\dot{z}_1(t_0)$, $z_2(t_0)$ and $\dot{z}_2(t_0)$ at $t = t_0$. The system will have solutions (assuming such solutions exist) of the form

$$\begin{aligned} z_1 &= f_1[z_1(t_0), z_2(t_0), \dot{z}_1(t_0), \dot{z}_2(t_0); t] \\ z_2 &= f_2[z_1(t_0), z_2(t_0), \dot{z}_1(t_0), \dot{z}_2(t_0); t] \end{aligned} \tag{2.2.4}$$

From Equations (2.2.4) we deduce that if the numerical values of the initial conditions are known at *any* time $t = t_0$ then z_1 and z_2 are known for all times $t > t_0$.[11] From the arbitrariness of the solutions it is obvious that given the initial conditions at t_0 we could compute z_1, z_2, $\dot{z}_1$, $\dot{z}_2$ at some time t_1 and hence by using t_1 as the initial time obtain a new solution of the system valid over the same range as the equation in t_0. The quantities z_1, z_2, $\dot{z}_1$, $\dot{z}_2$ are the definers of the state of the system which, if known as functions of time, specify the condition of the instrument completely. Accordingly, the four quantities form a set of *state variables* for the instrument.

The set is not unique. Any four permissible initial conditions (and in some instances four combinations of initial conditions) could be used. Thus z_1 and its first three derivatives could serve in place of z_1, z_2 and their first derivatives. In this book, however, we shall adopt a restricted definition of state variables which has been found to meet the needs of instrument designers.

Definition 2.2.1 LET A SYSTEM HAVE n OUTPUTS $z_1, z_2, z_3, \ldots, z_n$ AND LET THERE BE A SET OF n ORDINARY DIFFERENTIAL EQUATIONS

$$f_i(z_1, z_2, \ldots, \dot{z}_1, \dot{z}_2, \ldots, \ddot{z}_1, \ddot{z}_2, \ldots, \dddot{z}_1, \dddot{z}_2, \ldots, t) = 0$$

AND LET $d^{m_i}z_i/dt^{m_i}$ BE THE HIGHEST-ORDER DERIVATIVE OF A z_i VARIABLE APPEARING IN ANY OF THE EQUATIONS. THEN THE QUANTITIES

$$z_i, \frac{dz_i}{dt}, \ldots, \frac{d^{m_i-1}z_i}{dt^{m_i-1}} \qquad i = 1, 2, \ldots, n$$

SHALL BE CALLED THE STATE VARIABLES OF THE SYSTEM.

THE STATE VARIABLES SHALL BE LABELED u_k AND SHALL BE ASSIGNED SUBSCRIPTS ACCORDING TO THE FOLLOWING SCHEME:

$$\frac{d^p z_i}{dt^p} = u_{pn+i} \qquad \begin{aligned} i &= 1, 2, \ldots, n \\ p &= 0, 1, 2, \ldots, m_i - 1 \end{aligned} \tag{2.2.5}$$

Adoption of the scheme embodied in Definition 2.2.1 and Equation (2.2.5) will permit recognition of the system variable from an array of state variables by noting that

$$z_i = u_i \qquad i = 1, 2, \ldots, n$$

Substitution of the state variables into the differential equations will yield a set of ordinary first-order differential equations in the form

$$\dot{u}_k = g_k(u_1, u_2, \ldots, u_N; t) \qquad \begin{aligned} k &= 1, 2, \ldots, N \\ N &= \sum_{j=1}^{n} m_j \end{aligned} \tag{2.2.6}$$

Of the N equations, n will be obtained from the differential equations and $N - n$ from the definitions of Equation (2.2.5).

Example 2.2.3 Write the equations presented in Example 2.2.1 in state variable form.

Solution From Equation (2.2.5) we obtain $u_1 = z_1$, $u_2 = z_2$, $u_3 = \dot{z}_1$, $u_4 = \dot{z}_2$. The array of first-order Equations (2.2.6) becomes

$$\begin{aligned} \dot{u}_1 &= u_3 \\ \dot{u}_2 &= u_4 \\ \dot{u}_3 &= -u_1u_4 - u_1^2 + y_1(t) \\ \dot{u}_4 &= -u_1 - u_2^2 + y_2(t) \end{aligned}$$

in which the first two equations come from the definition [Equation (2.2.5)] and the last two from Equations (2.2.1) and (2.2.2) in Example 2.2.1.

It is possible to construct nonlinear differential equations in which the state variable form is not well defined. For example, the equation

$$\left(\frac{dz}{dt}\right)^4 + z = y(t)$$

has the state variable form

$$\dot{u} = (\dot{y}_1 - u_1)^{1/4}$$

There are four possible values of $(\dot{y}_1 - u_1)^{1/4}$ for each value of $\dot{y}_1 - u_1$, and thus the result is not unique. Mercifully such examples do not usually arise in engineering practice.

2.3 The Theorem of Cauchy-Lipschitz

The existence of solutions of ordinary differential equations is easily determined once the equations have been transformed to state variable form if they can be shown to satisfy certain criteria known as the *Lipschitz conditions*.

Definition 2.3.1 LET A SYSTEM BE REPRESENTABLE BY AN ARRAY

$$\dot{u}_k = g_k(u_1, u_2, u_3, \ldots, u_N; t) \qquad k = 1, 2, 3, \ldots, N$$

THEN THE SYSTEM IS SAID TO SATISFY A LIPSCHITZ CONDITION IN THE RESTRICTED FORM IF EACH g_k IS CONTINUOUS IN EACH u_k, AND EACH FIRST PARTIAL DERIVATIVE OF EACH g_k WITH RESPECT TO EACH STATE VARIABLE EXISTS AND IS CONTINUOUS.[12]

Nothing is said or implied about the variable t. The driving functions can be quite discontinuous in time without causing the solution to perish. Definition 2.3.1 is much more restrictive than the Lipschitz condition found in the literature.[13] For the purposes of this book and of instrument design the derivative condition is more than adequate. If a system is Lipschitzian, unique solutions of the equations exist.

Theorem 2.3.1 LET $g_k(u_1, u_2, \ldots, u_N; t)$ SATISFY A LIPSCHITZ CONDITION; THEN THERE EXISTS A UNIQUE SET OF SOLUTIONS u_k, $k = 1, 2, \ldots, N$, WHICH DEPEND ONLY ON THE STATE OF THE SYSTEM AT AN INITIAL TIME t_0 AND UPON THE TIME t.

Theorem 2.3.1 is a combination of the existence theorem of Cauchy-Peano and the uniqueness theorem of Lipschitz.[14] Although somewhat abstract, the theorem is the foundation of the so-called state variable technique now in vogue in control and measurement engineering practice. The test is extremely practical and can be applied fairly easily; for example, the state equations of Example 2.2.3 are clearly Lipschitzian at each fixed value of t. Actually the g_k in Example 2.2.3 have continuous derivatives of all orders in all u_k, and thus we can guarantee that the solutions will be analytic. This fact, which can be useful in handling nonlinear equations, is embodied in the following theorem.

Theorem 2.3.2 IF THE FUNCTIONS g_k ARE ANALYTIC IN ALL VARIABLES IN A NEIGHBORHOOD OF $t = t_0$, THEN THE SOLUTIONS OF THE DIFFERENTIAL EQUATIONS ARE ANALYTIC IN ALL VARIABLES IN THE SAME NEIGHBORHOOD.

2.4 Linear Ordinary Differential Equations

The previous discussion is quite general and can be applied to all lumped parameter systems representable by ordinary differential equations. A solution of the state equations is said to exist if the conditions discussed in Section 2.3 can be applied to the state functions g_k. That a solution is theoretically possible does not imply, however, that it can be found. In the set of all ordinary differential equations, only a minute subset is explicitly solvable. In general, a nonlinear system which satisfies a Lipschitz condition cannot be "solved" in a common sense that an explicit set of expressions in the initial conditions and time can actually be found.[15] However, the existence of a solution permits the scientist to construct approximate numerical solutions with every chance of success.[16]

One must agree, however, that explicit solutions can yield great insight and can simplify design. Thus the subset of solvable differential equations for which a routine solution exists, the linear ordinary differential equations with constant coefficients, is of particular interest to the designer. If the system can be represented by linear equations or if the nonlinear equations can be beaten into linear equations by limiting assumptions, then a solution of the system equations can be found and exhibited "fairly easily." But even in such a case the operative phrase is "fairly easily" rather than "easily." If the order of the system is higher than the fourth, the solution of the equations will depend on the possibility of extracting the roots of a high order polynomial, which, as is well known, cannot be accomplished exactly for any polynomial of order higher than the fourth.[17] In spite of all the difficulties, the known techniques for handling linear ordinary differential equations are so explicit and so useful that they form the mathematical foundation of most engineering and much science.

To connect the theory of linear equations to the discussion of Section 2.3 it is useful to convert to state variable notation. A set of ordinary linear differential equations (OLDE) with constant coefficients can be written in the notation of this book as

$$\sum_{i=1}^{n} \sum_{m=0}^{m_i} c_{im}^{j} \frac{d^m z_i}{dt^m} = y_i(t) \tag{2.4.1}$$

which, upon substitution of Equation (2.2.5), can be written in the form[18]

$$\dot{u}_k = \sum_{j=1}^{N} a_{kj} u_j + r_k(t) \tag{2.4.2}$$

in which r_k is a function of the various $y_i(t)$.

Example 2.4.1 Represent the following set of OLDE in state variable form.

$$\frac{d^2 z_1}{dt^2} + 3\frac{d^2 z_2}{dt^2} + 7z_1 = 4$$

$$\frac{dz_2}{dt} + \frac{d^2 z_1}{dt^2} + z_2 = e^t$$

Solution

$$u_1 = z_1 \qquad u_2 = z_2$$

$$u_3 = \dot{z}_1 \qquad u_4 = \dot{z}_2$$

Thus by direct substitution we obtain

$$\begin{aligned}
\dot{u}_1 &= u_3 \\
\dot{u}_2 &= u_4 \\
\dot{u}_3 &= -u_4 - u_2 + e^t \\
\dot{u}_4 &= -\frac{1}{3}\dot{u}_3 - \frac{7}{3}u_1 + \frac{4}{3}
\end{aligned}$$

The last equation is not in standard form. If we substitute from the third equation into the fourth, we obtain the answer

$$\begin{aligned}
\dot{u}_1 &= u_3 \\
\dot{u}_2 &= u_4 \\
\dot{u}_3 &= -u_4 - u_2 + e^t \\
\dot{u}_4 &= -\frac{7}{3}u_1 + \frac{1}{3}u_2 + \frac{1}{3}u_4 - \frac{1}{3}e^t + \frac{4}{3}
\end{aligned}$$

There exist well-known classical methods for the solution of Equation (2.4.2) which are simple and explicit. In spite of the elegant clarity of classical analysis it has been customary in the practice of instrument engineering to use Laplace transform techniques to manipulate and solve OLDE. Theoretically, nothing is gained through the use of transforms: practically, however, a few advantages can be discovered. In the first place, transforms present a routine, automatic procedure which is nearly foolproof. Secondly, transforms relate to the so-called "transfer functions" and to element impedance easily and directly. Finally, a number of design procedures have been developed, notably the "frequency plane" technique, which are easily applied through the transform calculus.[19]

Since the reader is presumably familiar with the relevant theorems of the Laplace transform, the theory of the $\mathfrak{L}$ calculus is not presented in this chapter. To meet the contingency that the author's opinion of the reader's knowledge is too sanguine, a self-contained introduction to Laplace transforms will be found in Appendix B.

The system variables and the state variables shall be identified by lowercase letters and the transforms of the variables by uppercase letters:

$$\mathfrak{L}y_i = Y_i$$
$$\mathfrak{L}z_i = Z_i$$
$$\mathfrak{L}u_i = U_i$$
$$\mathfrak{L}r_i = R_i$$

In the state variable presentation only the transform of the first derivative is needed

$$\mathfrak{L}\dot{u}_i(t) = sU_i - u_i(0^+)$$

in which $u_i(0^+)$ represents the initial state as a limit from the right

$$u_i(0^+) = \lim_{t \to 0} u_i(t) \qquad t > 0$$

The state equations can now be transformed into

$$sU_k - u_k(0^+) = \sum_{j=1}^{N} a_{kj} U_j + R_k \tag{2.4.3}$$

The summation in Equation (2.4.3) has a term U_k which can be collected with the similar term on the left. To permit a compact representation the Kronecker delta is introduced

$$\delta_{kj} = \begin{cases} 1 & \text{if } k = j \\ 0 & \text{if } k \neq j \end{cases}$$

and then Equation (2.4.3) becomes

$$\sum_{j=1}^{N} (s\delta_{kj} - a_{kj}) U_j = R_k + u_k(0^+) \tag{2.4.4}$$

The array of polynomials in Equation (2.4.4) can be solved by Cramer's rule in the form

$$U_i = \frac{B_i}{\Delta} \tag{2.4.5}$$

in which Δ is the characteristic polynomial of the system (2.4.2)

$$\Delta = (-1)^N \begin{vmatrix} a_{11} - s & a_{12} & \cdots & a_{1N} \\ a_{21} & a_{22} - s & \cdots & a_{2N} \\ \cdots & \cdots & \cdots & \cdots \\ a_{N1} & a_{N2} & \cdots & a_{NN} - s \end{vmatrix}$$

and B_i is the determinant Δ with the ith column replaced by the elements $R_k + u_k(0^+)$. By the rules for the addition of determinants,[20] B_k can be written

$$B_i = B_i^R + B_i^u$$

$$= (-1)^N \begin{vmatrix} a_{11} - s & \cdots & a_{1(i-1)} & R_1 & a_{1(i+1)} & \cdots & a_{1N} \\ a_{21} & \cdots & a_{2(i-1)} & R_2 & a_{2(i+1)} & \cdots & a_{2N} \\ \cdots & \cdots & \cdots & \cdots & \cdots & \cdots & \cdots \\ a_{N1} & \cdots & a_{N(i-1)} & R_N & a_{N(i+1)} & \cdots & a_{NN} - s \end{vmatrix}$$

$$+ (-1)^N \begin{vmatrix} a_{11} - s & \cdots & a_{1(i-1)} & u_1(0^+) & a_{1(i+1)} & \cdots & a_{1N} \\ a_{21} & \cdots & a_{2(i-1)} & u_2(0^+) & a_{2(i+1)} & \cdots & a_{2N} \\ \cdots & \cdots & \cdots & \cdots & \cdots & \cdots & \cdots \\ a_{N1} & \cdots & a_{N(i-1)} & u_N(0^+) & a_{N(i+1)} & \cdots & a_{NN} - s \end{vmatrix}$$

in which B_i^R, the first determinant in the expansion of B_i, is only a function of the transformed inputs R_k and the second determinant, B_i^u, depends only on the initial state. The transform of the system state variables can now be written as the sum of two terms, one of which describes the effects of the various inputs on the state variables, and another term which exists only as the system begins to operate from nonrest conditions.

$$U_i = \frac{B_i^R}{\Delta} + \frac{B_i^u}{\Delta}$$

The quantities B_i^R, B_i^u and Δ can be inverted into time functions by the methods described in Appendix B, and the resulting expressions can be introduced into the convolution to yield

$$\begin{aligned} u_i(t) = \int_{0^+}^{t} \left.\mathfrak{L}^{-1} B_i^R\right]_{t'=t-\tau} \left.\mathfrak{L}^{-1}\frac{1}{\Delta}\right]_{t'=\tau} d\tau \\ + \int_{0^+}^{t} \left.\mathfrak{L}^{-1} B_i^u\right]_{t'=t-\tau} \left.\mathfrak{L}^{-1}\frac{1}{\Delta}\right]_{t'=\tau} d\tau \end{aligned} \tag{2.4.6}$$

in which t' is defined by the inversion

$$\mathfrak{L}^{-1} F(s) = f(t')$$

The solution of the system equations is separable into two parts:

the first one which depends only on the input functions and the characteristics of the system, and a second one which depends only on the initial state and the characteristics of the system. The effects of forcing functions and of initial conditions can be now treated separately and the resulting effects added to yield a total solution. In instrument design and practice, a nonzero initial state is simply an error to be removed or minimized as rapidly as possible. If the design is to be accomplished by linear techniques, the instrument performance can be determined with a zero initial state and the initial condition considered later in the error analysis phase.

2.5 Observability

Equation (2.4.6) gives a clue to the formalization of the concept of observability. The determinant B_i^R can be expanded into the form

$$B_i^R = \sum_{j=1}^{N} R_j C_{ij}$$

in which C_{ij} is the cofactor of $a_{ij} - \delta_{ij} s$ in Δ. For zero initial state, Equation (2.4.6) becomes

$$u_i = \sum_{j=1}^{N} \int_{0^+}^{t} \mathcal{L}^{-1} R_j C_{ij} \Big]_{t'=t-\tau} \mathcal{L}^{-1} \frac{1}{\Delta} \Big]_{t'=\tau} d\tau \tag{2.5.1}$$

Equation (2.5.1) indicates the kth state variable is affected by the jth input function only if $\mathcal{L}^{-1} R_j C_{ij} \neq 0$. If $\mathcal{L}^{-1} R_j C_{ij} = 0$ the input $r_j(t)$ is said to be unobservable with respect to u_i.

For certain combinations of functions it may well be that the whole summation $\mathcal{L}^{-1} \sum_{j=1} R_j C_{ij} = 0$ or that the sum of a portion of the terms is zero,[21] say,

$$\mathcal{L}^{-1}(R_{a_1} C_{ia_1} + R_{a_2} C_{ia_2} + \cdots + R_{a_k} C_{ia_k}) = 0$$

Then the inputs $r_{a_1}, r_{a_2}, \ldots$ are said to be relatively unobservable with respect to u.

2.6 Impulsive Response

For zero initial state, Equation (2.4.5) can now be written

$$U_i = \sum_{j=1}^{N} R_j \frac{C_{ij}}{\Delta} = \sum_{j=1}^{N} R_j D_{ij} \tag{2.6.1}$$

in which $D_{ij} = C_{ij}/\Delta$. The inversion of D_{ij} is labeled $I_{ij}(t)$, from which it follows that

$$\mathcal{L}^{-1} U_i = u_i(t) = \sum_{j=1}^{N} \int_{0^+}^{t} r_j(t - \tau) I_{ij}(\tau)\, d \tag{2.6.2}$$

It also follows that if

$$R_j = 0 \quad \text{for } j \neq A \qquad R_j = 1 \quad \text{for } j = A$$

then Equation (2.6.1) reduces to $U_i = D_{iA}$. From this it can be concluded that if there exists a function δ ($\mathfrak{L}\delta = 1$) which can be used as the Ath input, then $I_{iA}(t)$ is the response of the ith state variable to the function δ as the Ath input. Thus

$$I_{ij}(t) = \int_{0^+}^{t} \delta(t - \tau) I_{ij}(\tau)\, d\tau$$

From the integral it can be deduced that $\delta(t - \tau) = 0$ at $t \neq \tau$ and is not zero at $t = \tau$, so that

$$I_{ij}(t) = I_{ij}(t) \int_{0^+}^{t} \delta(t - \tau)\, d\tau$$

and thus

$$\int_{0^+}^{t} \delta(t - \tau)\, d\tau = \int_{0^+}^{\infty} \delta(t - \tau)\, d\tau = 1$$

The function $\delta(t - \tau)$ is known as an impulse at time $t = \tau$, and $I_{ij}(t)$ is known as the impulsive response of the ith state variable with respect to the jth input. To understand the meaning of the word "impulse" as used in this instance, consider any function representable by a maximum height h at time $t = \tau$ and such that the function is essentially zero outside of an interval of width b about the time $t = \tau$ (Fig. 2.1). The relationship between h and b is selected so that the area under the curve is always one regardless of the size of b. If b is made as small as possible (that is, $b \rightarrow 0$), then to maintain the area, h will have to

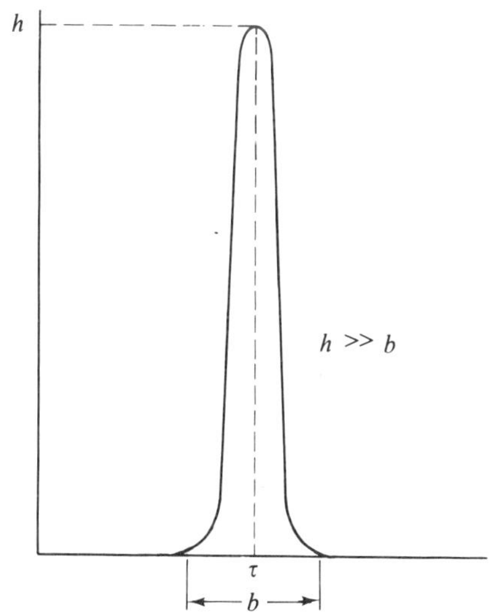

FIGURE 2.1 A typical approximation to an impulse

become very large ($h \longrightarrow \infty$). In a mechanical system $\delta(t-\tau)$ would be a mathematical model of the physical impulse, that is, a very large input of extremely short duration.

It is clear that the delta function is not a specific function but rather the limit of any of a set of functions which satisfy the integral property.[22] Three well-known examples of functions which approach δ as $b \longrightarrow 0$ are

$$\delta_1(t-\tau) = \lim_{b\to 0}\left\{\frac{u[t-(\tau-(b/2))]-u[t-(\tau+(b/2))]}{b}\right\}$$

$$\delta_2(t-\tau) = \lim_{b\to 0}\frac{1}{b\sqrt{\pi}}e^{-[(t-\tau)/b]^2}$$

$$\delta_3(t-\tau) = \lim_{b\to 0}\frac{b}{\pi}\frac{\sin^2[(t-\tau)/b]}{(t-\tau)^2}$$

Equation (2.6.2) acquires a remarkable meaning with respect to the impulse function. $I_{kj}(\tau)$ is a continuously occurring impulsive response, and the response $u_k(t)$ of the instrument is obtained by integrating or summing the response of the system to a sequence of impulses which approximate the driving function. In fact, if the impulsive response of a linear system is known, the response of the system to any input is also known. Thus $I_{ij}(t)$ is a set of functions which completely characterize the dynamics of the instrument.

From an experimental point of view it is difficult to apply an impulse to the instrument. In vibration measuring instruments, for example, any input which approximates an impulse will probably destroy the instrument. However, there is an alternate definition of $I_{ij}(t)$ which is occasionally realizable. Assume that the input functions are all zero but that the initial values of the state variables are not zero. Then the transformed response Equation (2.4.5) can be written as

$$U_i = \sum_{j=1}^{N} u_j(0^+)D_{ij}(s)$$

the inversion of which is

$$u_i = \sum_{j=1}^{N} u_j(0^+)I_{ij}(t) \qquad \textbf{(2.6.3)}$$

The ith output is thus a linear combination of all the impulsive outputs weighted by the initial states. If all initial conditions are set to zero except one, say, the $j = J$ condition, then the ratio

$$\frac{u_i(t)}{u_J(0)} = I_{iJ}(t) \qquad \textbf{(2.6.4)}$$

yields the Jth impulsive response.

Experimentally, $I_{ij}(t)$ can be obtained by disturbing (if possible) the jth state variable at time $t = 0$ and recording all the outputs and then by applying (2.6.4). The procedure is ideal for analogue computer study of instrument performance but is often difficult to achieve in the prototype instrument since it is required to disturb a high-order derivative to achieve the impulsive response of the system variables.

2.7 Computing Impulsive Response

The general linear ordinary differential equation with constant coefficients is given by

$$\dot{u}_i = \sum_{j=1}^{N} a_{ij} u_j + r_i(t)$$

in which $r_i(t)$ is the kth effective (or equivalent) input. Let $r(t) = a_i \delta(t)$ in which a is a set of constants. Then the transform of the ith state variable can be written as

$$U_i = \frac{1}{\Delta} \begin{vmatrix} s - a_{11} & \cdots & -a_{1(i-1)} & -a_1 & -a_{1(i+1)} & \cdots & -a_{1N} \\ -a_{21} & \cdots & -a_{2(i-1)} & -a_2 & -a_{2(i+1)} & \cdots & -a_{2N} \\ \cdots & \cdots & \cdots & \cdots & \cdots & \cdots & \cdots \\ -a_{N1} & \cdots & -a_{N(i-1)} & a_N & a_{N(i+1)} & \cdots & s - a_{NN} \end{vmatrix}$$

If all the $a_i = 0$ except for $i = J$, then we obtain

$$I_{iJ} = \mathcal{L}^{-1} \frac{U_i}{a_J}$$

Example 2.7.1 Find the impulsive response I_{11} of the OLDE

$$\frac{dz_1}{dt} + 3z_1 = y(t)$$

Solution $z_1 = u_1$, $r_1 = y(t) = a_1 \delta(t)$ and thus

$$\dot{u}_1 = --3u_1 + r_1$$

$$U_1 = \frac{a_1}{(s + 3)}$$

from which it immediately follows that

$$I_{11}(t) = \mathcal{L}^{-1}\left(\frac{U_1}{a_1}\right) = \exp\ -3t$$

Example 2.7.2 Find the impulsive responses I_{12} and I_{22} of the system represented by

$$\frac{d^2 z_1}{dt^2} + 8\frac{dz_1}{dt} + 15z_1 = y(t)$$

Solution

$$z = u_1$$

$$\dot{u}_1 = u_2$$

$$\dot{u}_2 = -15u_1 - 8u_2 + r_2(t)$$

$$\mathcal{L}I_{12} = \frac{1}{s^2 + 8s + 15} = \frac{1}{(s+3)(s+5)}$$

$$\mathcal{L}I_{22} = \frac{s}{(s+3)(s+5)}$$

Thus

$$I_{12} = \frac{1}{2}(e^{-3t} - e^{-5t})$$

$$I_{22} = -\frac{1}{2}(3e^{-3t} - 5e^{-5t})$$

I_{12} is the system response and I_{22} is the derivative of I_{12}.

Example 2.7.3 Obtain the expression for the impulsive response of the output z_1 with respect to the input function $y_2(t)$ if

$$\frac{d^2z_1}{dt^2} + 3\frac{dz_2}{dt} + 7\frac{dz_1}{dt} + z_1 = y_1(t)$$

$$\frac{d^2z_2}{dt^2} + \frac{dz_1}{dt} + 4z_2 = y_2(t)$$

Solution The state variable presentation of the performance equations is

$$\begin{aligned}
\dot{u}_1 &= u_3 \\
\dot{u}_2 &= u_4 \\
\dot{u}_3 &= -3u_4 - 7u_3 - u_1 + y_1(t) \\
\dot{u}_4 &= -4u_2 - u_3 + y_2(t)
\end{aligned}$$

with a characteristic polynomial

$$\Delta = s^4 + 7s^3 + 2s^2 + 28s + 4$$

The impulsive response I_{14} is required; hence

$$\mathcal{L}I_{14} = \frac{1}{\Delta}\begin{vmatrix} 0 & 0 & -1 & 0 \\ 0 & s & 0 & -1 \\ 0 & 0 & s+7 & 3 \\ 1 & 4 & 1 & s \end{vmatrix}$$

$$= \frac{-3s}{s^4 + 7s^3 + 2s^2 + 28s + 4}$$

The quartic can be factored[23] by exact methods; thus

$$\mathcal{L}I_{14} = \frac{-3s}{(s+0.28)(s+0.88)(s^2+7.96s+16.5)}$$

which can be inverted into

$$\begin{aligned}
I_{14} = {} & 0.074e^{-0.28t} - 0.180e^{-0.88t} \\
& + 0.73e^{-3.98t}\cos(0.78t + \theta) \qquad \theta = 1.72 \text{ radians}
\end{aligned}$$

2.8 Transfer Matrices

The transformed relationship between the several inputs and the output state variables is given, whenever the initial state is zero, by Equation (2.6.1):

$$U_i = \sum_{j=1}^{N} R_j D_{ij} \qquad i = 1, 2, \ldots, N$$

The first n transformed variables represent the system outputs z_1, $z_2, \ldots, z_n$, and the remaining quantities ($i > n$) are the transformed derivatives of the state variables. For design purposes, the knowledge of the $n \times N$ transforms D_{ij} is sufficient to fix the various outputs. If the initial state is known and if the system variables can be exhibited in a solution, then it is not necessary to compute all the state variables. In addition, it can be seen that there are only n functions $r_i(t)$ corresponding to the input quantities y_i in the form

$$r_N(t) = y_n(t) \qquad r_{N-1}(t) = y_{n-1}(t), \ldots, r_{N-n+1} = y_1(t)$$

and $r_i(t) = 0$ for $i < n$.

We can define a transformed system of the form

$$z_i = \sum_{j=1}^{n} (\mathcal{L} y_j) D_{ij} \tag{2.8.1}$$

which corresponds to the solution

$$z_i = \sum_{j=1}^{n} \int_{0^+}^{t} y_i(t - \tau) I_{ij}(\tau)\, d\tau \tag{2.8.2}$$

If the $n \times n$ quantities D_{ij} are known, and hence the $I_{ij}(\tau)$, then for any set of input functions y_i the system solutions z_i are known. If a matrix of the relevant transforms is given

$$\mathbf{T} = \begin{bmatrix} D_{11} & D_{12} & \cdots & D_{1n} \\ D_{21} & D_{22} & \cdots & D_{2n} \\ \cdots & \cdots & \cdots & \cdots \\ D_{n1} & D_{n2} & \cdots & D_{nn} \end{bmatrix}$$

then the system dynamic characteristics are known uniquely. The matrix $\mathbf{T}$ is known as the transfer matrix of the system and each D_{ij} is known as the transfer function between the ith input and the kth output.

Example 2.8.1 Obtain the transfer matrix of the following system:

$$\dot{u}_1 = u_3$$

$$\dot{u}_2 = u_4$$

$$\dot{u}_3 = -u_2 - 2u_3 + y_1(t)$$

$$\dot{u}_4 = -u_2 - 2u_3 + y_2(t)$$

Solution We need the transforms for u_1 and u_2. These are obtained directly

$$U_1 = \frac{\begin{vmatrix} 0 & 0 & -1 & 0 \\ 0 & s & 0 & -1 \\ Y_1 & 3 & s+2 & 0 \\ Y_2 & 1 & 2 & s \end{vmatrix}}{\begin{vmatrix} s & 0 & -1 & 0 \\ 0 & s & 0 & -1 \\ 0 & 3 & s+2 & 0 \\ 0 & 1 & 2 & s \end{vmatrix}}$$

$$= \frac{Y_1(s^2 + 1) - 3Y_1}{s(s^3 + 2s^2 + s - 4)}$$

By a similar procedure we obtain

$$U_2 = \frac{-2sY_1 + (s + 2)Y_2}{s(s^3 + 2s^2 + s - 4)}$$

and thus the elements of **T** are found

$$D_{11} = \frac{s^2 + 1}{s(s^3 + 2s^2 + s - 4)}$$

$$D_{12} = \frac{-3}{s(s^3 + 2s^2 + s - 4)}$$

$$D_{21} = \frac{-2s}{s(s^3 + 2s^2 + s - 4)}$$

$$D_{22} = \frac{s + 2}{s(s^3 + 2s^2 + s - 4)}$$

2.9 State and Measurement

On the somewhat infirm base of linear theory one may argue that the practical design of instruments is somewhat removed from the concept of state. As was demonstrated in the previous section, the useful design equations are the n relationships which connect the various inputs to the several outputs. In fact, since the state of the system is directly connected to the initial conditions and since in instrument design one usually sets the initial conditions to zero, considerable labor can be saved by ignoring the transformation to state variable form. In sophisticated design practice, however, it is usually necessary to consider both the state performance equations and the system performance equations. If for all valid inputs the instrument can be shown to obey a set of differential equations which are always linear with constant coefficients, it is easy to demonstrate that no real advantage is gained from the state variable representation. If, on the other hand, the system is modeled by a nonlinear set of ordinary differential equations, it can be demonstrated[24] that to discuss the stability of the system or to prove that the state of the system is uniquely predictable, the state variables presentation is almost necessary.

Consider the solution of the system in state form

$$u_k = u_k[u_1(0), u_2(0), \ldots, u_N(0); r_1(t), r_2(t), \ldots, r_N(t); t]$$

in which u_k is a set of unique relationships theoretically obtainable from the original set of differential equations. A necessary condition for measurability of a set of inputs is that the several outputs reach values which are independent of the initial conditions. Thus it is required that there exist a set of unique functions g_k such that the limit

$$\lim_{t\to\infty} u_k(t) = g_k[r_1(t), r_2(t), \ldots, r_N(t)]$$

exists irrespective of the initial state, and such that the form of g_k is independent of time.* In the study of linear systems a simple test can be applied to determine whether the limit exists and thus to discriminate between stable and unstable behavior.[25] In the study of nonlinear systems, however, such a clear decision cannot be made. In some systems a particular set of initial conditions can lead to a set of functions $g_k^{(1)}$, while another set of initial conditions can lead to a completely different ultimate state $g_k^{(2)} \neq g_k^{(1)}$. Occasionally a very small disturbance in one or more of the $u_k(t_0)$ can lead to a drastic error in one or more of the ultimate states. Meaningful examination of nonlinear states can best be carried out by first writing the system equations in state form.

2.10 The Block Diagram

The signal flow or energy path diagram (chap. 1) is a graphical model of a measurement system. Each block represents an element or a combination of elements with a single output and a single input. The directions of signals are marked by arrows and the functional behavior indicated in some fashion in the box (Fig. 2.2a). For a linear system it is simpler to use $\mathcal{L}$ transformed quantities with the transfer function $\mathcal{L}z_{12}/\mathcal{L}z_{11} = G(s)$ written in the box. With this convention the input transform is "transferred" to the output transform by the multiplication $(\mathcal{L}z_{11})G = \mathcal{L}z_{12}$ (Fig. 2.2b). If the relationship between input and output is nonlinear and describable only by a discontinuous function, a graph of the functional relationship is drawn in the box. For example, if the relationship is

$$z_{12} = \begin{cases} a & z_{11} \geq 0 \\ -a & z_{11} < 0 \end{cases}$$

*This means that if we wait long enough, say, a time t_1, we can make $|u_k - g_k|$ as small as we wish. Good design requires that for a preselected $|u_k - g_k|$, t_1 be made as small as possible.

the block diagram often appears as in Figure 2.2c. A triangle is used to indicate a summer or a summing junction both with transformed and untransformed functions (Fig. 2.2d). A triangle with an M in it is used to represent a multiplier (Fig. 2.2e). Other symbols will be introduced as needed.

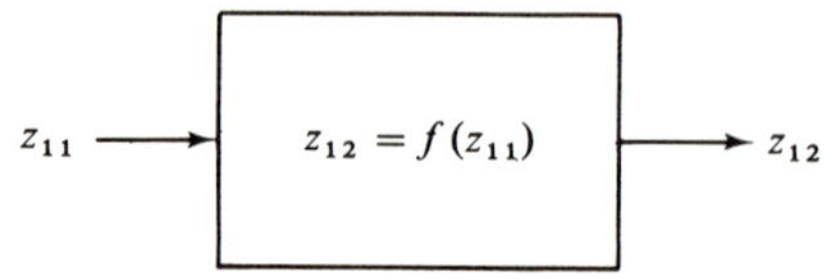

(a) Block element representation, general

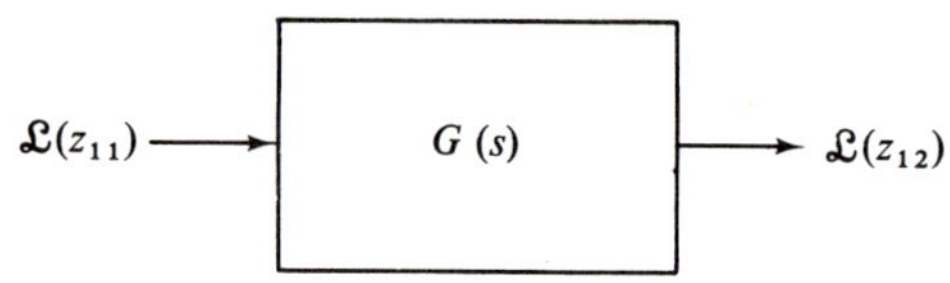

(b) Block element representation, linear transform

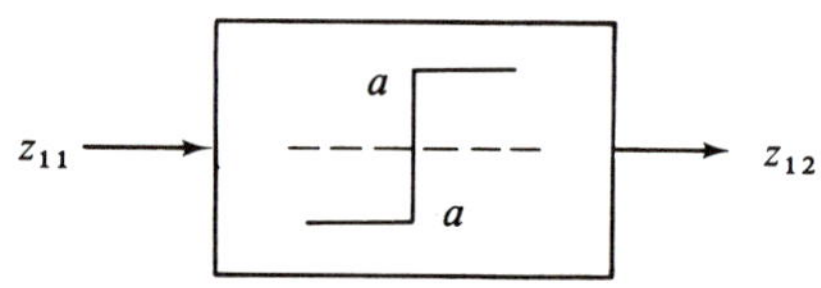

(c) Block element representation, discontinuous function

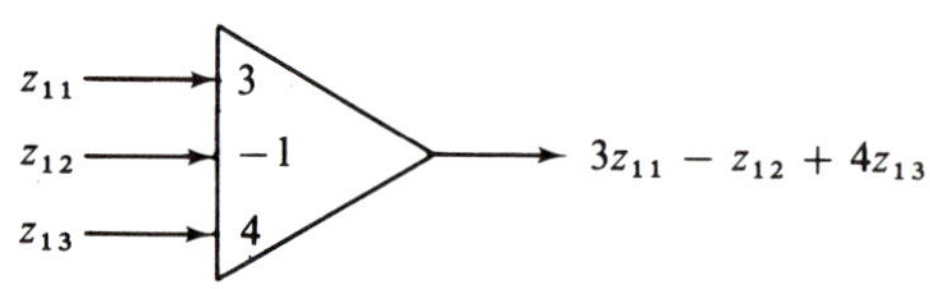

(d) Summer

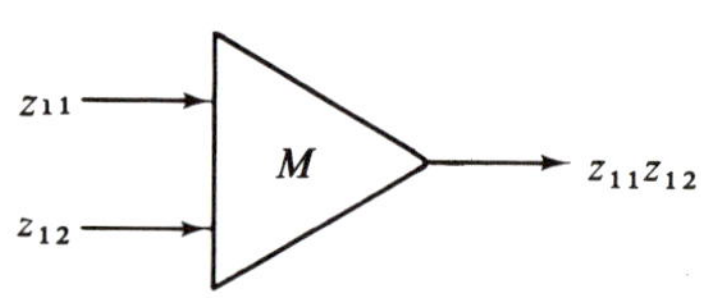

(e) Multiplier

FIGURE 2.2 Block diagram conventions

Example 2.10.1 A signal has been shown to follow the paths represented in Figure 2.3. Obtain an expression relating the input y to the output z.

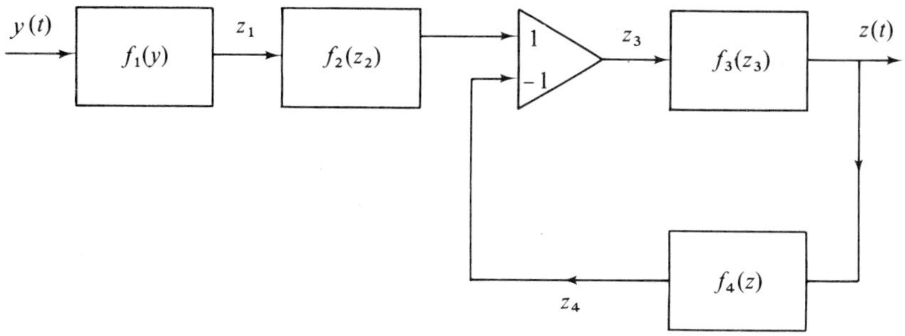

FIGURE 2.3 Block diagram for Examples 2.10.1 and 2.10.2

Solution The element equations are

$$z_1 = f_1(y) \qquad z_2 = f_2(z_1) \qquad z = f_3(z_3) \qquad z_4 = f_4(z)$$

The summing point equation is

$$z_3 = z_2 - z_4$$

We now eliminate the z_i and obtain the relationship between z and y

$$z = f_3\{f_2[f_1(y) - f_4(z)]\}$$

Example 2.10.2 In the system shown in Figure 2.3 the relationships are given by the differential equations

$$\frac{dz_1}{dt} + 3z_1 = y$$

$$\frac{dz_2}{dt} + z_2^2 = 4z_1$$

$$\frac{dz}{dt} + 7z = 2z_3$$

$$z_4 = 9z$$

Obtain the state variables expression in terms of z and its derivatives.

Solution We clear variables out of the first two equations

$$\frac{d}{dt}\left(\frac{1}{4}\frac{dz_2}{dt} + \frac{z_2^2}{4}\right) + \frac{3}{4}\left(\frac{dz_2}{dt} + z_2^2\right) = y$$

$$\frac{d^2z_2}{dt^2} + 2z_2\frac{dz_2}{dt} + 3\frac{dz_2}{dt} + z_2^2 = 4y$$

But $z_2 - z_4 = z_3$. Thus

$$z_2 = 9z + \frac{1}{2}\left(\frac{dz}{dt} + 7z\right) = 16z + \frac{1}{2}\frac{dz}{dt}$$

from which it immediately follows that

$$\frac{d^3z}{dt^3} + \left(\frac{dz}{dt} + 32z + 35\right)\frac{d^2z}{dt^2} + 32.5\left(\frac{dz}{dt}\right)^2 + (1056z + 96)\frac{dz}{dt} + 512z^2 = 8y$$

The state variables are $z = u_1$, $dz/dt = u_2$ and $d^2t/dt^2 = u_3$, whence the answer is

$$\begin{aligned}\dot{u}_1 &= u_2\\ \dot{u}_2 &= u_3\\ \dot{u}_3 &= -(u_2 + 32u_1 + 35)u_3 - 32.5u_1^2\\ &\quad - (1056u_1 + 96)u_2 - 512u_1^2 + 8y\end{aligned}$$

Example 2.10.3 Draw a block diagram for the system of Equation (2.8.1) under the assumption that the $D_{ij}(s)$ represent the transforms of the system equations of specific elements.

Solution The transformed equation has the form

$$\mathfrak{L}z_i = \sum_{j=1}^{N} (\mathfrak{L}y_j)D_{ij}(s)$$

If we now let $\mathfrak{L}z_i = Z_i$ and $\mathfrak{L}y_i = Y_i$, we obtain the block diagram shown in Figure 2.4.

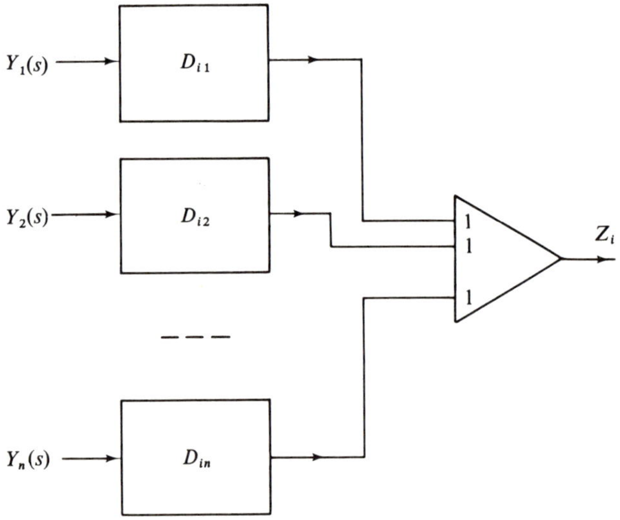

FIGURE 2.4 Block diagram for Example 2.10.3

Example 2.10.4 The components of a measuring system are interconnected as indicated by the equations. Obtain a signal flow graph for the system.

$$\frac{dz_1}{dt} + 0.06z_1 = 2y_1$$

$$\frac{dz_2}{dt} + 0.03z_2 = 1.25y_2$$

$$1.75z_1 - 3.11z_2 = z_3$$

$$\frac{d^2z_3}{dt^2} + 0.213\frac{dz_3}{dt} + 0.717z_3 = 2z$$

Label z, z_1, z_2, z_3 and y at the appropriate junctions.

Solution The signal flow is $y_1 \longrightarrow z_1$, $y_2 \longrightarrow z_2$, $(z_1, z_2) \longrightarrow z_3$, $z_3 \longrightarrow z$ and thus the block diagram is as shown in Figure 2.5 with

$$G_1(s) = \frac{2}{s + 0.06}$$

$$G_2(s) = \frac{1.25}{s + 0.03}$$

$$G_3(s) = \frac{2}{s^2 + 0.2135 + 0.717}$$

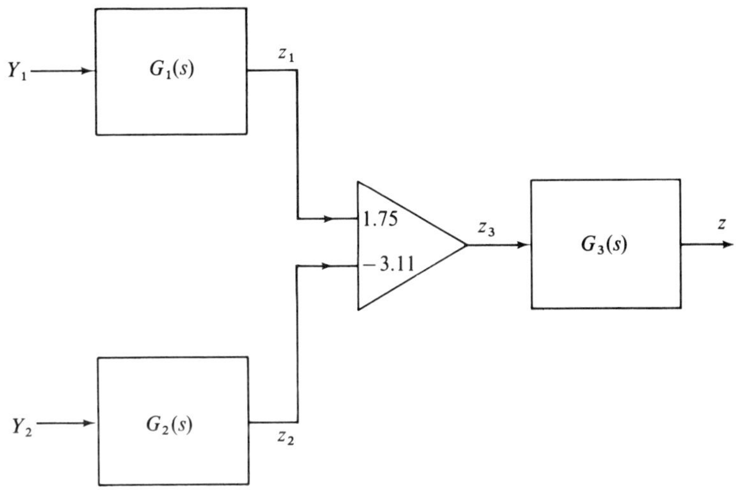

FIGURE 2.5 Block diagram for Example 2.10.4

Example 2.10.5 A system obeys the following set of component equations:

$$y = z \mid z_1$$

$$z_2 = \begin{cases} 0 & z_1 \leq 0 \\ 3 & z_1 > 0 \end{cases}$$

$$z = z_2 z_3$$

$$z = \frac{dz_3}{dt} + 0.293z_3$$

Draw a block diagram or signal flow graph of the system.

Solution See Figure 2.6.

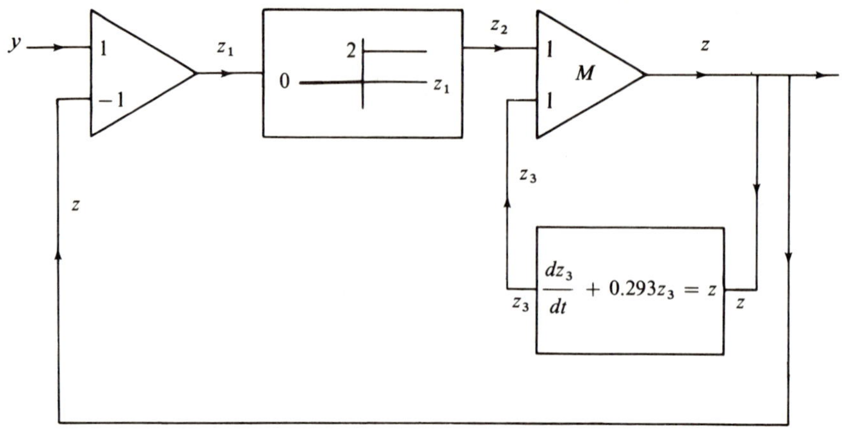

FIGURE 2.6 Signal flow path for Example 2.10.5

PROBLEMS

2.1 Reduce the following to the state variable form:

$$\ddot{z}_1 + \dot{z}_2\dot{z}_1 + 2.91\dot{z}_2 + z_1 z_2 = 2.77y_1$$
$$(1 - \dot{z}_1)\ddot{z}_2 + \dot{z}_2 + 7.34\dot{z}_1 + z_2^3 = 2.34y_2$$

2.2 Reduce to state variable form.

$$\dot{z}_1 + 0.702z_1 z_2 = y_1$$
$$\dot{z}_2 + 0.391z_2 z_3 = y_2$$
$$\ddot{z}_3 + z_1\dot{z}_3 + \dot{z}_2 + z_3 = y_3$$

2.3 Show how the system represented in problem 2.1 fails to meet a Lipschitz condition at least at one state.

2.4 In the system modeled in problem 2.2, let $z_1 = 1$, $\dot{z}_3 = 0$, $z_2 = 0$, $z_3 = 1$ at $t = 0$, and let $y_1 = 0$, $y_2 = 0$, $y_3 = 4$ for all times. Find the Taylor series solution for z_1, z_2 and z_3 valid near the initial state. Expand only to the term of order t^4.

2.5 Represent the following set of OLDE in state variable form:

$$\ddot{z}_1 + 2.71\dot{z}_2 + 3.14\dot{z}_1 + 0.711z_1 = 3.25y_1(t)$$
$$\ddot{z}_2 + 2.87\dot{z}_2 + z_1 + 9.76z_2 = y_2(t)$$

2.6 Obtain the differential equation for z_1 alone in problem 2.5 and exhibit it in state variable form.

2.7 Obtain the transforms I_{13} and I_{33} for the system in problem 2.5.

2.8 Obtain the transfer matrix for the system shown in problem 2.5.

2.9 Obtain the system equation $z = f(y)$ for the block diagram shown in Figure 2.7.

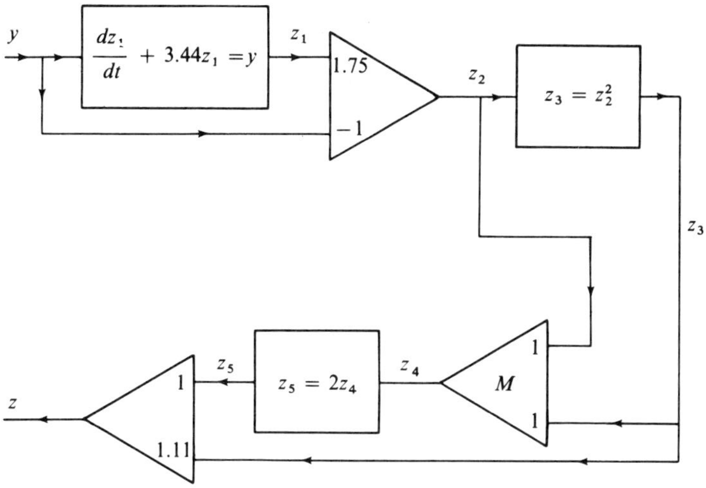

FIGURE 2.7 Block diagram for problem 2.9

2.10 A system is modeled by the following equations:

$$z_1 = y - z \qquad \dot{z}_2 + 2.21z_2 = z_1$$
$$2.66z_1z_2 = z_3 \qquad \dot{z}_4 + 0.002z_4 = z_3$$
$$z_4 + z_2z_3 = z$$

Draw a block diagram of the system.

NOTES

[1]Although the statistic used is often the mean of the state variables taken over time, other statics can be used as long as the numbers are uniquely derivable from the state and vice versa.

[2]"Stochastic" is a construct from the Greek word *stokasticos*, a skillful marksman. The implication is obvious and the suspicion evident.

[3]Ref. 57.

[4]Ref. 56, pp. 117–59.

[5]Perhaps we should say "has been exposed to."

[6]Ref. 45, pp. 185–89.

[7]Ref. 60.

[8]Ref. 53 is a single example of complete analysis. The author discusses a complicated distributed system and then reduces the scope of the problem by linearization. See also Ref. 54, 55.

[9]The coupling determines the form of the differential equation in one variable and is implicit in the "uncoupled" system as a pattern (q.v.).

[10]For proof that a solution exists see discussion in Section 2.3 and Ref. 14, p. 5.

[11]We arbitrarily exclude "negative time."

[12]Ref. 58, pp. 8, 9, 37, 41. Also Ref. 14, pp. 4–6.

[13]Ref. 58, p. 8. For restrictions weaker than the Lipschitz conditions, see Ref. 58, pp. 48–52.

[14]Ref. 14, pp. 6 and 10.

[15]Existence theorems are licenses to hunt for a solution.

[16]Existence proofs can be "constructive" in the sense that a solution can be constructed from elements of the proof. See, for example, Ref. 58, p. 8, theorem 2.1.

[17]Ref. 34.

[18]In recent practice matrix notation has come into use. The great compactness and the elegant theory connected with the matrix calculus makes the technique very attractive. In a book of this nature, however, matrix methods are superfluous. Matrix techniques relevant to our subject are discussed in Ref. 59, chapters 4 and 5.

[19]All the known linear techniques, such as the Nyquist-Bode plot method and the root-locus procedure can be derived elegantly without recourse to transforms. To neglect transforms, however, is a disservice to the reader since most modern systems literature is steeped in transform techniques.

[20]Ref. 34, p. 206.

[21]Ref. 57, pp. 128–51.

[22]Ref. 60, pp. 16–21.

[23]The method of Ferrari, Ref. 34, pp. 94–98.

[24]Refs. 14, 61, 62.

[25]Chap. 4.

3

SENSITIVITY

3.1 Introduction

It is commonly agreed that "sensitivity" is a measure of instrument effectiveness, a relationship between input and output which can be used to describe the performance of instruments. When it comes to a usable technical definition, however, no agreement exists. Some authors define sensitivity as the ratio of output to input,[1] others as the derivative of output with respect to input,[2] while still others simply do not define sensitivity at all. In commercial practice, especially in the electronics field, sensitivity has two completely different meanings. Some manufacturers define sensitivity as the ratio of output to input under static conditions, while others call sensitivity what we have called *threshold*.[3] The consensus seems to be in favor of the ratio of output deflection to input change, that is, the unit output corresponding to the unit input under static conditions.

The confusion of definitions is natural and will continue, since it is possible to show that there is no single parameter which can be used to indicate the relationship between output and input under all circumstances. The difficulties arise from the dynamics of the system and the nonlinearities inherent in the instrument configuration. It will be shown that in the case of time-varying inputs there exists no time-invariant function connecting the input and output. Indeed it will be shown that the output is a time-varying function of the input and all the derivatives of the input, and that the single sensitivity is at best a crude approximation and at worst an outright fraud. In instruments with strong nonlinearities it will be shown that even with steady inputs and outputs no single number can be given as the "sensitivity" of the instrument.

3.2 Definitions

In practice, it is hoped to design an instrument to give a response

$$\xi = Sy(t)$$

in which S is a constant. Unfortunately, such a response is not completely achievable and what is actually obtained (at least in a linear system) is

$$\xi(t) - S(t)y(t) + f(t) \tag{3.2.1}$$

The quantity S varies with time and the output response has a superposed dynamic error f.

In most well-designed systems f becomes less than a preassigned value ϵ_1, and S approaches a constant plus or minus a preassigned value ϵ_2 after a "response time" t_o, so that at any time $t \geq t_o$ we have

$$\xi(t) = [S(\infty) + \epsilon_2]y(t) + \epsilon_1$$

in which ϵ_1 and ϵ_2 approach zero as t approaches infinity. Clearly the smaller is t_0 the faster is the instrument response and the nearer to an ideal instrument is the measuring system.

If the quantity $y(t)$ is steady (not a function of time), then the time t_0 is usually not important. The instrument operator can wait a sufficient time for the instrument to settle to its ultimate state, that is, the final value of $\xi(t)$ under the input $y(t)$. If $y(t)$ is a function of time, the system would be designed in such a manner that all transients* vanish before $y(t)$ becomes greater than the design threshold. Part of the design procedure, therefore, is to select the instrument system parameters in such a fashion that within the limits of the accuracy and precision of the instrument

$$\xi = Sy(t)$$

in which $S = \lim_{t\to\infty} S(t)$ is a constant throughout the relevant portion of the measurement procedure.

**Transients* are those portions of the instrument output which vanish in time. The component of the output which persists after the suppression of the transients is called the *ultimate state.*

3.2.1 In commercial practice the "sensitivity" S is always presented as an ultimate state value. Usually a single number is given as the ratio of output indication to input. This number is often meaningless unless the "linearity" of the sensitivity parameter is indicated.

In commercial instruments an average linear relationship is obtained by defining a "best fit" straight line through the actual ultimate state output vs. input curve. The nonlinearity is then computed by drawing two parallel lines (Fig. 3.1a) between which the actual curve is contained. The distance between the two lines, stated in terms of a percentage of full scale output or full scale input, is called the *maximum nonlinearity* of the system. It can be seen, therefore, that a "best fit" straight line is the line which minimizes the nonlinearity.

Sensitivity is measured by introducing a constant input and recording the ultimate state output deflection as in Figures 3.1a and 3.1b. The

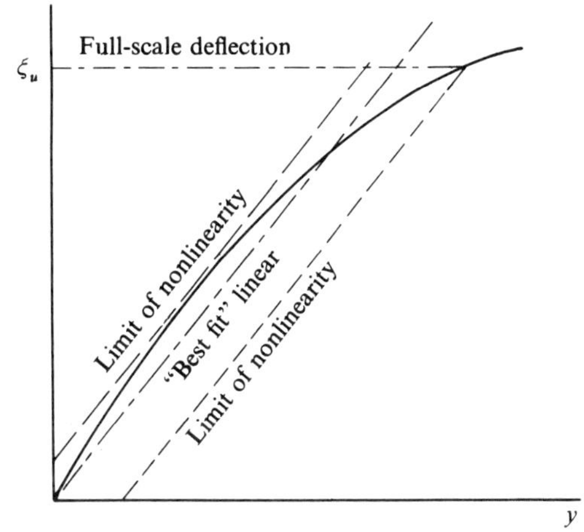

FIGURE 3.1a Nonlinear ultimate state response

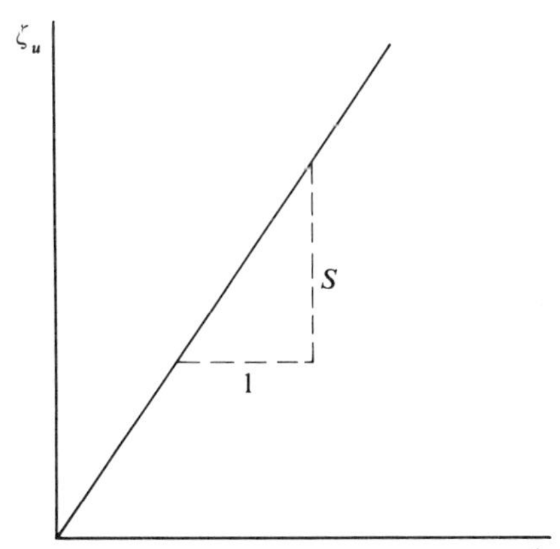

FIGURE 3.1b Linear ultimate state response

relationship may be linear (as in 3.1b) or nonlinear (3.1a). In the linear case there is little or no argument as to the meaning of ultimate state sensitivity $S = \xi_u/y$ in which ξ_u represents the ultimate state of ξ for a constant y. In the nonlinear case, S is given an interpretation largely according to whim. The honest way would be to give the ξ_u vs. y graph for the particular instrument: this unfortunately is almost never done. Instead, a single number is found which, in the practice of the firm, is called *sensitivity*. Commonly three definitions of sensitivity are found in the literature:[4]

(a) *Tangent* or "true" *sensitivity*: simply the slope of the output-input curve at a point. Plotted as a function of input, the tangent sensitivity is the best representation of local sensitivity. $S_L = \partial\xi/\partial y$.

(b) *Chord Sensitivity*: a representative, nominal sensitivity (Fig. 3.2) given by $S_c = \xi_{u1}/y_1$. Usually the slope of the straight line is given together with the maximum deviation of the sensitivity from the straight line.

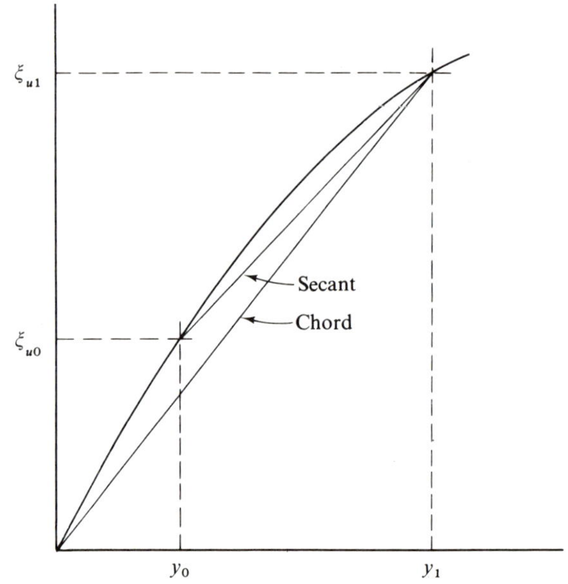

FIGURE 3.2 Static sensitivity

(c) *Secant Sensitivity*. If the relevant measurements are to be made in a region or band of acceptable inputs, say,

$$y_0 \leq y \leq y_1$$

it is occasional practice to define a secant sensitivity

$$S_s = \frac{\xi_{u1} - \xi_{u0}}{y_1 - y_0}$$

It is also possible to divide the range of the instrument into several subranges and to give different secant sensitivities and deviations from the linear in each range.

It is obvious that if the instrument is truly linear, $S_L = S_c = S_s = S$ and the ultimate state sensitivity is unique.[5]

In this book we shall always define the *static sensitivity* (ultimate state sensitivity) as

$$S = \lim_{t \to \infty} \left(\frac{\partial \xi}{\partial y} \right)$$

It will be shown that this quantity is a function of the instrument configuration alone and that under certain circumstances S can be a system constant.

Example 3.2.1 A pendulum is suspended from a pivot which generates a linear torque $-C\dot{\theta}$ in which θ is the angular displacement.

The complete assembly is restricted to translation in a vertical plane and is accelerated with an acceleration (**a**) on the same plane. Show that this device can be used to detect acceleration and obtain the static sensitivity of the system (Fig. 3.3).

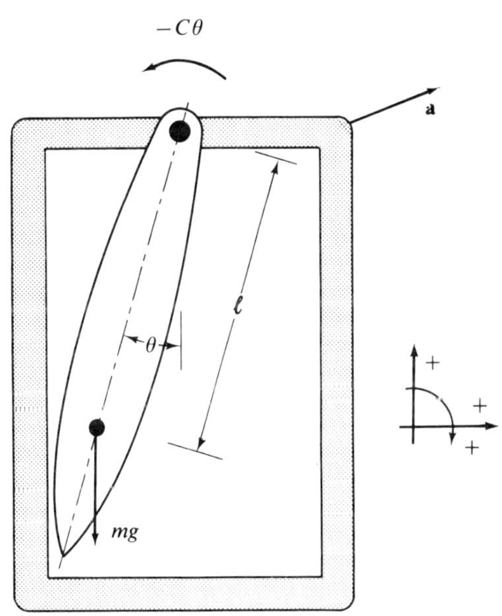

FIGURE 3.3 Pendulous accelerometer

Solution To obtain the equation of motion we use Lagrange's equation[6]

$$\frac{d}{dt}\left(\frac{\partial T}{\partial \dot{\theta}}\right) - \frac{\partial T}{\partial \theta} = Q$$

in which T is the kinetic energy of the system and Q represents the applied torques. From Figure 3.3 we obtain

$$T = \frac{1}{2}m(\dot{x}^2 + \dot{y}^2) + \frac{1}{2}I_0\dot{\theta}^2$$

where

I_0 = moment of inertia about center of gravity
m = mass of pendulum

From the geometry of the system it follows that

$$\dot{x} = \dot{u} - \ell\dot{\theta}\cos\theta$$

$$\dot{y} = \dot{v} + \ell\dot{\theta}\sin\theta$$

in which u and v are the horizontal and vertical displacements of the pivot, respectively. Thus we see that

$$T = \frac{1}{2}m[(\dot{u} - \ell\dot{\theta}\cos\theta)^2 + (\dot{v} + \ell\dot{\theta}\sin\theta)^2] + \frac{1}{2}I\dot{\theta}^2$$

Application of Lagrange's equation yields

$$(I_0 + m\ell^2)\ddot{\theta} + m\ell(\ddot{v}\sin\theta - \ddot{u}\cos\theta) = Q$$

Q is easily obtained by observation

$$Q = -C\dot{\theta} - mg\ell\sin\theta$$

If we let $I = I_0 + m\ell^2$, we obtain the performance equation of the system

$$I\ddot{\theta} + C\dot{\theta} + mg\ell\sin\theta = -m\ell(\ddot{v}\sin\theta - \ddot{u}\cos\theta)$$

It is obvious that the quantity $\theta(t)$ is a function of both the acceleration components $\ddot{u}$ and $\ddot{v}$.

The static sensitivity is obtained by setting $\ddot{\theta} = \dot{\theta} = 0$ and solving for $\theta = \theta_u$:

$$\tan\theta_u = \frac{\ddot{u}}{g + \ddot{v}}$$

This result shows the difficulties encountered in the use of this device. θ_u depends nonlinearly on $\ddot{v}$ as well as on $\ddot{u}$. Hence the pendulum is a device for detecting one component of acceleration only if $\ddot{v} \ll g$. Let us find the static sensitivity of the pendulum to $\ddot{u}$.

$$S_u = \frac{\partial\theta}{\partial\ddot{u}} = \frac{\cos^2\theta}{g\ddot{v}}$$

Thus

$$S_u = \frac{g + \ddot{v}}{(g + \ddot{v})^2 + \ddot{u}^2}$$

The sensitivity of the system is not a constant but is a function of $\ddot{v}$ and $\ddot{u}$. This static sensitivity is plotted in Figure 3.4 in a dimensionless form

$$gS_u = \frac{1 + (\ddot{v}/g)}{[1 + (\ddot{v}/g)^2 + (\ddot{u}/g)]^2}$$

At low horizontal acceleration $\ddot{u}$ the response is not single valued. At

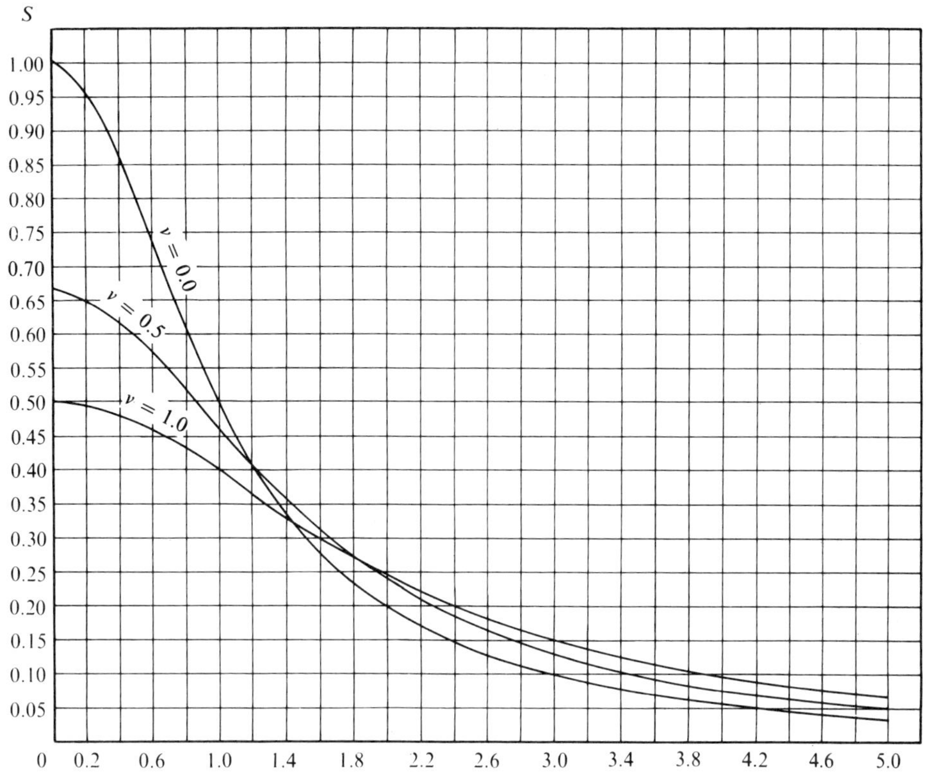

FIGURE 3.4 Sensitivity of pendulous accelerometer

high horizontal acceleration the "accelerometer" becomes insensitive to cross acceleration, but the system sensitivity goes down.[7] We must conclude that the pendulum is not a very good accelerometer.

3.3 Dynamic Theory of Sensitivity

Consider a system representable by a linear ordinary differential equation with constant coefficients in which ξ is the output indication and y is the input expected value.

$$\sum_{n=0}^{N} a_n \frac{d^n \xi}{dt^n} = Ky(t) \tag{3.3.1}$$

It is assumed that the instrument is at rest at time zero, that ξ and all its derivatives are zero initially and that $y(t)$ is "switched on" at time zero. This last assumption is made to avoid discussion of a measurement made on an instrument initially not at rest. It will be shown later that the effects of initial nonzero conditions can be treated separately without affecting the conclusions of this section.

As was shown in the previous chapter, we can write the solution of the differential equation in the form

$$\xi(t) = \int_{0^+}^{t} I(\tau)y(t - \tau)\, d\tau \tag{3.3.2}$$

in which $I(t)$ is the impulsive response, the solution of Equation (3.3.1) for $y(t) = \delta(t)$.

Let us suppose that $y(t)$ is continuous and possesses continuous derivatives of all orders [$y(t)$ is analytic]. Then it is possible to expand $y(t - \tau)$ into a uniformly convergent series in τ[8]

$$y(t - \tau) = \sum_{n=0}^{\infty} (-1)^n \frac{d^n y}{dt^n}\bigg|_{\tau=0} \frac{\tau^n}{n!}$$

so that upon substitution into (3.3.2) we obtain

$$\xi = \sum_{n=0}^{\infty} S_n(t) \frac{d^n y(t)}{dt^n} \tag{3.3.3}$$

where

$$S_n(t) = (-1)^n \int_0^t I(\tau)\frac{\tau^n}{n!}\, d\tau \tag{3.3.4}$$

The functions $S_n(t)$ depend only on the instrument parameters and are completely independent of the input function. If we let

$$y^{(n)} = \frac{d^n y}{dt^n}$$

we see immediately that

$$S_n = \frac{\partial \xi}{\partial y^{(n)}}$$

and that the S_n represents the *sensitivity* of the instrument to the nth derivative of the input function. It appears, therefore, that the expected value of the output ξ is a function not only of the input quantity but also of the instrument parameters and of all the derivatives of the input function.

3.3.1 If we now compare Equations (3.2.1) and (3.3.3), it is evident that

$$\xi(t) = S_0(t)y(t) + \sum_{n=1}^{\infty} S_n(t)y^{(n)}(t)$$

and thus that the conditions for a perfect instrument are

$$S_0(t) = S = \text{constant}$$

$$S_n(t) = 0 \qquad n > 0$$

Unfortunately such conditions are never realizable since $I(t)$ has a solution of the form (see chapter 2 and Appendix B)

$$I(\xi) = \sum e^{\alpha_j t} (A_j \cos \beta_j t + B_j \sin \beta_j t)$$

and that at best the $S_n(t)$ vanish exponentially. All that can be done from a design point of view is to pick system parameters such that $I(t)$ vanishes in a negligible time with respect to the speed of the input.

Example 3.3.1 A system obeys the differential equation

$$\tau_0 \frac{d\xi}{dt} + \xi = 2y(t)$$

Select the relevant parameter τ_0 to make $I(t) < 0.01$ after 0.1 second.

Solution From chapter 2 we see that

$$I(t) = \frac{2}{\tau_0} e^{-\tau/\tau_0}$$

Substitution of the design conditions gives

$$\tau_0 \leq 200e^{-0.1/\tau_0}$$

By trial and error we find that $\tau_0 \leq 0.010107$ second. If we pick $\tau_0 =$ 0.01 we obtain a reasonable solution.

Example 3.3.2 Determine the sensitivity functions for the system considered in the previous example with $\tau_0 = 0.01$. If $y = 3t$, determine the error from the ideal response at 0.1 second and at 1 second (to the fifth decimal).

Solution

$$S_n = (-1)^n \int_{0^+}^{t} 200e^{-100\tau} \frac{\tau^n}{n!}\, d\tau$$

For our problem we need only evaluate S_0 and S_1:

$$S_0 = 2(1 - e^{-100t})$$

$$S_1 = 2[e^{-100t}(t + 0.01) - 0.01]$$

To five decimals we can compute

t	S_0	S_1	ξ
0.01	1.26624	−0.02000	0.02215
1.0	2.00000	−0.02000	5.94000

in which $\xi = 3tS_0 + 3S_1$.

In an ideal system $\xi_I = 2(3t)$ and thus, writing the relative error

$$\delta = \left| \frac{\xi_I - \xi}{\xi_I} \right|$$

we obtain

t	ξ	ξ_I	δ
0.01	0.02215	0.06000	0.63083
1.0	5.94000	6.00000	0.01000

At $t = 0.01$ s the relative error is 63.1% of the expected value of the output while at $t = 1.0$ s the error has gone down to 1%. It is easy to show that after $t = 1$ s, $\xi_I - \xi$ approaches 0.06 with increasing accuracy and the relative error goes down. For example, after 10 seconds the relative error will be only 0.1%. From a measurement point of view we have presented a fairly slow system.

Such a first-order system can represent the performance equation of a thermometer; and if we consider the rather slow configuration changes in thermal measurement, a system constant of 0.01 is faster than is normally needed. It is not possible, however, to decide if the instrument is fast enough without first discussing the speed range of the measurand. If, for example, it were necessary to use a thermal transducer to detect the passage of a shock wave in a shock tube, a time constant of order 10^{-2} s would be insufficient.

3.4 Sensitivity Test Functions

With the exception of $S_0(t)$ it is not easy to make direct measurements of the sensitivity functions. The difficulty is inherent in the functions themselves. If we write the S_n in a form which exhibits the input functions necessary to make a particular $S_n(t)$ the output of the instrument, the difficulty will become evident. We can show that

$$(-1)^n\tau^n = (-\tau)^n = [(t-\tau)-t]^n$$
$$= \sum_{m=0}^{n} (-1)^m (t-\tau)^{n-m} t^m \frac{n!}{(n-m)!\,m!}$$

Thus it follows from Equation (3.3.4) that

$$S_n(t) = \sum_{m=0}^{n} \frac{(-1)^m t^m}{m!} A_{n-m}(t) \qquad \textbf{(3.4.1)}$$

in which the $A_{n-m}(t)$ are called the sensitivity *test* functions

$$A_{n-m} = \int_{0^+}^{t} I(\tau)\frac{(t-\tau)^{n-m}}{(n-m)!}\,d\tau \qquad \textbf{(3.4.2)}$$

Since the $A_{n-m}(t)$ represent the response of the instrument to the input function $t^{n-m}/(n-m)!$ and since the t^m appears outside the integral, it follows, in all instances in which $n \neq 0$, that there exists no single input function which will make $S_n(t)$ an output of the instrument.[9] The functions $A_n(t)$, on the other hand, are well-defined outputs to equally well-defined inputs and hence the $S_n(t)$ can be computed from A_{n-m}.

For $n = 0$ we obtain $S_0(t) = A_0(t)$, which indicates that the sensitivity to the input function can be obtained by subjecting the instrument to a unit step input. All other sensitivity functions are obtained by combinations of the appropriate sensitivity test functions. Consideration of the use of S_n and A_n in instrument design is deferred until the next chapter. It is not remiss, however, to examine a simple instrument system and to compute the system characteristics from experimental evidence.

Example 3.4.1 A "fast response" temperature-measuring system is subjected to a unit input ($y = 1$), a unit rate input ($y = t$) and a unit acceleration input ($y = t^2/2$) with the following results:

t (s)	0.0	0.1	0.2	0.3	0.4	0.5
A_0	0.000	0.272	0.495	0.677	0.827	0.948
A_1	0.000	0.014	0.052	0.111	0.186	0.276
A_2	0.000	0.000	0.004	0.011	0.027	0.049
t (s)	0.6	0.8	1.0	1.5	2.0	3.0
A_0	1.049	1.197	1.293	1.425	1.473	1.497
A_1	0.375	0.481	0.851	1.538	4.526	11.25
A_2	0.082	0.149	0.284	0.918	1.863	4.875

Compute S_0, S_1, S_2 and determine (to the order of S_2) the response of the system to the input $y = e^t$ in the interval $0 \leq t \leq 3$ s.

Solution From Equation (3.4.1) it follows that

$$S_0 = A_0$$
$$S_1 = A_1 - tA_0$$
$$S_2 = A_2 - tA_1 + \frac{t^2}{2}A_0$$

and from the Definition 3.3.3 we can write

$$\begin{aligned}\xi_1 &= y(t)S_0 + y'(t)S_1 + y''(t)S_2 + \cdots \\ &= e^t(S_0 + S_1 + S_2) + \cdots\end{aligned}$$

With the equations given above we can now evaluate the required parameters.

t (s)	0.0	0.1	0.2	0.3	0.4	0.5
S_0	0	0.272	0.495	0.677	0.827	0.948
S_1	0	−0.013	−0.046	−0.092	−0.144	−0.198
S_2	0	0.000	0.000	0.008	0.018	0.029
ξ_1	0	0.286	0.548	0.800	1.046	1.284
t (s)	0.6	0.8	1.0	1.5	2.0	3.0
S_0	1.049	1.197	1.293	1.425	1.473	1.497
S_1	−0.254	−0.356	−0.441	−0.600	−0.709	−0.741
S_2	0.046	0.057	0.079	0.214	0.288	0.361
ξ_1	1.532	1.998	2.530	4.657	7.773	22.41

It is instructive to notice that up to three seconds the approximation

$$\xi_1 = e^t(S_0 + S_1 + S_2)$$

works quite well. If the system is subjected to the input e^t, the time response ξ obtained experimentally agrees well with ξ_1 as can be seen from the following table and from Figure 3.5. At $t = 10$ seconds, however, we obtain $\xi = 22{,}026$ and $\xi_1 = 24{,}669$. The difference between the two quantities is of the order of 11% and remains fairly constant after $t = 0.4$ second.

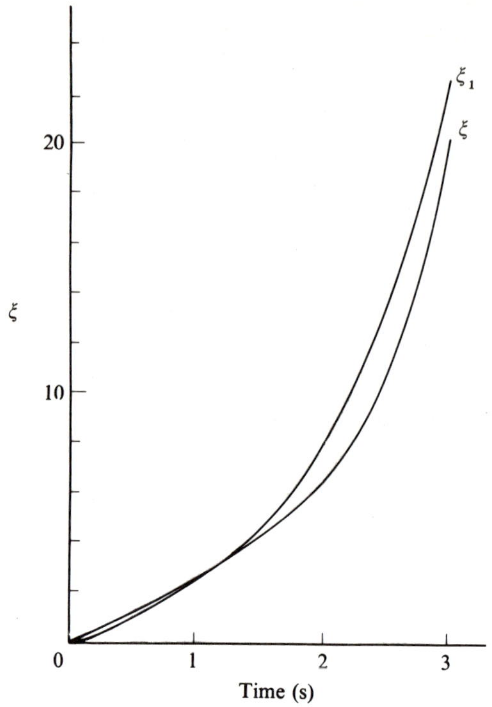

FIGURE 3.5 Response of "fast response" thermometer

t (s)	0.1	0.2	0.3	0.4	0.5	0.6
ξ	0.286	0.551	0.801	0.943	1.271	1.521
ξ	0.286	0.548	0.800	1.046	1.284	1.532

t (s)	0.8	1.0	1.5	2.0	3.0
ξ	2.023	2.573	4.432	7.379	20.084
ξ_1	1.998	2.530	4.657	7.773	22.41

3.5 Static Sensitivity of Selected Instruments

It is necessary to consider the dynamic sensitivity theory discussed above whenever an instrument is to be used to present an output signal proportional to a varying input. Such a dynamical analysis of instruments is delayed until the next chapter. In the design of many instruments it is often sufficient to discuss the static sensitivity

$$S = \lim_{t \to \infty} S_0(t)$$

if the input function is time-independent or else is such that $d^n y/dt^n \ll y$ for all n. Such steady or quasi-steady inputs can usually be treated by the equation

$$\xi = \frac{K}{a_0} y = Sy \tag{3.5.1}$$

in which K and a_0 are given by Equation (3.3.1):

$$a_n \frac{d^n \xi}{dt^n} + a_{n-1} \frac{d^{n-1} \xi}{dt^{n-1}} + \cdots + a_1 \frac{d\xi}{dt} + a_0 \xi = Ky(t)$$

It must be understood that a complete analysis of an instrument requires some consideration of dynamics, if nothing else, to determine the filtering characteristics (chap. 5) of the instrument. If the instrument is subjected to sinusoidal vibrations, typically small-amplitude, high-frequency inputs, such that

$$y = y_0 + \sum_{n=1}^{\infty} \delta_n \cos(\omega_n t + \phi_n)$$

in which $\delta_n \ll y_0$, it is necessary to consider the dynamical behavior even though the δ_n are so small that no significant effects are noticed through the use of Equation (3.5.1). It is evident that since

$$\frac{dy}{dt} = -\sum_{n=1}^{\infty} \delta_n \omega_n \sin(\omega_n t + \phi_n)$$

if at least one of the ω_n is large, the contribution of S_1 may become significant. The ability of the instrument to suppress the unwanted sinusoidal terms, i.e., to filter, has to be considered even though the input is assumed to be steady. It is good practice, therefore, to obtain the complete performance equation during the static sensitivity analysis of the system and to use the complete equations in filtering and stability studies.

In general the analysis proceeds as follows:

(1) determine the general structure of the system;
(2) separate into elements;
(3) obtain performance equation of each element;
(4) determine overall performance equation;
(5) determine the ultimate state sensitivity;
(6) adjust the parameters to meet the sensitivity specifications as closely as possible.

3.5.1 The dc galvanometer is essentially a spring-restrained, damped dc electric motor so constructed that in the ultimate state the torque generated by the current through the armature is balanced by the spring torque. For analysis the system can be divided into a mechanical torque meter, an electrical circuit containing the armature or field coils (or both), the source of input current and some intermediate circuitry. The electrical and mechanical systems are connected to each other by a magnetic element. The structure of the system, therefore, is quite simple (Fig. 3.6).

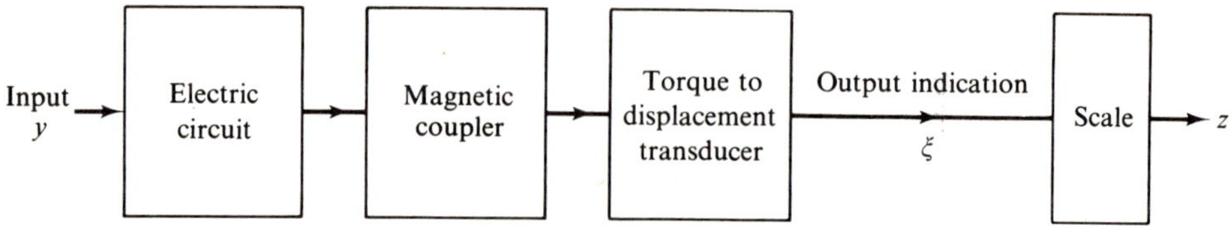

FIGURE 3.6 Elements of the dc galvanometer

The mechanical system normally can take one of two forms. The most common structure consists of a pivoted armature mounted on low friction bearings. The armature is located between the pole pieces of a source of magnetic flux, either a permanent magnet or an electromagnet (Fig. 3.7).[10] Such pivoted galvanometers are usually used for steady state measurements and on time-varying measurands with relevant frequencies not in excess of 10 cycles per second (10 Hz).

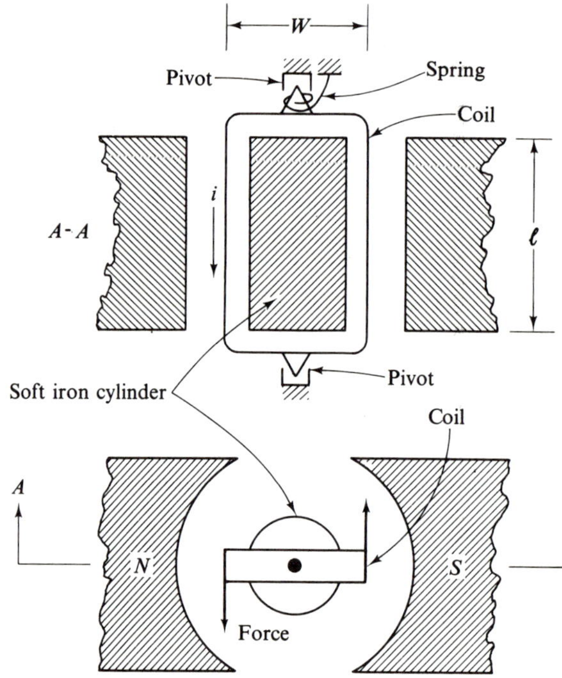

FIGURE 3.7 dc moving coil galvanometer

For high-frequency measurements (up to 5,000 Hz) the customary suspension is "filar."[11] The armature, usually very light and of low moment of inertia, is suspended between two segments of tightly stretched wire (Fig. 3.8). The wires serve both as spring and as nearly frictionless pivots. A mirror is mounted on the armature so that the angular displacement of the coil is measured by the motion of a reflected light spot on a screen. Filar galvanometers have very fast response

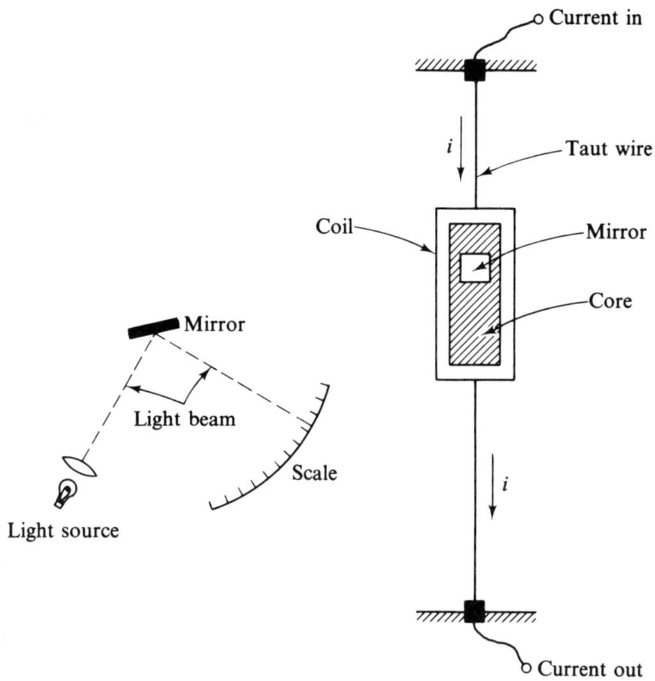

FIGURE 3.8 Filar suspension galvanometer

as a result of the high spring constant of the wire, low moment of inertia of the armature and negligible damping.

Let us now analyze the elements of the system individually.

(a) A free-body diagram of the mechanical system is shown in Figure 3.9. The coil, shaft and other moving parts are assumed to be properly balanced and to have a moment of inertia J. (The general analyses of the pivoted and the filar galvanometers are obviously identical. Only the numbers differ.) The spring can be assumed to generate a torque T_S and the suspension (bearings or wire) introduces an additional torque T_B. The electromagnetic signal produces a torque T. Clearly we have identified an element of the galvanometer with torque as input signal and angular displacement as output signal.

The performance equation is easily obtained from the free-body diagram

$$J\ddot{\phi} = T - T_S - T_B$$

in which ϕ is the angular deflection of the coil.

What T_S and T_B are is a matter of experiment and conjecture. Usually it is possible to design torsion springs in which the torsion torque is proportional to the angular displacement, $T_S = k\phi$. The bearing resistance is due to energy dissipation in the bearing. There

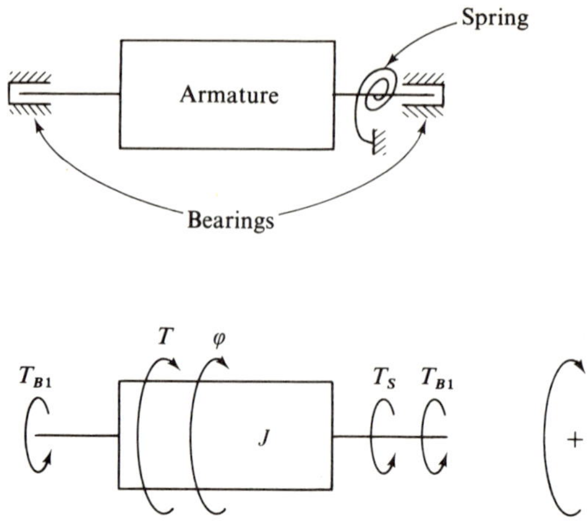

FIGURE 3.9 The mechanical element of the dc moving coil galvanometer

are three possible mechanisms: friction, viscous losses in lubricated bearings and internal (molecular) damping in an elastic system. All three phenomena are present in each galvanometer design, although it is evident that friction is dominant in dry jewelled bearings; viscous resistance in lubricated bearings; and internal molecular resistance in filar suspensions. Viscous suspension is rarely used, friction suspension is common in most modern low-frequency galvanometers and filar suspension is universally applied to high-speed galvanometers.[12] In most instances it can be shown experimentally that $T_B = T_B(\phi, \dot{\phi})$. Although T_B is a nonlinear function of both variables, it can be demonstrated, again by experiment, that $\dot{\phi}$ is the dominant parameter and the effects of ϕ reduce to small change in the spring constant. This is equivalent to saying that in the MacLaurin expansion of T_B

$$T_B = T_0 + T_\phi \phi + T_{\dot{\phi}} \dot{\phi} + T_{\dot{\phi}\phi} \phi \dot{\phi} + T_{\phi\phi} \phi^2 + \cdots$$

only $T_{\dot{\phi}}$ is relevant. Thus the assumed linear performance differential equation of the mechanical system is

$$J \frac{d^2\phi}{dt^2} + f \frac{d\phi}{dt} + k\phi = T \qquad \textbf{(3.5.2)}$$

(b) The torque T applied to the galvanometer mechanical element is a function of the current impressed on the armature. The electromagnetic element, therefore, has an input of current and an output of torque. It is well known that the force acting on one side of the coil (Fig. 3.1) is given by[13]

$$F = nB\ell i$$

in which

F = force
n = number of active turns
ℓ = length of active coils
B = flux density in the vicinity of the coils
i = current

The torque on the coil is

$$T = wnB\ell i = K_1 i \tag{3.5.3}$$

in which w is the active width of the coil and $K_1 = wnB\ell$ is a constant dependent entirely on the physical characteristics of the galvanometer system.

(c) The combination of Equations (3.5.2) and (3.5.3) indicates that the output displacement of the mechanical movement is a function of the current flowing through the armature. If the galvanometer is connected into an electrical circuit, it will draw a current which will depend on its own electrical characteristics and on the circuit elements external to the armature. Thus the galvanometer output is indeterminate and must be calibrated on each circuit. To avoid this indeterminacy, an isolating circuit is connected across the galvanometer in such a fashion that the galvanometer does not disturb the measurand (does not draw appreciable energy from the input configuration) and such that the galvanometer output is a function of a given input property regardless of the combination of elements in the input configuration.

Two major effects must be noted.

(1) As the armature turns, the galvanometer acts as a generator and creates a back electromotive force, thereby affecting the circuit current.

(2) The current flowing through the armature is affected by the coil resistance, the coil inductance and by some coil capacitance.

The resistance, however, is the dominant effect, and as a first approximation we shall assume that the galvanometer is represented in the equivalent circuit by a voltage source (back emf generator) and a series resistance (the armature resistance). The coupling network shall consist of a "shunt" resistance R_D, an input series resistance R_S and a "swamping" resistance R_I (Fig. 3.10).

According to the theorem of Thevenin[13] the circuit to which the galvanometric system is attached can be replaced by a voltage source and a series impedance. In many galvanometric systems the galvanometer isolating circuit "looks into" a resistive system. Thus we shall assume that the input configuration consists of an equivalent voltage source E with an equivalent series resistance R. We now

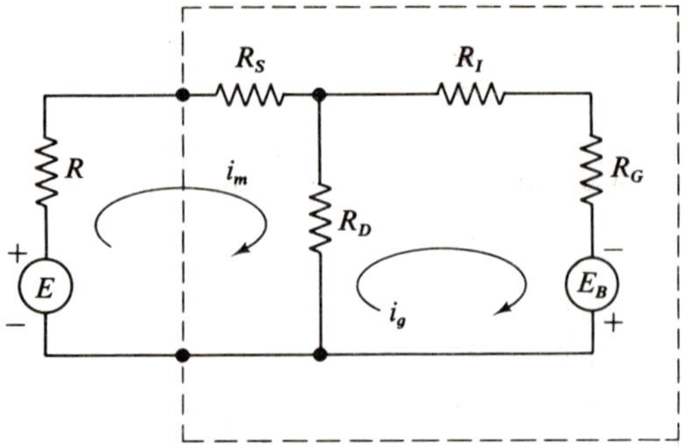

FIGURE 3.10 Equivalent circuit for dc galvanometer. Galvanometer equipment is inside broken line box; equivalent resistive load, outside box.

establish the following nomenclature:

E_B = back emf
E = input equivalent voltage
R = input equivalent resistor
R_S = galvanometer series resistance
R_D = galvanometer shunt resistance
R_I = galvanometer "swamping" resistance
R_G = resistance of galvanometer armature coil
i_m = input current
i_G = current through armature

We can now write the Kirchhoff equations and solve for i_G

$$i_G = -a_1 E_B + a_2 E$$

in which

$$a_1 = \frac{R + R_S + R_D}{R_G(R + R_S + R_D) + R_I(R + R_S + R_D) + R_D(R + R_S)}$$
$$= \frac{1}{R_G + R_I + R_D(R + R_S)/(R + R_S + R_D)}$$

and

$$a_2 = \frac{1}{[(R + R_S)/(R_D + 1)]R_G + R_I + R_D(R + R_S)/(R + R_S + R_D)}$$

It is well known that the emf created by a generator with the general construction of the galvanometer (Fig. 3.7) is[14]

$$E_B = nDB\frac{d\phi}{dt}$$

in which D is the planform area of the armature and the other nomenclature was previously defined. It follows, therefore, that the performance equation of the electrical circuit is

$$i_G = -a_1 nDB\frac{d\phi}{dt} + a_2 E \qquad \textbf{(3.5.4)}$$

Equations (3.5.2), (3.5.3), and (3.5.4) can be combined to yield

$$J\frac{d^2\phi}{dt^2} + (f + K_1 a_1 S B_n)\frac{d\phi}{dt} + k = k_1 a_2 E \qquad \textbf{(3.5.5)}$$

Equation (3.5.5) is the performance equation of the galvanometer measuring an equivalent Thevenin circuit with a resistance in series with a voltage. The quantities R_I, R_S and R_D are chosen to adjust the coefficient of $d\phi/dt$ to meet the dynamic specifications of the system, to create a static sensitivity

$$S = \frac{K_1 a_2}{k}$$

according to specifications and to minimize the current i_m.

3.5.2 The accelerometer consists of a cylindrical mass free to move in a cylindrical case and attached to a strain column (Fig. 3.11). To the column is bonded a strain gage of nominal resistance R connected to a resistance bridge which is supplied by a voltage source E. The output is indicated by a galvanometer G. Application of acceleration is resisted by the inertia of the mass M and hence by a change in length of the strain column which, in turn, is reflected in a change of resistance in the strain gage and thus a current change in the bridge. We can descry five elements of the instrument (Fig. 3.12). The input transducer, which consists of the mass M and its supporting column, converts the component of acceleration a_x along the mass major axis x into a change of linear dimensions of the column δ_x. The first modifier is a strain gage which converts the linear displacement δ_x into a resistance change ΔR. The second modifier is the resistance bridge which converts the resistance change ΔR into a current i. A secondary energy source E provides the Wheatstone Bridge excitation. The output of the bridge actuates

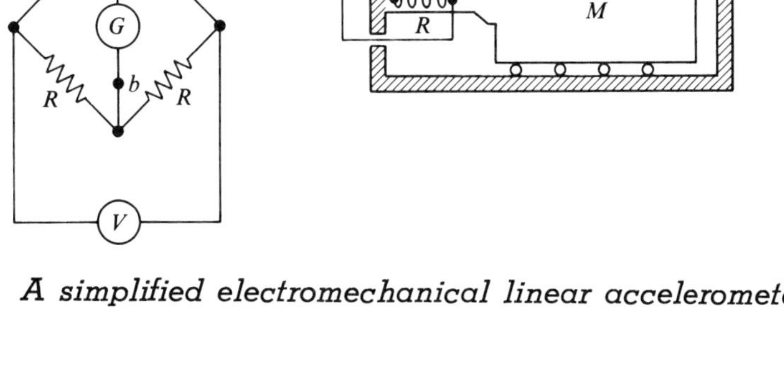

FIGURE 3.11 A simplified electromechanical linear accelerometer

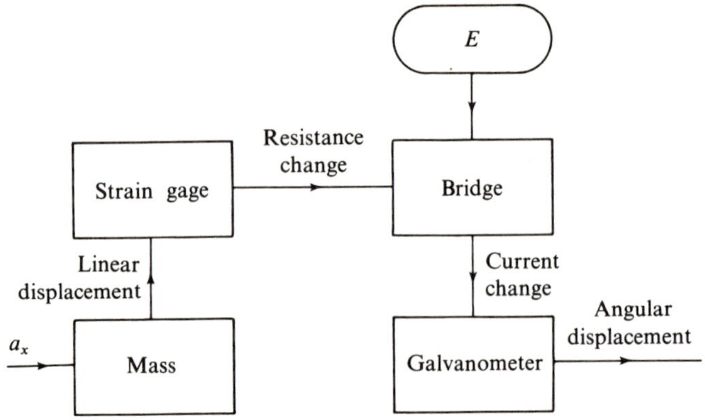

FIGURE 3.12 Signal flow diagram for a simple accelerometer

the galvanometer display. The sensitivity of the galvanometer should be computed in terms of the associated bridge circuit.

(a) Under an acceleration, the mass M resists motion inertially, thereby compressing or expanding the strain column. The column is a spring element and we shall assume that it can be treated as a linear spring of modulus k. Some damping exists in the system, through friction in the contacts between the mass and the case and through the internal energy losses in the strain column. The inherent damping in such a system is low but by no means negligible. We shall assume that the damping is linear with a constant coefficient f. The accelerometer case is assumed to be attached to a vehicle which is accelerating in some arbitrary direction. Since the mass M is constrained to move only along its major axis, it is obvious that (ideally, at least) the accelerometer will respond only to accelerations along its major axis. Thus we consider only the input component of acceleration a_x. The case moves parallel to a_x and its coordinate is labeled x. The mass moves relative to the case through a distance δ_x. Thus a free-body diagram can be drawn (Fig. 3.13) and the equation of motion found:

$$\ddot{\delta}_x + \frac{f_0}{M}\dot{\delta}_x + \frac{k_0}{M}\delta_x = -\ddot{x} = -a_x \tag{3.5.6}$$

The strain column is loaded by the inertia element M. If the length of the column is L, there is created in the column a strain $\epsilon = \delta_x/L$. This last expression is rather approximate in the sense that the column is assumed to be inertialess and to respond instantaneously to the load. If the acceleration a_x changes slowly, the assumption is quite good. For high-frequency acceleration changes the assumption must be examined with care. Under normal circumstances $\epsilon = \delta_x/L$ is quite legitimate and we now can write Equation (3.5.6) in a form which relates input acceleration to column strain.

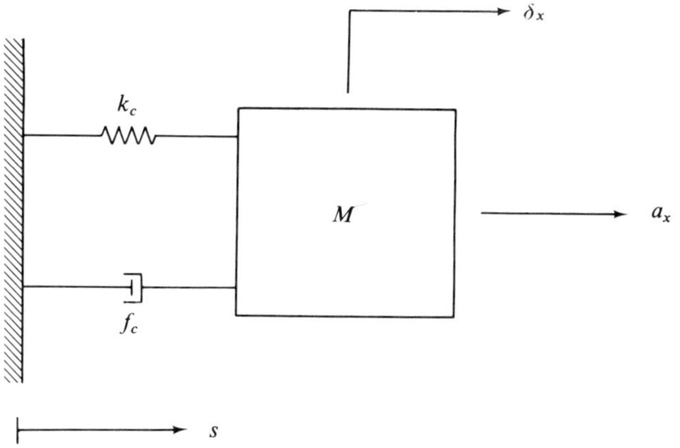

FIGURE 3.13 Acclerometer as a lumped-parameter system

$$\ddot{\epsilon} + \frac{f_0}{M}\dot{\epsilon} + \frac{k_0}{M}\epsilon = -\frac{a_x}{L} \tag{3.5.7}$$

(b) The detection of ϵ is accomplished by means of a resistance strain gage. The theory of the electrical strain gage has an extensive literature of its own.[15] If a resistance wire of uniform cross-sectional area A and of length ℓ is composed of a material of uniform specific resistance ρ, then the wire will exhibit a resistance given by

$$R = \rho\frac{\ell}{A}$$

If the wire is subjected to strain, the length and the area will change under the load. The resistivity may change under the strain but, in addition, may change as a result of a change in ambient conditions, especially temperature. Thus there occurs a change in resistance

$$dR = \frac{\rho A\,d\ell + A\,d\rho - \rho\,dA}{A^2}$$

from which we obtain

$$\frac{dR}{R} = \frac{d\ell}{\ell} + \frac{d\rho}{\rho} - \frac{dA}{A}$$

Since $d\ell$ represents a change in length, $d\ell/\ell$ is obviously the strain ϵ. It is well known[16] that the transverse strain of a wire subjected to uniform stress changes in terms of the Poisson ratio μ

$$\epsilon_T = -\mu\epsilon$$

We consider a plane element of area $dx\,dy$ and note that a stress perpendicular to it will cause the area to deform into $dx\,(1 + \epsilon_x)$ $dy\,(1 + \epsilon_y)$. Thus the change in the cross-sectional area due to the load will be

$$\begin{aligned} dA &= dx\,dy(1 + \epsilon_x + \epsilon_y + \epsilon_x\epsilon_y) - dx\,dy \\ &= dx\,dy(-2\mu\epsilon + \mu^2\epsilon^2) \end{aligned}$$

under the assumption $\epsilon_x = \epsilon_y = -\mu\epsilon$. Therefore

$$-\frac{dA}{A} = (2\mu\epsilon - \mu^2\epsilon^2)$$

and hence

$$\frac{dR}{R} = \left(1 + 2\mu - \mu^2\epsilon + \frac{d\rho}{\rho}\right)\epsilon$$

In a well-designed gage[17]

$$\frac{d\rho}{\rho} \approx 0$$

Furthermore $\mu \approx 0.5$ and ϵ is of the order of 10^{-6} whence (subject to errors of the order of 10^{-13}) we obtain the performance equation of the strain wire

$$\frac{dR}{R} = (1 + 2\mu)\epsilon$$

In general, the gage consists of a grid of wires all mounted parallel to the strain axis and interconnected in series. The interconnection does not materially alter the gage equation. However, since R is larger for the grid than for a single wire as long as the gage, dR will also be larger. As is customary in strain gage practice we define a strain gage factor F

$$\frac{\Delta R}{R} = F\epsilon \qquad \textbf{(3.5.8)}$$

which is a function of gage material, gage dimensions and gage temperature. Equation (3.5.8) is the performance equation of the strain transducer.

(c) The bridge and the galvanometer are interconnected as in Figure 3.11. A means of adjusting the bridge and zeroing the galvanometer at zero applied strain should be provided in the circuit but will not be discussed here. Insofar as the galvanometer is concerned, the bridge is represented by a two-terminal network connected to the galvanometer at a and b (Figs. 3.11 and 3.14). We apply Thevenin's theorem to compute the equation of the bridge as seen from the galvanometer. The equivalent voltage, under the assumption that the voltage source has no internal resistance, is given by

$$E_e = \frac{E}{2}\left[\frac{1}{1 + (\Delta R/R)} - 1\right] = -\frac{E}{4}\frac{\Delta R}{R} + \cdots$$

in which the last equality is obtained by expansion into the binomial series. The higher-order terms can be neglected since $\Delta R/R$ is

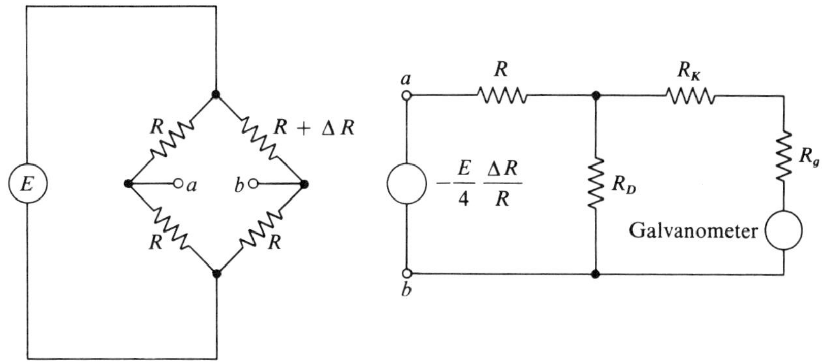

FIGURE 3.14

of order 10^{-6}. The equivalent resistance (as seen by the galvanometer) is

$$\begin{aligned} R_e &= \frac{R}{2}\left[\frac{1+(\Delta R/R)}{1+(\Delta R/2R)}+1\right] \\ &= \frac{R}{2}\left(2+\frac{\Delta R}{R}\cdots\right) \\ &= R+\cdots \end{aligned}$$

The equivalent galvanometer circuit can now be drawn as in Figure 3.14b. From the figure we can write the galvanometer equation (3.5.5) which, when combined with Equation (3.5.8), yields the performance equation of the strain gage accelerometer

$$\begin{aligned} &\frac{d^4\phi}{dt^4}+\left[\frac{f_0}{M}+\frac{f+k_1a_1SBn}{J}\right]\frac{d^3\phi}{dt^3} \\ &+\left[\frac{k_0}{M}+\frac{(f+k_1a_1SBn)f_0}{JM}+\frac{k}{j}\right]\frac{d^2\phi}{dt^2} \\ &+\left[\frac{(f+k_1a_1SBn)k_0}{JM}+\frac{kf_0}{JM}\right]\frac{d\phi}{dt}+\frac{kk_0}{JM}\phi \\ &=\frac{k_1a_2EFa_x}{4LJ} \end{aligned} \tag{3.5.9}$$

This combination shows the incredible number of parameters that the designer must cope with even when studying a "simple" problem. The static sensitivity of the instrument is quite messy:

$$S=\frac{k_1EF}{4Lkk_0[(R+R_S/R_D)+1]R_G+R_I+[R_D(R+R_S)/(R+R_S+R_D)]}$$

In general, only E, R_S, R_D and R_I are available to the designer. The other parameters are usually fixed by the available commercial galvanometers and strain gages. L is usually fixed by structural and size considerations.

3.5.3 It must not be assumed that it is always possible to study the dynamics of a system theoretically. Quite often, even in simple devices, the complete theory is so involved that elementary analysis is not feasible. A good example of such a physically simple yet difficult to analyze system is the thermocouple thermometer.

If two conductors of different material are joined in a loop, and if one junction is placed in a region of uniform temperature T_1 and the other junction in a region of uniform temperature $T_2 \neq T_1$, a potential difference is developed in the loop. If one of the conductors is cut (Fig. 3.15) and connected to a very high impedance voltmeter (so that almost no energy is drawn from the loop), an open circuit potential is measured at the voltmeter which is a function of $T_2 - T_1$. This phenomenon, known as the Seebeck effect, has been known since 1821[18] and is the basic principle of the thermocouple thermometer. If a junction is kept at a known reference temperature, then the voltage will vary with the temperature of the other junction, and the voltmeter can be calibrated in temperature units.

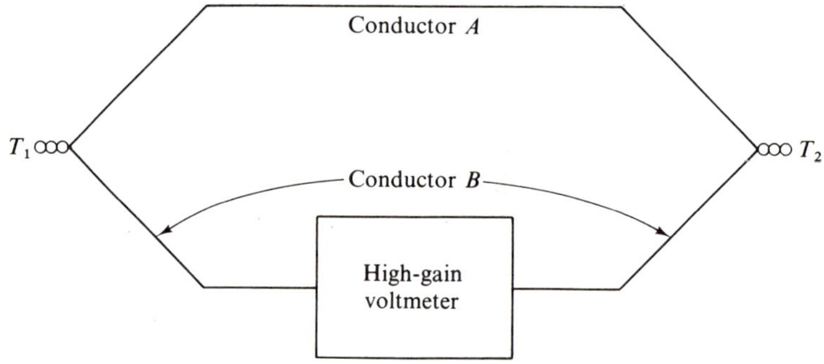

FIGURE 3.15 Elementary thermocouple circuit

A complete analysis of the system requires five integro-differential equations to be solved simultaneously. Physically, the investigator must consider the overall energy balance, the junction energy balances and the differential equation for the temperature distribution in the conductors. These equations have not been solved generally although special linearized instances have been examined in the literature.[19] Under such circumstances the engineer must have recourse to experiment.

A system for testing thermocouple dynamics can be built fairly easily. Two chambers can be built with a small aperture between them. A thermocouple is mounted on a spring mechanism in line with the aperture so that upon release of the mechanism the junction will move rapidly through the hole from one chamber to the other.[20] If the chambers are heated at different temperatures T_0 and T, the response of the thermocouple as recorded in a high performance recorder will be $(T_0 - T)S_0(t)$. From the results it will be possible to obtain a differential equation of motion for the system.

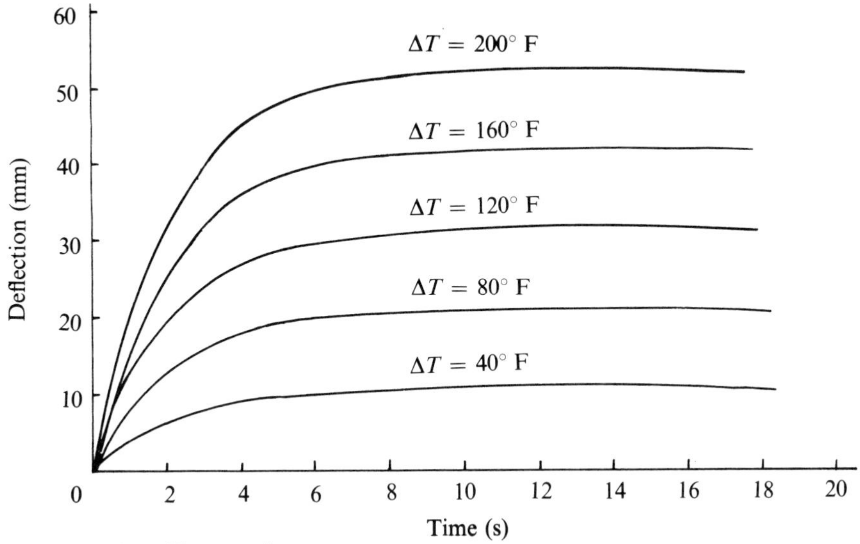

FIGURE 3.16 Typical response curves

A typical set of response curves was obtained by the author in his laboratory (Fig. 3.16). The linear displacement of a recording galvanometer (measured in mm) was obtained as a function of time for five different temperature jumps. (See Table 3.1.)

The data have been smoothed (see chapter 6) by statistical means and are, in this fashion, somewhat "ideal." The curves indicate that S_0 reaches an ultimate state as $t \longrightarrow \infty$. Thus it is possible to obtain the ultimate state sensitivity S by plotting the deflection for large t versus the temperature input (Fig. 3.17) to obtain $S = 0.260$ mm/°F.

TABLE 3.1 Temperature jump °F

Time (s)	60°	80°	120°	160°	200°
0.000	0.000	0.000	0.000	0.000	0.000
0.382	1.893	3.786	5.678	7.571	9.464
0.764	3.432	6.026	10.30	13.73	17.32
1.146	4.690	9.381	14.07	18.76	23.45
1.528	5.730	11.46	17.19	22.92	28.65
1.910	6.573	13.14	19.72	26.29	32.86
2.865	8.081	16.16	24.24	32.32	40.40
3.820	8.996	17.99	26.99	35.98	44.98
4.775	9.547	19.09	28.64	38.19	47.74
5.730	9.880	19.76	29.64	39.52	49.40
7.640	10.21	20.42	30.64	40.85	51.06
9.550	10.33	20.65	30.98	41.31	51.64
13.37	10.39	20.70	31.17	41.56	51.95
17.19	10.40	20.80	31.20	41.60	52.00

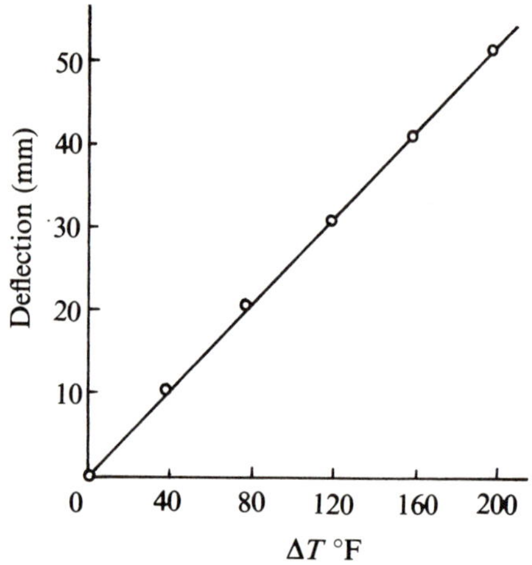

FIGURE 3.17 Ultimate state deflection as a function of temperature

Let us now try to fit a linear ordinary differential equation to the system. We write

$$\sum_{n=1}^{N} \frac{d^n\delta}{dt^n} a_n = a_0 A \, \Delta T u(t)$$

which in the ultimate state becomes

$$\delta_{us} = S\Delta T$$

We need to discover N and all the a_n. Since we start with the system at rest, we know that we have initial conditions

$$\frac{d^n\delta}{dt^n} = 0 \qquad n = 0, \ldots, N-1$$

Thus the $d^N\delta/dt^N$ derivative is undefined.

Examination of the curves shows us that $d\delta/dt \neq 0$ and thus the system can only be of the first order. If it were of the second order or higher, $d\delta/dt = 0$. Thus we write

$$a_1 \frac{d\delta}{dt} + a_0\delta = a_0 S \, \Delta T$$

Let us set $a_1/a_0 = \tau$ and now we obtain

$$\tau \frac{d\delta}{dt} + \delta = S \, \Delta T$$

We can solve this equation subject to the requirement $\delta(0) = 0$ to obtain

$$\delta(t) = S \, \Delta T(1 - e^{-t/\tau})$$

It remains to determine τ. When $t = \tau$ we have

$$\delta(\tau) = (S \, \Delta T)0.632$$

Thus τ appears to be a constant which is equivalent to the time $t = \tau$

at which the deflection equals 63.2% of the difference between initial value and final value. From the table we compute 63.2% of $S\,\Delta T$.

ΔT	40.	80.	120.	160.	200.
$S\,\Delta T$	10.40	20.80	31.20	41.60	52.00
$.632\,S\,\Delta T$	6.573	13.14	19.72	26.29	32.86

All the responses $0.632\ S\,\Delta T$ occur at $t = 0.191$ second and thus we have obtained the equation

$$0.191\frac{d\delta}{dt} + \delta = 0.260\,\Delta T$$

as a representation of the system.

One might wonder about this experimentally achieved equation. The system consists of a thermocouple, an amplifier and a recording galvanometer. With the galvanometer representable by a second-order ordinary differential equation and the amplifier being at least of the first order (see Chapter 4), we would expect the overall system to be at least of the fourth order (assuming a minimum of a first-order system for the thermocouple). However, we meet here for the first time the practical application of an instrumentation rule. We selected an amplifier and galvanometer so fast that their transients vanished many times faster than the transients of the thermocouple, so that effectively, the recorded response is the response of the thermocouple alone.

3.5.4 The measurement of pressure is universally effected by balancing the pressure force against a known load. Most common is the balancing of a pressure force against an elastic force. The construction variations are infinite[21] and have a specialized subliterature of their own. Theoretical static sensitivities cannot be found and recourse must be had to experimental tests. The bourdon tube, for example, is a theoretical nightmare, the elastic theory of which has not been solved completely today.[22] There exists, however, an enormous mass of experimental information collected over a period of 150 years[23] on the sensitivity, construction and care of bourdon tubes. No one needs to design bourdon tubes any more—one simply buys one.

On the other hand, electrical pressure transducers, although still in a state of development, have an approximate theory available. Electrical pressure transducers consist of a metallic diaphragm, usually circular (Fig. 3.18) backed by a mechanical-to-electrical energy transducer, say, a strain column, somewhat in the same fashion as the linear accelerometer described previously. The diaphragm is either a "thin membrane," with thickness (t) very much less than the diameter (d), or else a "true" diaphragm with a substantial thickness. The second case is the most common and will, consequently, be described here.

The elastic theory needed for the description of the motion is too extensive to repeat. It is, however, well known. A good derivation is found in Reference 63.[24]

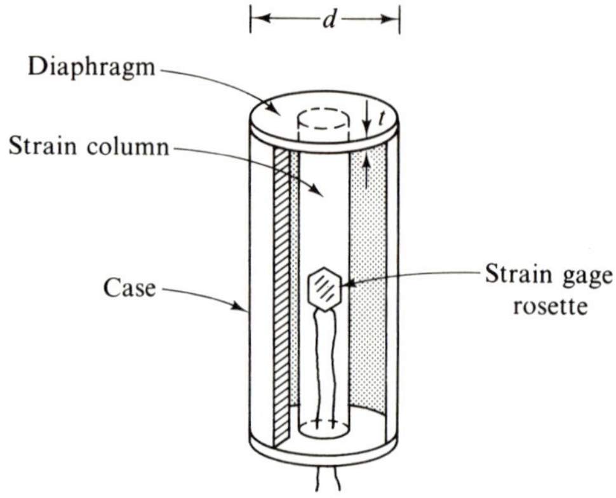

FIGURE 3.18 Electrical pressure transducer

Consider a circular plate clamped rigidly at its outer edge free to deform in the interior (Fig. 3.19). Assume that a uniform pressure (p) is applied above the plate and that the diaphragm deforms elastically downward through a distance (ξ_c). By some unspecified means, let (ξ_c) be detected without unduly loading the plate; the displacement (ξ_c) can then be easily shown to be a function of the pressure, the time and the physical characteristics of the diaphragm. If the physical characteristics of the diaphragm are completely uniform, it can be assumed that the displacement is symmetrical about the center of the plate. It follows, then, that the diaphragm must satisfy at all times the following

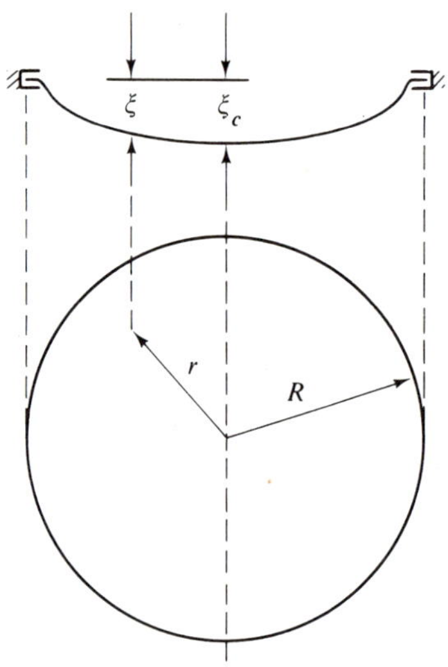

FIGURE 3.19 Clamped diaphragm

boundary conditions:

$$\frac{\partial \xi}{\partial r} = 0 \quad \text{at } r = R \quad \text{and} \quad r = 0$$

$$\xi = 0 \quad \text{at } r = R$$

It is assumed that the shape of the diaphragm can be described by the equation

$$\xi = \xi_c\left(1 - \frac{r^2}{R^2}\right)^2$$

which demonstrably satisfies the boundary conditions. Assuming such a solution is no guarantee of correctness—after all, many other forms can be used to satisfy the boundary conditions and the equations of elasticity. It can be shown by experiment, however, that the kinetic energy and the potential energy of the plate are relatively insensitive to the form of ξ as long as the boundary conditions are satisfied and that the lowest mode, at least, can be computed sufficiently well for practical applications.[25] Accordingly we obtain the kinetic energy (T) and the potential energy (V) for the plate and then compute the performance equation with our assumed (ξ). In Reference 63 we find

$$V = \frac{\pi E h^3}{12(1 - \nu^2)} \int_0^R \left(\frac{\partial^2 \xi}{\partial r^2} + \frac{1}{2}\frac{\partial \xi}{\partial r}\right)^2 r\, dr$$

$$T = \pi \rho h \int_0^R \dot{\xi}^2 r\, dr$$

in which

ρ = mass density of diaphragm material
E = Young's modulus
h = thickness of plate
r = radial coordinate from center of plate
R = radius of plate
ν = Poisson's ratio
ξ = vertical displacement of plate from the unloaded position

Substitution of the expression for ξ and integration yield

$$T = \left(\frac{\rho h \pi R^2}{5}\right)\frac{1}{2}\dot{\xi}_c^2 = \frac{1}{2} m_e \dot{\xi}_c^2$$

$$V = \left[\frac{16}{9}\frac{\pi E h^2}{(1 - \nu^2)R^2}\right]\frac{1}{2}\xi_c^2 = \frac{1}{2} k_e \xi^2$$

Application of Lagrange's equation to $L = T - V$ gives the equation of motion

$$m_e \ddot{\xi}_c + k_e \xi_c = F$$

in which m_e is the effective mass and k_e the effective spring constant of the diaphragm with respect to the displacement of the center. F is the external force acting on the plate.

The driving force (F) is a function of pressure and may be computed in terms of the work performed on the diaphragm. Under a pressure (p) the plate is deformed through a distance (ξ) and thus the work can be written as

$$W = \int_0^{2\pi} \int_0^R p\xi r \, dr \, d\theta$$

which, upon integration, becomes

$$W = \frac{\pi R^3}{3} p\xi_c$$

which can be considered an equivalent force acting through a displacement ξ_c. Thus we obtain the ideal equation of the circular clamped diaphragm

$$m_e\ddot{\xi}_c + k_e\xi_c = A_e p$$

in which

$$m_e = \frac{m_0}{5} = \text{effective mass}$$

$$m_0 = \text{mass of diaphragm} = \rho h\pi R^2$$

$$k_e = \frac{16}{9}\frac{\pi E h^3}{(1 - \nu^2)R^2} = \text{effective spring constant}$$

$$A_e = \frac{A}{3} = \text{effective area}$$

$$A = \text{area of diaphragm} = \pi R^2$$

The effective quantities as developed are only approximations to the measured values. It is found, also, that the usual pressure transducer diaphragm is not a lossless system and that there exists a small measurable damping due partially to the resistance of the surrounding medium and partially to internal losses in the vibrating plate. Thus the equation of the pressure transducer is given by

$$m_e\ddot{\xi}_c + c_e\dot{\xi}_c + k_e\xi_c = A_e p$$

in which m_e, c_e, k_e and A_e are determined experimentally. As a first design approximation, however, the theoretical values developed in this section are quite usable.

It is now possible to compute the ultimate state sensitivity of the diaphragm

$$S = \frac{A_e}{k_e} = \frac{3}{16}\left(\frac{R}{h}\right)^3\left(\frac{R}{E}\right)(1 - \nu^2)$$

Proper design of diaphragms involves the selection of materials to satisfy sensitivity requirements, dynamic response requirements and strength (allowable stresses) requirements.[26]

3.6 Feedback

Employment of feedback to maintain favorable sensitivity characteristics has become a major technique in aerospace instrumentation. The principal use is in the correction of attitude sensor characteristics, especially gyroscope and pendulum devices.[27] The procedure consists of determining the system configuration for maximum sensitivity and then supplying a force or moment to the system in such a fashion that the transducer is forced to operate in the optimum region. Since the reset force or moment must change with the input, then the torque or moment becomes the transducer output.

Consider, for example, the pendulous accelerometer discussed in Example 3.2.1. It was shown that near the output angle $\theta = 0$, the static sensitivity was at a maximum and that for $\theta > 0$, the sensitivity became smaller rapidly with increasing θ. In addition, it was shown that the pendulum was quite sensitive to cross accelerations $\ddot{v}$. The device was shown to be nearly useless as an accelerometer. With feedback, however, it becomes an excellent acceleration-measuring device. Consider the setup shown in Figure 3.20. The pendulum shaft is connected at one end to a potentiometer, a device which generates an electric signal proportional to the angular position of the shaft. The output of the potentiometer is connected to a controller which generates a voltage that bears a functional relationship to the shaft position and that is then used as the control voltage, say, the field voltage, of an electric motor connected to the pendulum shaft. The motor is controlled in such a fashion that the pendulum is forced to return to the $\theta = 0$ position. Here we have a typical feedback device; a function of the output is fed back as an input torque in opposition to the applied torque. A signal flow chart demonstrates the feedback relationship graphically (Fig. 3.21)

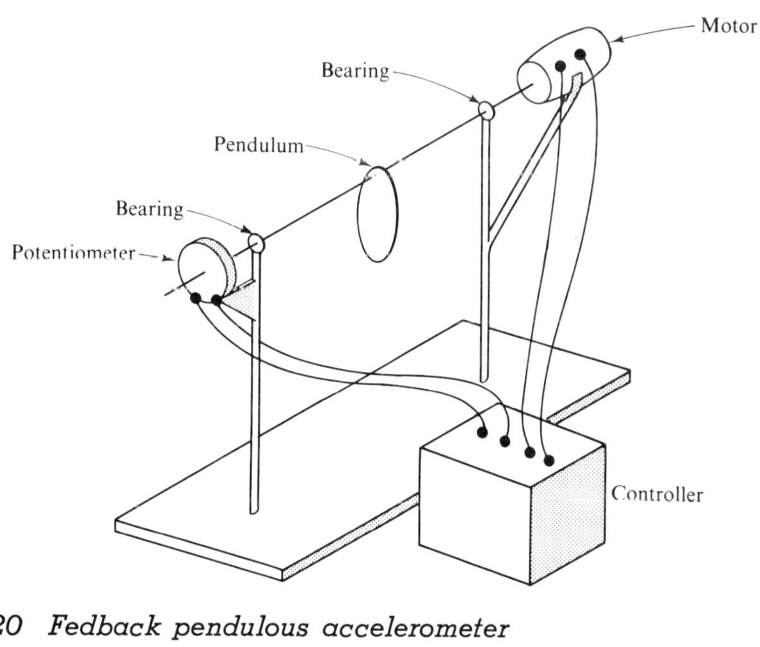

FIGURE 3.20 Fedback pendulous accelerometer

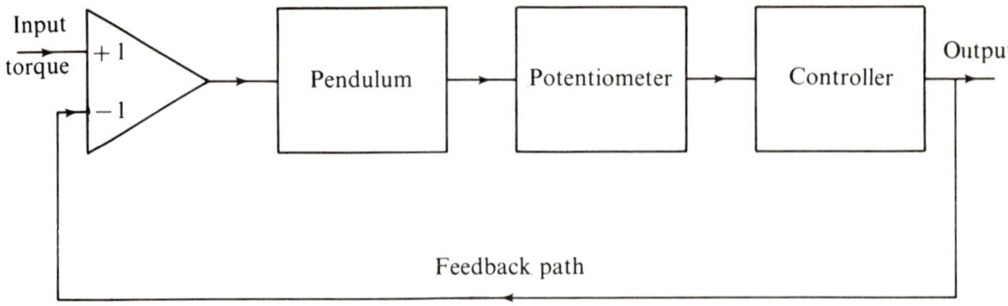

FIGURE 3.21 Signal flow chart

The system equation is now

$$I\ddot{\theta} + c\dot{\theta} + mg\ell \sin\theta = -m\ell(\ddot{v}\sin\theta - \ddot{u}\cos\theta) - T_F$$

in which T_F is the feedback torque. The controller is designed to operate with the performance equation

$$T_F = k_1\theta + k_2\dot{\theta} + k_3\int_0^t \theta\, dt$$

that is, the torque is proportional to a linear combination of the angular deflection, its first derivative and its integral.[28] The system equation now becomes

$$\begin{aligned} I\dddot{\theta} + c\ddot{\theta} + mg\ell\cos\theta\,\dot{\theta} = &-m\ell(\ddot{v}\sin\theta - \ddot{u}\cos\theta) \\ &- m\ell(\ddot{v}\cos\theta + \ddot{u}\sin\theta)\,\dot{\theta} \\ &- k_1\dot{\theta} - k_2\ddot{\theta} - k_3\theta \end{aligned}$$

Let a steady acceleration $[(\ddot{u})^2 + (\ddot{v})^2]^{1/2}$ be introduced and let us assume that the system is stable so that an ultimate state is reached. Then all the derivatives of θ will in time equal zero, which implies, by virtue of $\ddot{u}$ and $\ddot{v}$ being zero, that θ ultimately becomes zero. From that it follows that in the ultimate state

$$T_{F_u} = m\ell\ddot{u}$$

Thus, since the torque is proportional to the motor field voltage, the output voltage of the controller is proportional, with constant sensitivity to the horizontal acceleration. The resulting instrument is essentially linear and, in the ultimate state, insensitive to cross acceleration. Obviously, as soon as an angle θ exceeding the threshold is cranked into the system, the controller will begin to crank the pendulum back below the threshold. Thus the angle θ will not diverge much from zero and can be treated as small. Hence $\theta \approx \sin\theta$, $1 \approx \cos\theta$ and the equation of motion becomes

$$I\ddot{\theta} + (c + k_2)\dot{\theta} + (mg + m\ell\ddot{v} + k_1)\theta + k_3\int_0^t \theta\, dt = m\ddot{u}$$

For constant v and u the equation can be written as

$$I\dddot{\theta} + (c + k_2)\ddot{\theta} + (mg + k_1)\dot{\theta} + k_3\theta = 0$$

which obviously has the ultimate state $\theta = 0$.

The equation for the torque is now

$$I\dddot{T} + (c + k_2)\ddot{T} + (mg\ell + m\ell\ddot{v} + k_1)\dot{T} + k_3 T$$
$$= k_3 m\ell\ddot{u} + k_1 m\ell\ddot{u} + k_2 m\ell\dddot{u}$$

which, for steady u, yields the result $T = m\ell\ddot{u}$. The steady ultimate state is, therefore, insensitive to steady cross acceleration.

PROBLEMS

3.1 Compute S_0 and S_1 and plot them as functions of time if the system equation has the form

$$\frac{d^2\xi}{dt^2} + 6\frac{d\xi}{dt} + 5\xi = 2.7y(t)$$

3.2 Let $y(t)$ in problem 3.1 be sin 0.25 t. Compute ξ using only S_0 and S_1 and also obtain the exact solution. At what time will the difference between the two solutions exceed 0.1% of the exact solution?

3.3 Show, with the aid of Equation (3.3.3), that if the input to a system is $y(t) = \sin \omega t$, the output is given by $\xi = S_s \sin \omega t + S_c \cos \omega t$ in which S_s and S_c are functions of time representable as infinite series in the sensitivity functions. Show that

$$S_s = \int_{0^+}^{t} I(\tau) \cos \omega\tau \, dy$$

3.4 A thermocouple is connected into a galvanometer to give a deflection ξ (in radians) which corresponds to a temperature input θ (°C). The system obeys the equation

$$\frac{d^3\xi}{dt^3} + 2.91\frac{d^2\xi}{dt^2} + 5.82\frac{d\xi}{dt} + 6.21\xi = 0.195\theta(t)$$

(a) If the maximum deflection of the galvanometer needle is 90, determine the temperature range in °C.
(b) What is the ultimate state sensitivity in radians/°C?
(c) What is the scale factor in °C/radians?

3.5 Determine S_0 and S_1 for the system in problem 3.4.

3.6 Prove that $A_1 = S_1 + tS_0$. Use the results of problem 3.5 to compute A_1.

3.7 Compute and plot A_2 for the system of problem 3.4.

3.8 Prove that

$$A_n = \sum_{m=0}^{n} a_m S_m(t) t^{n-m}$$

and determine a_m as a function of m and n.

3.9 The pendulous accelerometer has been analyzed under the assumption that the axis of rotation is maintained parallel to the earth's surface. Assume that the rotation axis forms a fixed nonzero angle with the horizon and determine the ultimate state sensitivity of the system.

3.10 A galvanometer is connected as shown in Figure 3.22. The voltage E is switched on at time $t = 0$.

(a) Determine the circuit behavior without the galvanometer.

(b) How does the galvanometer affect the circuit behavior?

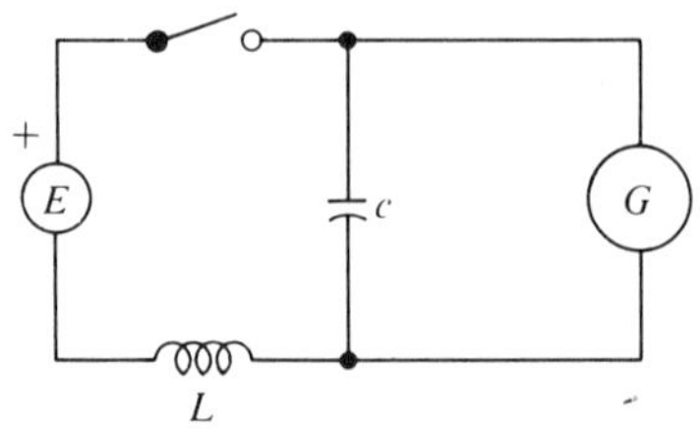

FIGURE 3.22 Galvanometer circuit for problem 3.10

3.11 Assume that a strain column attached to a flat diaphragm has no effect upon the diaphragm. Establish the performance equation of a pressure transducer which uses a strain column, strain gage bridge and galvanometer to give an output deflection in functional relationship with the input pressure. What is the ultimate state sensitivity?

3.12 Show that the pressure transducer responds to acceleration as well as pressure. Propose a method for canceling the acceleration effects.

NOTES

[1]Ref. 5, p. 13.

[2]Ref. 2, p. 3; Ref. 1, Vol. 3, p. 16. Compare with Ref. 38.

[3]Ref. 4, p. 9.

[4]Ref. 1, Vol. 3, p. 16.

[5]Much engineering effort is lost in making instruments truly linear and in defining a constant sensitivity. It would be better for instrument designers to accept nonlinearity and let it go at that. Since such a radical departure from practices would require engineers to learn about nonlinear systems, it is not likely that the millenium will ever come.

[6]Ref. 45, p. 472.

[7]A simple analysis of the instrument for $\ddot{v} = 0$ is given in Ref. 15, pp. 35–38.

[8]Ref. 30, p. 458.

[9]This argument is not a mathematical proof but simply an exercise in faith.

[10]Ref. 46, pp. 160–85. See also Ref. 50.

[11]Ref. 39 and 48.

[12]The magnitude of the friction of galvanometer pivots was investigated by Stott in 1931 (Ref. 49). J. Miller developed practical criteria, based on Stott's measurements, for the determination of bearing friction (Ref. 6, pp. 8–86).

[13]Ref. 44, pp. 430–36.

[14]Ref. 44, p. 594.

[15]Ref. 7, passim. For elastic basis see Ref. 72.

[16]Ref. 52, p. 50. See also Ref. 16.

[17]Ref. 7, p. 53 and passim.

[18]Ref. 53, p. 2.

[19]A fairly complete analysis is found in Ref. 53 for linearized "small" disturbance inputs. This simplified analysis, though not particularly difficult, is quite lengthy and cannot be introduced here. The static theory has been examined with some care in References 54 and 55.

[20]Ref. 1, Vol. 3, pp. 199–204.

[21]Refs. 1, 2, 3, 4, 5, 6, 11, 51, 86, 102.

[22]Ref. 66, pp. 1072–77.

[23]Ref. 6, pp. 3–16 to 3–26. See also Refs. 3, 4.

[24]Ref. 63, pp. 441–455.

[25]An integral assumption in the application of the method of Rayleigh-Ritz. Ref. 63, p. 446.

[26]Details can be found in Ref. 1, Vol. 3, pp. 407–10.

[27]Ref. 65, p. 120, Fig. 7–2.

[28]This is not difficult to do. See, for example, Ref. 67, pp. 1–9 to 1–17. Complete design details for feedback servomechanisms can be found in Ref. 16, 29, 71, 74.

4

DYNAMICAL ERROR

4.1 Introduction

It was established in chapter 3 that the output of a linear system is of the form

$$\xi = S_0(t)y(t) + f(t) \tag{4.1.1}$$

and that the design conditions should be $S_0 = S =$ constant, $f(t) = 0$. Unfortunately the best one can do is

$$\lim_{t\to\infty} S_0(t) = S$$
$$\lim_{t\to\infty} f(t) = \epsilon_D \tag{4.1.2}$$

in which the design parameters are selected so as to minimize ϵ_D and stabilize S as quickly as possible.

In many instances, even the relaxed conditions (4.1.2) cannot be met for all input functions. Consider, for example, a recording galvanometer connected to a voltage source. If the source were instantaneously* shut

*Engineering definition of *instantaneously*: as rapidly as possible using fastest available equipment.

off, the galvanometer would effectively be subjected to a negative step. The galvanometer will return to the zero position as rapidly as possible but not instantaneously. Inertia, friction and electrical losses all conspire to retard the action so that the actual galvanometer output trace will look like the broken line in Figure 4.1. The ultimate state, time t_2, is the time at which the system "freezes up" as a result of static friction or some other threshold mechanism. In this instance the expected output does not match the expected input (within an acceptable error) until time t_2 is reached. The voltage recorded between t_1 and t_2 is a dynamical error. Ideal design of this instrument would be achieved by making $t_2 - t_1 = 0$, whereas practical design is achieved by making $t_2 - t_1$ as small as possible, or at least small enough to be unimportant in the measurement situation for which the instrument is designed.

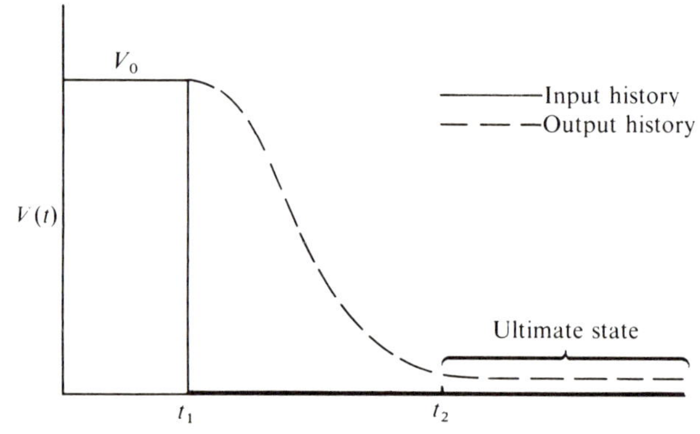

FIGURE. 4.1 Response of a galvanometer to a negative step

The statement at the end of the last paragraph is a major principle of economic design: it can be shown that cost increases rapidly with decreasing response time, and thus the engineer should not build into an instrument a response capability greater than is needed. A galvanometer used as a power station monitor can have a 10-second response lag without endangering its mission, whereas a galvanometer with a response lag of 15 milliseconds would be unable to respond correctly to a thermocouple used to detect the passage of a pneumatic shock wave in a shock tube.

The response of an instrument to a step function is characterized by the sensitivity coefficient $S_0(t)$. The response time to the unit step is the time T_R in which $S_0(t)$ becomes $S \pm \epsilon_S$ where ϵ_S is the design sensitivity tolerance. If, for example, the designer were willing to accept a 1% error in the output of the system, then T_R would be defined by the time such that for all subsequent times $t(t > T_R)$ the inequality

$$\frac{S_0(t)}{S} - 1 \leq 0.01$$

is satisfied.

What happens to the error $f(t)$ is usually a cause of unhappiness to the designer. To minimize $f(t)$ it is required to minimize $S_n(t)$ for $n > 0$. The transient part of $S_n(t)$ can usually be shown to vanish at least as rapidly as $S_0 \longrightarrow S$ so that optimization of S_0 usually is sufficient to guarantee the rapid decay of the transient response. Unfortunately the $S_n(t)$ have a nonzero ultimate state and hence the derivatives of a time-varying input function can be expected to contribute an output error at all times. Minimization of this error often runs counter to requirements for the satisfaction of other criteria. In most instances, extremization becomes a trade-off between available amplification and required dynamical response.

It is evident that dynamic design is accomplished by performing the following operations:

(1) determine the performance equations;
(2) determine the speed of the inputs;
(3) determine the design requirements;
(4) compute the $S_n(t)$ and S;
(5) make $S_0(t)$ meet the design requirements;
(6) determine $f(t)$ for the allowable inputs and attempt to minimize $f(t)$ to meet design requirements.

4.2 Stability

To establish a set of design parameters the designer is required to compute the sensitivity functions and consequently to compute the characteristic roots of the characteristic polynomial of the performance equation. Although the computation is quite straightforward it is laborious, and whenever the order of the performance equation exceeds the fourth order, must be performed by approximate procedures. Design is accomplished by varying the coefficients of the characteristic polynomial, computing the corresponding roots and selecting the design from the plotted family of root loci.

The designer does not enter the root computation derby blindly. He has absolute limits imposed on his computation by the equipment and material at his disposal, and he has a mathematical limit enforced by the needs of system stability. By "stability" is meant the requirement that the instrument reach an ultimate state, or, as will be shown later, that the various $S_n(t)$ become constants in time. Thus it is required that the first derivative of $S_n(t)$ becomes zero in time, which implies that

$$\lim_{t \to \infty} I(t) = 0$$

faster than t^n approaches infinity for any value of n. This last requirement can be translated to mean that $I(t)$ vanishes exponentially. It was shown in chapter 2 that if the roots of the characteristic equations are given by $r_1, r_2, \ldots, r_m$, then

$$I(t) = f_1(t)e^{r_1t} + f_2(t)e^{r_2t} + \cdots + f_m(t)e^{r_3t}$$

in which the $f_i(t)$ are either constants or polynomials in t. (The latter condition corresponds to the possible existence of repeated roots.) Thus it immediately follows that a necessary and sufficient condition for $I(t)$ to become zero in time is that the real part of the characteristic roots be always less than zero. A bound on the design parameters can be found, therefore, by determining the crossover points for the roots of the characteristic equation.

For a second-order system, in which $s^2 + as + b = 0$ is the characteristic equation, the crossover point is found directly. The roots are

$$s_{1,2} = -\frac{a}{2} \pm \sqrt{\left(\frac{a}{2}\right)^2 - b}$$

In the case where $0 < (a/2)^2 < b$, the real part of the roots is given by $-a/2$, and thus $a > 0, b > 0$ guarantee stability. For $(a/2)^2 > b$ it is obvious that there will be one positive root if $b < 0$. Thus, under all circumstances, $a > 0, b > 0$ is necessary and sufficient to guarantee the stability of a second-order system.

In systems of higher order the situation is not so straightforward. Although it is required that the coefficients of the characteristic equation be of the same sign, the condition is necessary but not sufficient to guarantee stability. Given the difficulty (and in many instances, the undesirability) of extraction of the roots, the generation of sufficiency conditions is not directly approachable.

An indirect theorem, known as the Routh-Hurwitz theorem, is used to do the job. Proof of this theorem is rather lengthy and will be omitted here.[1] There are two forms of the theorem—the Routh form, popular in the United States, and the Hurwitz form, which is most commonly used in Europe. The Routh criterion gives more information but is more complicated since it involves special treatment of certain singular cases. The Hurwitz criterion, on the other hand, is extremely straightforward.[2] Since, in instrument design, it is simply required to reject unstable systems, the Hurwitz criterion, which does just that, will be used.

Hurwitz Criterion

IN ORDER THAT THE ROOTS OF THE POLYNOMIAL

$$s^n + a_{n-1}s^{n-1} + a_{n-2}s^{n-2} + \cdots + \cdots a_1s + a_0 = 0$$

HAVE NEGATIVE REAL PARTS, IT IS NECESSARY AND SUFFICIENT THAT $a_i > 0$ AND THAT[3]

$$\begin{vmatrix} a_{n-1} & a_{n-3} \\ 1 & a_{n-2} \end{vmatrix} > 0$$

$$\begin{vmatrix} a_{n-1} & a_{n-3} & a_{n-5} \\ 1 & a_{n-2} & a_{n-4} \\ 0 & a_{n-1} & a_{n-3} \end{vmatrix} > 0$$

$$\begin{vmatrix} a_{n-1} & a_{n-3} & a_{n-5} & a_{n-7} \\ 1 & a_{n-2} & a_{n-4} & a_{n-6} \\ 0 & a_{n-1} & a_{n-3} & a_{n-5} \\ 0 & 1 & a_{n-2} & a_{n-4} \end{vmatrix} > 0$$

THE PROCESS OF CONSTRUCTION OF DETERMINANTS IS TO CONTINUE UNTIL THE TERM a_0 APPEARS IN THE BOTTOM ROW AND THE LAST COLUMN.

$$\begin{vmatrix} a_{n-1} & a_{n-3} & a_{n-5} & \cdots & 0 \\ 1 & a_{n-2} & a_{n-4} & \cdots & 0 \\ 0 & a_{n-1} & a_{n-3} & \cdots & 0 \\ 0 & 1 & a_{n-2} & \cdots & 0 \\ \cdots & \cdots & \cdots & \cdots & \cdots \\ 0 & 0 & 0 & \cdots & a_0 \end{vmatrix} > 0$$

Example 4.2.1 Obtain criteria which guarantee roots with negative real parts in the polynomial equations

$$s + a_0 = 0 \tag{4.2.1}$$

$$s^2 + a_1 s + a_0 = 0 \tag{4.2.2}$$

$$s^3 + a_2 s^2 + a_1 s + a_0 = 0 \tag{4.2.3}$$

$$s^4 + a_3 s^3 + a_2 s^2 + a_1 s + a_0 = 0 \tag{4.2.4}$$

Solution

$$a_0 > 0 \tag{4.2.1}$$

$$a_1 > 0 \qquad \begin{vmatrix} a_1 & 0 \\ 1 & a_0 \end{vmatrix} > 0$$

$$\text{Hence } a_1 > 0,\ a_0 > 0 \tag{4.2.2}$$

$$a_2 > 0 \qquad \begin{vmatrix} a_2 & a_0 \\ 1 & a_1 \end{vmatrix} > 0$$

$$\begin{vmatrix} a_2 & a_0 & 0 \\ 1 & a_1 & 0 \\ 0 & a_2 & a_0 \end{vmatrix} > 0$$

$$\text{Hence } a_1 > 0,\ a_2 > 0,\ a_0 > 0 \text{ and } a_2 a_1 - a_0 > 0 \tag{4.2.3}$$

$$a_0 > 0,\ a_1 > 0,\ a_2 > 0,\ a_3 > 0,\ a_e a_2 - a_1 > 0$$

$$a_1(a_3 a_2 - a_1) - a_0 a_3^2 > 0 \tag{4.2.4}$$

Example 4.2.2 Does the polynomial

$$s^{10} + s^9 - s^8 - 2s^7 + s^6 + 3s^5 - 2s^4 + s^3 + 2s^2 + s + 1 = 0$$

represent the characteristic polynomial of a stable system?

Solution No. Not all $a_i > 0$.

Example 4.2.3 Show that the roots of

$$s^6 + s^5 + 6s^4 + 5s^3 + 11s^2 + 7s + 6 = 0$$

satisfy the conditions of stability.

Solution

$$\begin{vmatrix} 1 & 5 \\ 1 & 6 \end{vmatrix} = 1 \qquad \begin{vmatrix} 1 & 5 & 7 \\ 1 & 6 & 11 \\ 0 & 1 & 5 \end{vmatrix} = 1$$

$$\begin{vmatrix} 1 & 5 & 7 & 0 \\ 1 & 6 & 11 & 6 \\ 0 & 1 & 5 & 7 \\ 0 & 1 & 6 & 11 \end{vmatrix} = 192 \qquad \begin{vmatrix} 1 & 5 & 7 & 0 & 0 \\ 1 & 6 & 11 & 6 & 0 \\ 0 & 1 & 5 & 7 & 0 \\ 0 & 1 & 6 & 11 & 6 \\ 0 & 0 & 1 & 5 & 7 \end{vmatrix} = 1{,}308$$

$$\begin{vmatrix} 1 & 5 & 7 & 0 & 0 & 0 \\ 1 & 6 & 11 & 6 & 0 & 0 \\ 0 & 1 & 5 & 7 & 0 & 0 \\ 0 & 1 & 6 & 11 & 6 & 0 \\ 0 & 0 & 1 & 5 & 7 & 0 \\ 0 & 0 & 1 & 6 & 11 & 6 \end{vmatrix} = 7{,}848$$

Since all the Hurwitz determinants are greater than zero, the polynomial can be the characteristic polynomial of a stable linear system.

Example 4.2.4 Determine the range of K for which the system represented by

$$\frac{d^3\xi}{dt^3} + 6\frac{d^2\xi}{dt^2} + K\frac{d\xi}{dt} + 9\xi = 7.2y(t)$$

is stable.

Solution From Equation (4.2.3) we find that $6K - 9 > 0$ is both necessary and sufficient. Thus stability is assured if $1.5 < K < \infty$.

4.3 Sensitivity Functions

The roots of a characteristic polynomial which represents a stable linear system can only be real negative roots or complex conjugate roots with negative real parts. As a consequence, factorization of the polynomial into the product of a set of lowest-order polynomials with real coefficients will yield factor of the form $(s + r_i)^{p_i}$ and $(s^2 + 2\zeta_i\,\omega_i s + \omega_i)^{q_i}$ in which the former is a consequence of the real roots and the latter the consequence of a system with pairs of conjugate complex roots. With the convention that complex pairs are left in quadratic form, the factorization of the impulsive transfer function

$$I_m(t) = \frac{K}{s^m + a_{m-1}s^{m-1} + a_{m-2}s^{m-2} + \cdots + a_0}$$

which corresponds to the system equation

$$\frac{d^m\xi}{dt^m} + a_{m-1}\frac{d^{m-1}\xi}{dt^{m-1}} + \cdots + a_0\xi = K\,\delta(t)$$

is given by

$$I_m(t) = \frac{K}{[(s+r_1)\ldots(s+r_\mu)][(s^2+2\zeta_1\omega_1 s+\omega_1^2)\ldots(s^2+2\zeta_\nu\omega_\nu s+\omega_\nu^2)]}$$

$$\mu + 2\nu = m \qquad \textbf{(4.3.1)}$$

whenever there are no repeated roots. (The treatment of repeated roots is a separate question.) Inversion is straightforward:

$$I_m(t) = \sum_{i=1}^{\mu} A_i' e^{-r_i t} + \sum_{i=1}^{v} B_i' e^{-\zeta_i\omega_i t} \sin(\omega_i\sqrt{1-\zeta_i^2}\,t) \qquad \textbf{(4.3.2)}$$

in which the A_i' and B_i' are independent of time and can be computed in terms of r_i, ζ_i and ω_i and K. Computation of the A_i' and B_i' shows that under all circumstances

$$A_i' = KA_i \quad \text{and} \quad B_i' = \frac{KB_i}{\omega_i\sqrt{1-\zeta_i^2}}$$

and that A_i and B_i are independent of K. The explicit form of the impulsive response is

$$I_m(t) = K\left[\sum_{i=1}^{\mu} A_i e^{-r_i t} + \sum_{i=1}^{\nu} \frac{B_i e^{-\zeta\omega_i t}}{\omega_i\sqrt{1-\zeta_i^2}} \sin(\omega_i\sqrt{1-\zeta_i^2}\,t)\right] \qquad \textbf{(4.3.3)}$$

The quantities A_i, B_i depend on the order of the differential equation and can be computed by direct inversion. Some values of A_i and B_i have been computed for the first four systems (first through fourth order) and are tabulated in Table 4.1.

TABLE 4.1 Impulsive response coefficients

1.	
Equation	$\frac{d\xi}{dt} + a_0\xi = Ky(t)$
Special conditions	$a_0 > 0$
Factored form	$s + a_0 = 0$
Coefficients	$A_1^{(1)} = 1, \quad B_1^{(1)} = 0$
	$A_i^{(1)} = B_i^{(1)} = 0, \quad i > 1$
2.	
Equation	$\frac{d^2\xi}{dt^2} + a_1\frac{d\xi}{dt} + a_0\xi = Ky(t)$
Special conditions	$\left(\frac{a_1}{2}\right)^2 - a_0 > 0$
Factored form	$(s + r_1)(s + r_2) = 0$

TABLE 4.1 (*Continued*)

Coefficients	$A_1^{(2)} = \dfrac{1}{r_2 - r_1}$
	$A_2^{(2)} = \dfrac{1}{r_1 - r_2}$
	$A_i^{(2)} = 0, \quad i > 2$
	$B_i^{(2)} = 0, \quad i \geq 1$
3.	
Equation	$\dfrac{d^2\xi}{dt^2} + a_1\dfrac{d\xi}{dt} + a_0\xi = Ky(t)$
	$a_1 = 2\zeta\omega, \quad a_2 = \omega^2$
Special conditions	$\left(\dfrac{a_1}{2}\right)^2 - a_0 < 0$
Factored form	$s^2 + 2\zeta\omega s + \omega^2 = 0$
Coefficients	$A_i^{(2)} = 0, \quad i \geq 1$
	$B_i^{(2)} = 1$
	$B_i^{(2)} = 0, \quad i > 1$
4.	
Equation	$\dfrac{d^3\xi}{dt^3} + a_2\dfrac{d^2\zeta}{dt^2} + a_1\dfrac{d\xi}{dt} + a_0\xi = Ky(t)$
Special conditions	$p = a_1 - \dfrac{a_2}{3}, \quad q = a_0 - \dfrac{a_1 a_2}{3} + \dfrac{2a_2^3}{27}$
	$\Delta = 4p^3 + 27q^2, \quad \Delta \leq 0$
Factored form	$(s + r_1)(s + r_2)(s + r_3) = 0$
Coefficients	$A_1^{(3)} = \dfrac{1}{(r_2 - r_1)(r_3 - r_1)}$
	$A_2^{(3)} = \dfrac{1}{(r_1 - r_2)(r_3 - r_1)}$
	$A_3^{(3)} = \dfrac{1}{(r_1 - r_3)(r_2 - r_3)}$
	$A_i^{(3)} = 0, \quad i > 3$
	$B_i^{(3)} = 0, i \geq 1$
5.	
Equation	$\dfrac{d^3\xi}{dt^3} + a_2\dfrac{d^2\xi}{dt^2} + a_1\dfrac{d\xi}{dt} + a_0 = Ky(t)$
Special conditions	$\Delta > 0$ (See 4, above.)
Factored form	$(s + r_1)(s^2 + 2\zeta\omega s + \omega^2)$
Coefficients	$A_1^{(3)} = \dfrac{1}{(\zeta\omega - r_1)^2 + \omega^2(1 - \zeta^2)}$

TABLE 4.1 (Continued)

	$A_i^{(3)} = 0, \quad i > 1$
	$B_1^{(3)} = \dfrac{1}{[(\zeta\omega - r_1)^2 + \omega^2(1 - \zeta^2)]^{1/2}}$
	$B_i^{(3)} = 0, \quad i > 1$
6.	
Equation	$\dfrac{d^4\xi}{dt^4} + a_3\dfrac{d^3\xi}{dt^3} + a_2\dfrac{d^2\xi}{dt^2} + a_1\dfrac{d\xi}{dt} + a_0\xi = Ky(t)$
Special conditions	$p = a_3a_1 - \dfrac{4}{3}a_0 - \dfrac{1}{3}a_2^2$
	$q = -a_3^2a_0 + \dfrac{1}{3}a_3a_2a_1 + \dfrac{8}{3}a_2a_0a_1^2 - \dfrac{2}{27}a_2^2$
	$\Delta = 4p^3 + 27q^2, \quad \Delta \leq 0$
Factored form	$(s + r_1)(s + r_2)(s + r_3)(s + r_4) = 0$
Coefficients	$A_1^{(4)} = \dfrac{1}{(r_2 - r_1)(r_3 - r_1)(r_4 - r_1)}$
	$A_2^{(4)} = \dfrac{1}{(r_1 - r_2)(r_3 - r_2)(r_4 - r_2)}$
	$A_3^{(4)} = \dfrac{1}{(r_1 - r_3)(r_2 - r_3)(r_4 - r_3)}$
	$A_4^{(4)} = \dfrac{1}{(r_1 - r_4)(r_2 - r_4)(r_3 - r_4)}$
	$A_i^{(4)} = 0, \quad i > 4, \quad B_i^{(4)} = 0, \quad i \geq 1$
7.	
Equation	$\dfrac{d^4\xi}{dt^4} + a_3\dfrac{d^3\xi}{dt^3} + a_2\dfrac{d^2\xi}{dt^2} + a_1\dfrac{d\xi}{dt} + a_0\xi = Ky(t)$
Special conditions	$\Delta > 0$ (See 6, above.)
Factored form	$(s + r_1)(s + r_2)(s^2 + 2\zeta\omega s + \omega^2) = 0$
Coefficients	$A_1^{(4)} = \dfrac{1}{(r_2 - r_1)(r_1^2 + 2\zeta\omega r_1 + \omega^2)}$
	$A_2^{(4)} = \dfrac{1}{(r_1 - r_2)(r_2^2 + 2\zeta\omega r_2 + \omega^2)}$
	$A_i^{(4)} = 0, \quad i > 2$
	$B_1^{(4)} = \dfrac{1}{u}$
	$u = [(r_2 - \zeta\omega)^2 + \omega^2(1 - \zeta^2)]$ $[(r_1 - \zeta\omega)^2 + \omega^2(1 - \zeta^2)]$
	$B_i^{(4)} = 0, \quad i > 1$

TABLE 4.1 (Continued)

8.

Equation	$\frac{d^4\xi}{dt^4} + a_3\frac{d^3\xi}{dt^3} + a_2\frac{d^2\xi}{dt^2} + a_1\frac{d\xi}{dt} + a_0\xi = Ky(t)$
Special conditions	$\Delta < 0$ (See 6, above.)
Factored form	$(s^2 + 2\zeta_1\omega_1 s + \omega_1^2)\cdot(s^2 + 2\zeta_2\omega_2 s + \omega_1^2) = 0$
Coefficients	$A_i^{(4)} = 0, \quad i \geq 1$
	$B_1^{(4)} = B_2^{(4)} = \frac{1}{u}$
	$B_i^{(4)} = 0, \quad i > 2$
	$u^2 = [(\zeta_2\omega_2 - \zeta_1\omega_1)^2 + (\omega_2\zeta_2^2 - \omega_1\zeta_1^2) + (\omega_1 - \omega_2)]^2 + 4\omega_2^2(1 - \zeta_2^2)(\zeta_2\omega_2 - \zeta_1\omega_1)$

Equation (4.3.3) can be introduced into the definition of the sensitivity parameters

$$S_n = (-1)^n \int_{0^+}^{t} I_m(\tau)\frac{\tau^n}{n!}\, d\tau$$

$$= (-1)^n \frac{K}{n!} \sum_{i=1}^{\mu} A_i \int_{0^+}^{t} e^{-r_i\tau}\,\tau^n\, d\tau$$

$$+ \sum_{i=1}^{\nu} \frac{B_i}{\omega_i\sqrt{1-\zeta_i^2}} \int_{0^+}^{t} \tau^n e^{-\zeta_i\omega_i\tau} \sin(\omega_i\,\tau\sqrt{1-\zeta_i})\, d\tau \qquad \textbf{(4.3.4)}$$

Equation (4.3.4) demonstrates that all sensitivity functions are compounded from only two types of integrals:

$$U_{n_i}^{(1)} = \frac{(-1)^n A_i^{(m)}}{n!} \int_{0^+}^{t} e^{-r_i\tau}\tau^n\, d\tau$$

$$U_{n_i}^{(2)} = \frac{(-1)^n B_i^{(m)}}{n!\,\omega_i\sqrt{1-\zeta_i^2}} \int_{0^+}^{t} e^{-\zeta_i\omega_i\tau} \sin(\omega_i\sqrt{1-\zeta_i^2}\,\tau)\tau^n\, d\tau$$

Both integrals can be evaluated by using the well-known formula[4]

$$\int_{0^+}^{t} \tau^n e^{-a\tau}\, d\tau = \frac{n!}{a^{n+1}}\, 1 - e^{-at} \sum_{i=1}^{n} \frac{(at)^n}{(n-1)!}$$

After considerable manipulation the integrals can be shown to be

$$U_{n_i}^{(1)} = (-1)^n \frac{A_i^{(m)}}{r_i^{n+1}} \left[1 - e^{-r_k t} \sum_{k=0}^{n} \frac{(r_i t)^{n-k}}{(n-k)!}\right] \qquad \textbf{(4.3.5)}$$

$$U_{n_i}^{(2)} = \frac{(-1)^n B_i^{(m)}}{\omega_i^{n+2}\sqrt{1-\zeta_i^2}}\Big\{\sin(n+1)\phi_i$$

$$+\, e^{-\zeta_i\omega_i t}\sin(\omega_i\sqrt{1-\zeta_i^2}\,t)\Big[\sum_{j=0}^{n}\frac{(\omega_i t)^{j+1}\cos(j+1)\phi_i}{(j+1)!}\Big]$$

$$-\, e^{-\zeta_i\omega_i t}\cos(\omega_i\sqrt{1-\zeta_i^2}\,t)\Big[\sum_{j=0}^{n}\frac{(\omega_i t)^{j+1}\sin(j+1)\phi_i}{(j+1)!}\Big]\Big\} \tag{4.3.6}$$

in which

$$\sin\phi_i = \sqrt{1-\zeta_i^2} \qquad \cos\phi_i = \zeta_i$$

Although Equation (4.3.6) is rather messy in appearance, it is fairly easy to compute. With the equation

$$S_n = K\Big(\sum_{i=1}^{\mu} U_{n_i}^{(1)} + \sum_{i=1}^{\nu} U_{n_i}^{(2)}\Big) \tag{4.3.7}$$

$$\mu + 2\nu = m$$

the sensitivity functions are completely defined in terms of the system characteristic roots.

4.4 Structure of the Solution

Equation (4.3.7) can be written in a structural form

$$S_n = (-1)^n K\Big[\sum_{i=1}^{\mu}\frac{A_i^{(m)}}{r_i^n} + \sum_{i=1}^{\nu}\frac{B_i^{(m)}}{\omega_i^{n+2}}M_{in}(\zeta_i)\Big]$$

$$+\,(-1)^{n+1}K\Big[\sum_{i=1}^{\mu} e^{-r_i t}R_i^{(1)}(t) + \sum_{i=1}^{\nu} e^{-\zeta_i\omega_i t}R_i^{(2)}(t)\Big]$$

$$\mu + 2\nu = m \tag{4.4.1}$$

in which M_i, $R_i^{(1)}$, $R_i^{(2)}$ are obvious functions.

In a stable system $r_i > 0$ and $\zeta_i\omega_i > 0$, whence it can be established that the elements in the second brackets will vanish in time. The summations within the first brackets are independent of time and contribute to the solution at all times. Thus the ultimate state sensitivity can be obtained immediately

$$\lim_{t\to\infty} S_0(t) = S = K\Big(\sum_{i=1}^{\mu}\frac{A_i^{(m)}}{r_i} + \sum_{i=1}^{\nu}\frac{B_i^{(m)}}{\omega_i^2}\Big)$$

$$\mu + 2\nu = m$$

If it is remembered that the solution of the system equation has the form

$$\xi = S_0(t)y(t) + S_1(t)y^{(1)}(t) + \cdots$$

it becomes evident that the output ξ will *always* respond to the input function and all its derivatives. Even in the limit, as $t \longrightarrow \infty$, the S_n ($n > 0$) do not vanish, so that ξ may, for a very rapidly changing input, respond more nearly to some derivatives of the input than to the input itself.

Example 4.4.1 A system has ultimate state sensitivities

$$S = 1 \qquad S_{1u} = -0.5 \qquad S_{2u} = 10^{-7} \qquad S_{nu} = 10^{-7(n-1)} \qquad (n \geq 3)$$

The instrument is required to respond to an input $y(t) = \sin 10^3 t$. Show that the system responds mostly to $\cos 10^3 t$.

Solution

$$\frac{d^n y}{dt^n} = \begin{cases} (-1)^{n/2} 10^{3n} \sin 10^3 t & (n \text{ even}) \\ (-1)^{(n+1)/2} 10^{3n} \cos 10^3 t & (n \text{ odd}) \end{cases}$$

Thus

$$\begin{aligned} \xi &= \sin 10^3 t - 0.5 \cdot 10^3 \cos 10^3 t - 10^{-7} \cdot 10^6 \sin 10^3 t \\ &\quad + 10^{-14} \cdot 10^9 \cos 10^3 t + 10^{-21} \cdot 10^{12} \sin 10^3 t + \ldots \\ &= (1.01 - 10^{-1} + \ldots) \sin 10^3 + (50 - 10^{-5} + \ldots) \cos 10^3 t \\ &\approx 1.01 \sin 10^3 t - 50 \cos 10^3 t \end{aligned}$$

The expected output is only 2% of the unwanted first derivative output. In this instance (fortunately very unlikely in linear systems) the information signal is only a minor perturbation of the dynamical error.

4.5 The Ultimate State Dynamic Error for a First-order System

Both minimization of the ultimate state dynamic error and rapid decay of the transient dynamic error are design requirements. It is not possible to treat the two separately for the simple reason that both conditions are minimized by variation of the same parameters. Consider, for example, a simple first-order system

$$\frac{d\xi}{dt} + a_0 \xi = Ky(t) \qquad \textbf{(4.5.1)}$$

Optimization of the system requires that an infinite number of sensitivity functions be set at zero and that $S_0 = S$ as rapidly as possible. But only two parameters are available to the engineer, a_0 and K; evidently absolute optimization is out of the question.

The sensitivity coefficients for the first-order system are given by Equations (4.3.5), (4.3.6), and Table 4.1 as

$$S_n = (-1)^n \frac{K}{a_0^{n+1}} \left[1 - e^{-a_0 t} \sum_{k=0}^{n} \frac{(a_0 t)^{n-k}}{(n-k)!} \right]$$

so that the system response is represented by

$$\xi = K \left\{ \sum_{n=0}^{\infty} \left[\frac{(-1)^n}{a_0^{n+1}} \right] y^{(n)} - e^{-a_0 t} \sum_{n=0}^{\infty} \left[\frac{(-1)^n}{a_0^{n+1}} \right] y^{(n)} \sum_{k=0}^{\infty} \frac{(a_0 t)^{n-k}}{(n-k)!} \right\} \qquad \textbf{(4.5.2)}$$

It is obvious that if we select $a =$ (maximum a_0) > 0, then a is the design element which will make the transient vanish as rapidly as possible.

No further optimization can be performed. $a = a_0$ fixes thc relative amplitude of ultimate state error. K, which is multiplied into all the terms, only provides a scale adjustment. As luck would have it, maximization of a_0 minimizes the dynamic ultimate state error. If, as is often done, we set $K = K_0 a_0$, $K_0 \geq 1$, the ultimate state has the form

$$\xi_u = K_0 y(t) - \frac{K_0}{a_0}\frac{dy}{dt} + \frac{K_0}{a_0^2}\frac{d^2y}{dt^2} - \cdots$$

The larger a_0 is, the closer the ultimate state is to the ideal response $\xi = K_0 y(t)$.

Example 4.5.1 A first-order system is represented by $a_0 = 1.41$ and $K = 3.56$. Compare the ultimate state responses to the functions t, t^5, e^t, $u(t)$.

Solution (a) $y = t \qquad \dfrac{dy}{dt} = 1 \qquad \dfrac{d^2y}{dt^2} = 0$

We need only S and S_{1u}. From Equation (4.5.2) we obtain

$$\xi_u = \frac{3.56}{1.41}\left[t - \frac{1}{1.41}\right] = 2.52(t - 0.709)$$

We can compute a relative error

$$e_{u1} = \frac{\xi_u - 2.52t}{2.52t} = -\frac{0.709}{t}$$

The actual error is a constant; hence the percentile error decreases with time (Fig. 4.2).

(b) $y = t^5$. From Equation (4.5.2) we obtain

$$\xi_u = 2.52t^5 - 8.95t^4 + 25.4t^3 - 54.1t^2 + 76.8t - 54.4$$

The ultimate state relative dynamic error is

$$e_{u2} = \frac{\xi_u - 2.52t^5}{2.52t^5} = -\frac{3.55}{t} + \frac{10.1}{t^2} - \frac{21.5}{t^3} + \frac{30.5}{t^4} - \frac{21.6}{t^5}$$

The relative error is plotted in Figure 4.2.

(c) For $y = e^t$, the steady part of Equation (4.5.2) yields

$$\xi_u = \left[\frac{3.56}{1.41}\sum_{n=0}^{\infty}(-1)^n\left(\frac{1}{1.41}\right)^n\right]e^t$$

The summation is recognized as the binomial series expansion of 0.585 and thus

$$\xi_u = 1.48e^t$$

The relative ultimate state error is

$$e_{u3} = \frac{k_u - 2.52e^t}{2.52e^t} = -0.413$$

The relative error e_{u3} is a constant $e_{u3} = -0.413$.

(d) For $y = u(t)$, $\xi_u = 2.52\,u(t)$ and thus the relative ultimate state error is zero.

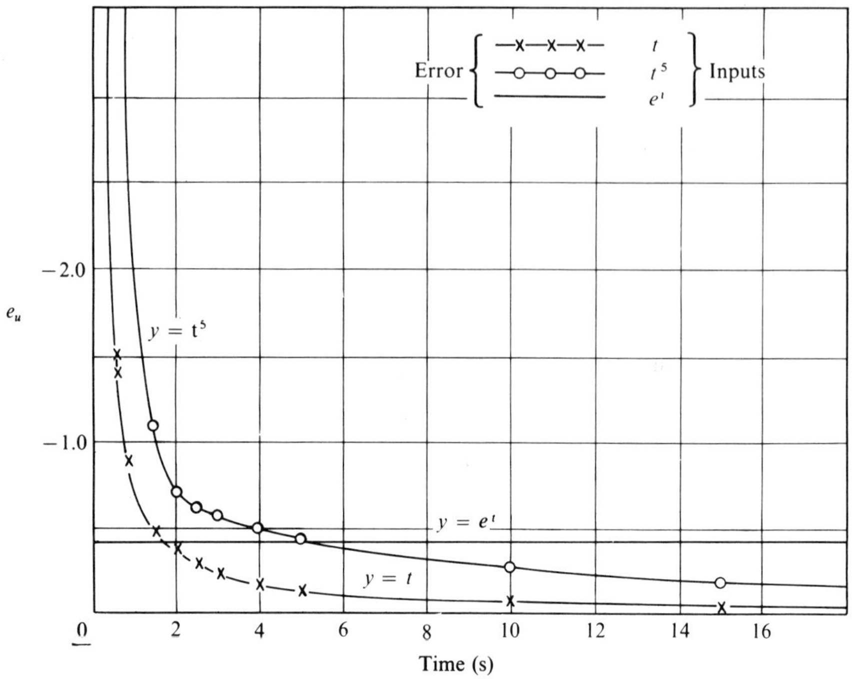

FIGURE. 4.2 Ultimate state error for various inputs, first-order system

It is clear, from Equation (4.5.2), that choosing a_0 as large as possible is the sole optimization criterion for a first-order system. If a set of specifications is given, for example, that the instrument should respond to a given input function so that at a given time the ultimate state relative dynamic error should be less than a given amount, it is possible to work backward and find the quantity a_0 which satisfies the requirement.

Example 4.5.2 A thermocouple is to be used to measure a system in which the temperature varies according to the input $\theta = 20 + 7.22t^2$. Select an a_0 such that the absolute value of ultimate state dynamic error is less than 2% at $t = 0.1$ second.

Solution According to Equation (4.5.2)

$$\xi_u = \frac{K}{a_0}\left[(20 + 7.22t^2) - \frac{1}{a_0}14.44t + \frac{1}{a_0^2}14.44\right]$$

and thus the dynamic ultimate state relative error becomes

$$e_u = \frac{a_{0u}/K - (20 + 7.22t^2)}{20 + 7.22t^2} = \frac{-14.4t/a_0 + 14.4/a_0^2}{20 + 7.22t^2}$$

Thus we require that

$$\frac{14.4}{a_0}\left|\frac{1}{a_0} - t\right| \leq 0.02(20 + 7.22t^2)$$

or

$$\frac{14.4}{a_0^2}\,|1 - 0.1a_0| \leq 0.02(20 + 0.072)$$

We note that $a_0 > 0$. Let us assume at this moment that $a_0 < 10$. Then we can remove the absolute value signs to obtain

$$a_0^2 + 3.59a_0 - 35.9 \geq 0$$

from which it follows that $a_0 \geq 4.46$ satisfies the specifications.

Example 4.5.3 Determine a_0 for a first-order system with $K = a_0$ to satisfy the specification

$$|e_u| \leq 0.01 \quad \text{at } t = 0.1 \quad \text{if } y(t) = e^t$$

Compare to Example 4.5.2.

Solution We follow Example 4.5.1 to obtain

$$e_u = \frac{1}{a_0 + 1} \leq 0.01 \quad \text{or} \quad a_0 \geq 99$$

Let us choose $a_0 = 100$. If the input function were $20.0 + 7.22t^2$ as in Example 4.5.2, the error would be only 0.065%. Obviously selecting e^t as a test function has a most salubrious effect on the design procedure.

4.6 USDE for Second-order Systems

The ultimate state dynamic error (USDE) for a second-order system either may be obtained from the superposition of two first-order systems (Table 4.1, item 2), whereupon the solution developed in Section 4.6 applies, or else may be described by an oscillatory impulsive response function (Table 4.1, 3). In the latter case the ultimate state sensitivities are found to be [from Equation (4.4.1) and Table 4.1, 3]

$$S_{nu} = \frac{(-1)^n K}{\omega^{n+2}} M_n(\zeta) \qquad \textbf{(4.6.1)}$$

in which

$$M_n = \frac{\sin (n + 1)\phi}{\sin \phi}$$

and

$$\sin \phi = \sqrt{1 - \zeta^2}$$

The performance equation for the oscillatory case has the form

$$\frac{d^2\xi}{dt^2} + 2\xi\omega\frac{d\xi}{dt} + \omega^2\xi = Ky(t)$$

and has complex characteristic roots. It follows, therefore, that $(\xi\omega)^2 - \omega^2 < 0$ and thus $\xi^2 < 1$. For stability to exist it is required that $\xi > 0$ and thus Equation (4.6.1) implies $0 < \xi < 1$.

It is fairly easy to compute the quantities M_n. The first seven are tabulated in Table 4.2. Subsequent expressions for M_n, $n > 6$, can be computed by the recursion formula

$$M_n = 2\xi M_{n-1} - M_{n-2} \quad \textbf{(4.6.2)}$$

which can be obtained from the performance equation by simple analysis (see problem 4.19). The ultimate state can be written as

$$\begin{aligned} \xi_u &= S_0 y(t) - S_1 y'(t) + S_2 y''(t) - \cdots \\ &= \frac{K}{\omega_2}\left[y(t) - \frac{M_1}{\omega}\frac{dy}{dt} + \frac{M_2}{\omega^2}\frac{d^2y}{dt^2} - \frac{M_3}{\omega^3}\frac{d^3y}{dt^3} \cdots \right] \end{aligned}$$

so that the ultimate state relative dynamic error can be written in the form

$$e_u = \frac{\omega^2\xi/K - y(t)}{y(t)} = -\frac{M_1}{\omega}\frac{y^{(1)}}{y} + \frac{M_2}{\omega^2}\frac{y^{(2)}}{y} - \frac{M_3}{\omega^3}\frac{y^{(3)}}{y} + \cdots \quad \textbf{(4.6.3)}$$

Minimization of the error requires minimization of $M_1, M_2, \ldots$ and maximization of ω. Ideally it would be best to select ξ in such a fashion that all the $M_n = 0$. Computation of the positive roots of each M_n demonstrates that there exists no single value of ζ which reduces all M_n to zero. (See Tables 4.2 and 4.3.) It is possible, however, to select ζ so as to suppress any M_n for $n > 1$. In any design situation in which some higher-order disturbance needs to be minimized, the choice of ζ can help. Unfortunately, the most important source of error, M_1, is a direct function of ζ and cannot be suppressed.

TABLE 4.2

n	M_n	Real Positive Roots of M_n
0	1	none
1	2ζ	none
2	$4\zeta^2 - 1$	0.500
3	$4\zeta(2\zeta^2 - 1)$	0.707
4	$10\zeta^4 - 12\zeta^2 + 1$	0.310, 0.809
5	$2\zeta(16\zeta^4 - 16\zeta^2 + 3)$	0.5, 0.866
6	$64\zeta^6 - 80\zeta^4 + 24\zeta^2 + 1$	0.222, 0.624, 0.901

Examination of the tables can soon convince us that a small ζ is probably the most practical solution. Actual tests of modern high-speed equipment indicate that in practice $0.100 \leq \zeta \leq 0.750$. Electrical pressure transducers usually have a low damping,[5] with a maximum value of $\zeta = 0.40$. High-speed galvanometers may be designed with $\zeta = 0.10$, while laboratory type galvanometers may have a $\zeta \approx 0.90$ or even larger.

TABLE 4.3 *Numerical values of M_n $(0 \leq \zeta \leq 1.00)$*

ζ	M_0	M_1	M_2	M_3	M_4	M_5	M_6	M_7	M_8
.05	1.000	0.100	−0.990	−0.100	0.980	0.090	−0.971	−0.187	0.952
.10	1.000	0.200	−0.960	−0.392	0.882	0.568	−0.868	−0.394	0.789
.15	1.000	0.300	−0.930	−0.579	0.756	0.806	−0.514	−0.652	0.318
.20	1.000	0.400	−0.840	−0.736	0.676	1.006	−0.274	−1.116	−0.172
.25	1.000	0.500	−0.750	−0.875	0.312	1.031	0.203	−0.924	−0.261
.30	1.000	0.600	−0.640	−0.984	0.050	0.954	0.523	−0.640	−0.907
.35	1.000	0.700	−0.510	−1.057	−0.230	0.996	0.927	−0.347	−1.170
.40	1.000	0.800	−0.360	−1.088	−0.510	0.680	1.054	0.163	−0.924
.45	1.000	0.900	−0.190	−1.052	−0.652	0.530	1.076	0.331	−0.811
.50	1.000	1.000	0.000	−1.000	−1.000	0.000	1.000	1.000	0.000
.55	1.000	1.100	0.210	−0.869	−1.187	−0.437	0.706	1.214	0.629
.60	1.000	1.200	0.440	−0.672	−1.246	−0.823	0.258	1.132	1.100
.65	1.000	1.300	0.690	−0.403	−1.214	−1.175	−0.313	0.768	1.311
.70	1.000	1.400	0.990	−0.014	−1.010	−1.400	−0.546	0.636	1.436
.707	1.000	1.414	1.000	0.000	−1.000	−1.414	−1.000	0.000	1.000
.75	1.000	1.500	1.250	0.375	−0.687	−1.405	−1.420	−0.725	0.332
.80	1.000	1.600	1.560	0.896	−0.126	−1.098	−1.633	−1.515	−0.821
.85	1.000	1.700	1.890	1.513	0.682	−0.354	−1.284	−1.829	−1.825
.90	1.000	1.800	2.240	2.232	1.776	0.965	−0.039	−1.035	−3.600
.95	1.000	1.900	2.610	3.059	3.202	3.025	2.545	1.810	0.414
1.00	1.000	2.000	3.000	4.000	5.000	6.000	7.000	8.000	9.000

In control engineering practice $\zeta = 0.707$ is often called an *optimum*. The reason for this selection depends on the filtering properties of the measurement system and on the band width requirements of control systems. Discussion of the falsity of this criterion is deferred until chapter 5.

Example 4.6.1 A second-order system is subjected to the input function $y = e^t$. The quantity ω^2 is fixed at 100 radian²/second², whereas ζ is at the disposal of the designer. Find the ultimate state relative dynamic error by direct integration and by use of the first five terms of Equation (4.6.3). Is there an optimum?

Solution Direct integration of the equation

$$\xi'' + 20\zeta\xi' + 100\xi = Ke^t$$

yields the result (as $t \longrightarrow \infty$)

$$\xi_u = \frac{Ke^t}{1 + 20\zeta + 100}$$

Thus

$$e_{u_1} = -\left(\frac{1 + 20\zeta}{101 + 20\zeta}\right)$$

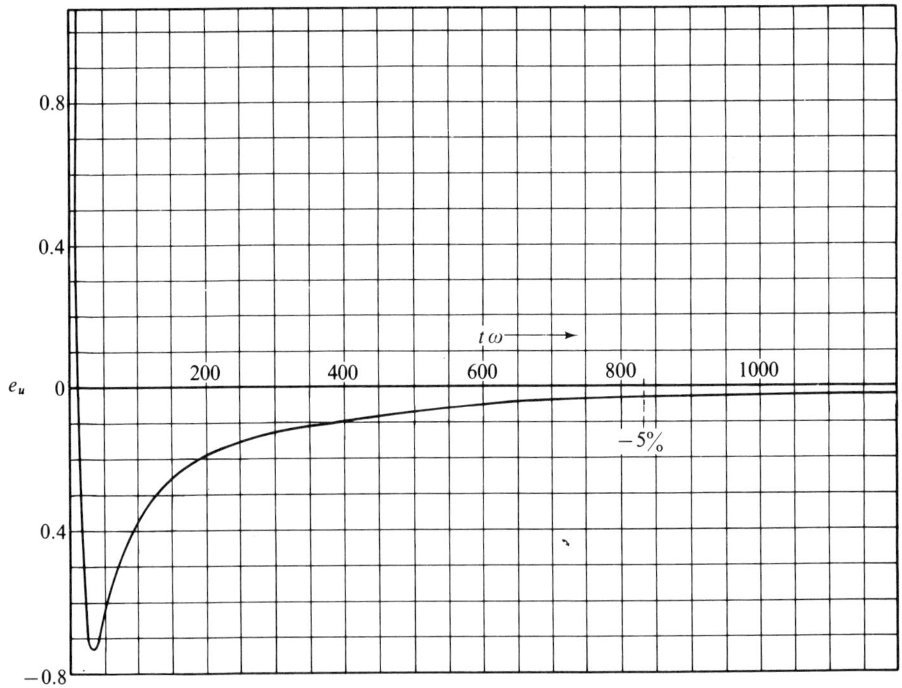

FIGURE 4.3 Dimensionless error as a function of dimensionless time in Example 4.6.2

The approximate solution is given by

$$e_{u_2} = -0.1M_1 + 0.01M_2 - 0.001M_2 + 0.0001M_4$$

Direct computation of e_{u_1} and computation of e_{u_2} with the help of Table 4.3 yields results identical to four significant figures for e_{u_1} and e_{u_2}:

$-e_u$	0.029	0.048	0.066	0.083	0.100	0.116	0.131	0.147	0.161
ζ	0.1	0.2	0.3	0.4	0.5	0.6	0.7	0.8	0.9

There is no optimum. Minimization of the error can be achieved only by selecting the smallest acceptable value of $\zeta > 0$.

Example 4.6.2 Let $\zeta = 0.707$ and $y(t) = t^3$. Find the value of ω for which $|e_u| \leq 0.05$ for all $t \geq 0.1$ second.

Solution From Table 4.3 and Equation (4.6.3) we obtain

$$e_u = -\frac{42.42}{(t\omega)} + \frac{600}{(t\omega)^2}$$

We plot e_u as a function of $t\omega$ (Fig. 4.3) and note that $|e_u| \leq 0.05$ for $t\omega > 832$. The solution of our problem then is

$$\omega > 8{,}320 \text{ rad/s}$$

4.7 Setting the Specifications

Instruments designed to measure time-varying inputs are actually measuring two measurands, the input function and time. Two methods are usually employed to record the quantities simultaneously: (a) timed-recording traces and (b) interval sampling.

The timed-recording trace procedure is typified by the strip chart galvanometer, the traveling pen galvanometer and the oscilloscope. The writing instrument is constrained to respond to the input function along a single axis and then either the whole writing assembly is driven along an axis vertical to the function axis by an accurate clock mechanism or the recording paper is moved along the time axis. In either case the allowable error is a function of the clock speed.

Consider, for example, a common recording oscillograph. The input is recorded on a continuous strip of paper which is driven at 5 mm/s.[6] The trace varies in thickness but can be as wide as 1 mm. It is possible to read 0.5-mm changes with fair accuracy so that a change of 0.25 mm is, in effect, zero change. For this instrument, therefore, anything which occurs before $t_0 = 0.05$ s is largely immaterial to the user. In setting local error, the "time-change relevance-threshold time" t_0 is then to be used, as was done in Example 4.6.2.

There are times, however, in which the time t_0 is determined by reference to the input function, rather than by the time axis speed. Consider again the recording galvanometer. In one particular instrument the chart is 50 mm wide and can be read to the nearest 0.5 mm. Let us suppose that the system is to record the function $y = 2.5\,t$ volts and that the ultimate state sensitivity is 1 mm/V. Since the first 0.25 mm cannot be distinguished from zero mm, we obtain

$$0.25 = 2.5t_0$$

and thus the reference time is $t_0 = 0.10$ s. Whatever the recorder shows before $t = 0.10$ is immaterial to the record.

Interval sampling is used as a means of digitizing the instrument output. The instrument output is recorded only at preprogrammed times, usually at fixed time intervals. What happens between zero time and the first sample is not noticed by the recorder. It is completely unimportant how wildly the instrument behaves during the first sample interval as long as the error is reduced to the specified minimum at the end of the interval.

Up to this point it has been assumed that the error is specified to be less than a given fraction of the local value. In such a case a 5% maximum error would simply mean 5% or less of the recorded output at any time and for all values of the output indication. In much current instrumentation promotional literature, however, the relative error is stated in terms of a fraction of full scale deflection. The local relative error would then vary from the specified value at the maximum range up to infinity near zero. A good example is given by the attempt to read a graph scaled in millimeters. If the peak value on the graph corresponds to 100 mm and if the visual resolution is 0.25 mm (a fairly good figure), the reading error is 0.25% of the full scale. Near zero, however, an output of 0.25 mm can be read as either zero mm or as 0.5 mm so that the error is 100%. It follows that local value specification is more stringent than full scale value specification.

As long as the instrument manufacturer states clearly how the error is to be computed, there is no difficulty in understanding. A 5% of full scale error means that there is a band of constant size delineated by a pair of lines parallel to the recorded tracing. The actual expected value will lie somewhere between those two lines.

Example 4.7.1 A second-order system is designed to record the input function $y = at$, in which a can have any value between 0.60 and 3.21. Select ζ and ω such that $|e_u| \leq 0.10$ if the system relevance threshold is $y(t_0) = 0.12$.

Solution From Equation (4.6.3) and Table 4.3 we find

$$e_u = -\frac{M_1}{\omega t}$$

This becomes, upon substitution of the error requirement,

$$\frac{2\zeta}{\omega t_0} \leq 0.1$$

The threshold time is found from

$$3.21t_0 = 0.12$$
$$t_0 = 0.0373 \text{ s}$$
$$0.6t_0 = 0.12$$
$$t_0 = 0.200 \text{ s}$$

Choosing the smaller threshold times gives us the design criterion

$$\zeta \leq 0.06\omega$$

Any combination of ζ and ω which satisfies the last inequality is a solution of the problem.

4.8 Transient Dynamic Error (First-order)

The transient portion of the dynamic solution for a first-order system has the form [Equations (4.4.1), (4.5.2)]

$$\xi_T = -e^{-at} \sum_{n=0}^{\infty} \sum_{k=0}^{\infty} \frac{(-1)^n K}{a^{n+1}} \frac{d^n y}{dt^n} \frac{(at)^{n-k}}{(n-k)!} \tag{4.8.1}$$

Substitution of

$$\frac{d^n y}{dt^n} = \left.\frac{d^n y}{dt^n}\right|_{y=0} + \left.\frac{d^{n-1} y}{dt^{n-1}}\right|_{y=0} t + \left.\frac{d^{n-2} y}{dt^{n-2}}\right|_{y=0} \frac{t^2}{2!} + \cdots$$

yields the simple result

$$\xi_T = -e^{-at} \sum_{n=0}^{\infty} S_{nu} y^{(n)}(0^+)$$

from which it immediately follows that

$$\xi_T = -e^{-at} \xi_u(0^+)$$

in which $\xi_u(0^+)$ is the *ultimate state dynamic error* evaluated at $t = 0^+$. Stability requires that $a > 0$ so that the maximum value of ξ_T must occur at the threshold, $t - t_0$. Thus the transient dynamic error is computed from the known ultimate state of the system and the threshold time.

Example 4.8.1 Find the maximum relative transient error connected with the system discussed in Example 4.5.3.

Solution We had $y = e^t$, $a = 100$, $t_0 = 0.10$, $S = K/100$, $\xi_u = K/101e^t$. Thus the relative transient error will be

$$e_T = \frac{\xi_T}{Se^t} = -\frac{100}{101} \exp - 101t$$

At $t = 0.1$ we find that

$$\max(e_T) = 0.990e^{-10.1} = 0.0000411$$

This error is, at worst, 0.004% and can be neglected in comparison to the 1% error introduced by the ultimate state component.

Example 4.8.2 Select a in such a fashion that the total dynamic error of a system subjected to the input $y = t^3$ is less than 5% for all $t \geq 0.2$ second.

Solution

$$\xi_u = St^3 + S_{1u}3t^2 + S_{2u}6t + S_{3u}6$$
$$\xi_T = -e^{-at}\xi_u(0^+) = -e^{-at}6S_{3u}$$
$$e_D = \frac{\xi_u + \xi_T - St^3}{St^3}$$
$$e_D = \frac{-3}{at} + \frac{6}{(at)^2} - \frac{6}{(at)^3}(1 - e^{-at})$$

As before, we plot e_D as a function of at and find the value of at for which $e_D \leq 0.05$ (Fig. 4.4). We find $at = 60$ and therefore $a = 60/0.2 = 300$ seconds^{-1}.

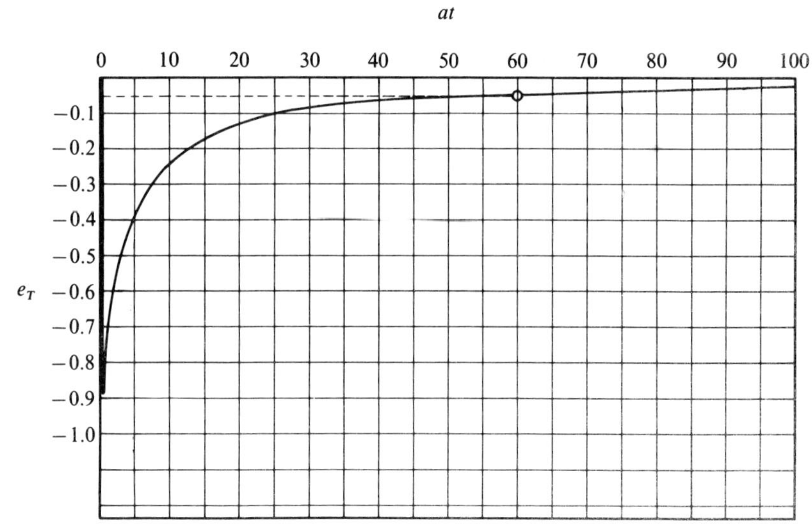

FIGURE. 4.4 Dimensionless error as a function of dimensionless time for Example 4.8.2

An interesting point about the quantity a is that it must have the dimensions of the reciprocal of time. Thus $1/a = T$ is a characteristic time parameter of the first-order system. T is often called the *response time* of a system. At time $t = T$ the transient error has become $e^{-1}\xi_u(0^+)$ or $0.3679\ \xi_u(0^+)$, so that the transient error has been reduced to 36.79% of the initial value. In control engineering practice, T is discussed as the value of the time at which a first-order system subjected to a constant input (step input) reaches an output which is 63.2% of the ultimate state.[7] This is an obvious result. Under a step input, the first-order response is

$$\xi = S_0(t)C$$

in which C is the amplitude of the step. From Equation (4.4.1) with

$r_i = 1/T$, one obtains

$$S_0(t) = K(1 - e^{-t/T})$$

so that

$$\xi(T) = KC0.632$$

4.9 Transient Dynamic Error (Second-order)

Let the performance equation have the form

$$\frac{d^2\xi}{dt^2} + 2\zeta\omega\frac{d\xi}{dt} + \omega^2\xi = Ky(t)$$

with the restriction $\zeta^2 < 1$. The transient portion of the solution [Equation (4.4.1)] is

$$\xi_T = -Ke^{\zeta\omega t}\sum_{n=0}^{\infty}(-1)^n y^{(n)}(t)R_n(t) \tag{4.9.1}$$

in which R_n is defined by Equation (4.3.6). If the $y^{(n)}(t)$ are expanded into MacLaurin series and substituted into (4.9.1) and if it is noted that

$$\cos(n+1)\phi = \zeta M_n - M_{n-1}$$

then Equation (4.9.1) can be written as

$$\begin{aligned}\xi_T = \frac{Ke^{-\zeta\omega t}}{\omega^2}\Bigg[&\sin bt \sum_{n=0}^{\infty}(-1)^n\frac{(\zeta M_n - M_{n-1})}{\omega^n\sqrt{1-\zeta^2}}y^{(n)}(0^+)\\ &- \cos bt\sum_{n=0}^{\infty}(-1)^n M_n y^{(n)}(0^+)\Bigg]\end{aligned} \tag{4.9.2}$$

Through use of the definition of M_n, ξ_T can now be written in terms of the initial values of the ultimate state solution:

$$\begin{aligned}\xi_T = -e^{-\zeta\omega t}\Bigg\{&\left[\frac{\sin\omega\sqrt{1-\zeta^2}\,t}{\sqrt{1-\zeta^2}}\right]\left[\zeta\xi_u(0^+) + \omega\dot{\xi}_u(0^+)\right]\\ &- [\cos\omega\sqrt{1-\zeta^2}\,t][\xi_u(0^+)]\Bigg\}\end{aligned} \tag{4.9.3}$$

If the instrument is stable, ζ_T will vanish with a damped oscillation of frequency $\omega\sqrt{1-\zeta^2}$ and characteristic time $T = 1/\zeta\omega$. From a design point of view, only ω is available as an optimizing parameter. For a given ζ, if it be required that the amplitude of ξ_T reach a certain value at a certain time, there is always a theoretical value of ω which will yield the proper result. Whether it is possible for such a value of ω to be achieved physically cannot be answered at this point.

A clearer physical picture, and a useful formula, can be obtained by defining a "phase shift" angle

$$\tan\phi_T = \frac{\xi_u(0^+)\sqrt{1-\zeta^2}}{\zeta\xi_u(0^+) + \omega\dot{\xi}_u(0^+)}$$

Then ξ_T can be written as

$$\xi_T = -e^{-\zeta\omega t}\left[\frac{\xi_u^2(0^+) + 2\zeta\omega\xi_u(0^+)\dot{\xi}_u(0^+) + \dot{\xi}_u^2(0^+)\omega^2}{1-\zeta^2}\right]^{1/2} \cdot \sin(\omega\sqrt{1-\zeta^2}\,t - \phi_T) \qquad \textbf{(4.9.4)}$$

When plotted, ξ_T will have the form shown in Figure 4.5. The transient error will oscillate between the boundaries of the envelope indicated by

$$\pm\, e^{-\zeta\omega t}\left[\frac{\xi_u^2(0^+) + 2\zeta\omega\xi_u(0^+)\dot{\xi}_u(0^+) + \dot{\xi}_u^2(0^+)}{1-\zeta^2}\right]^{1/2}$$

whereupon the expression for the relative transient error becomes

$$e_T \leq \frac{\omega^2 e^{-\zeta\omega t}}{Ky(t)}\left[\frac{\xi_u^2(0^+) + 2\zeta\omega\xi_u(0^+)\dot{\xi}_u(0^+) + \dot{\xi}_u^2(0^+)}{1-\zeta^2}\right]^{1/2} \qquad \textbf{(4.9.5)}$$

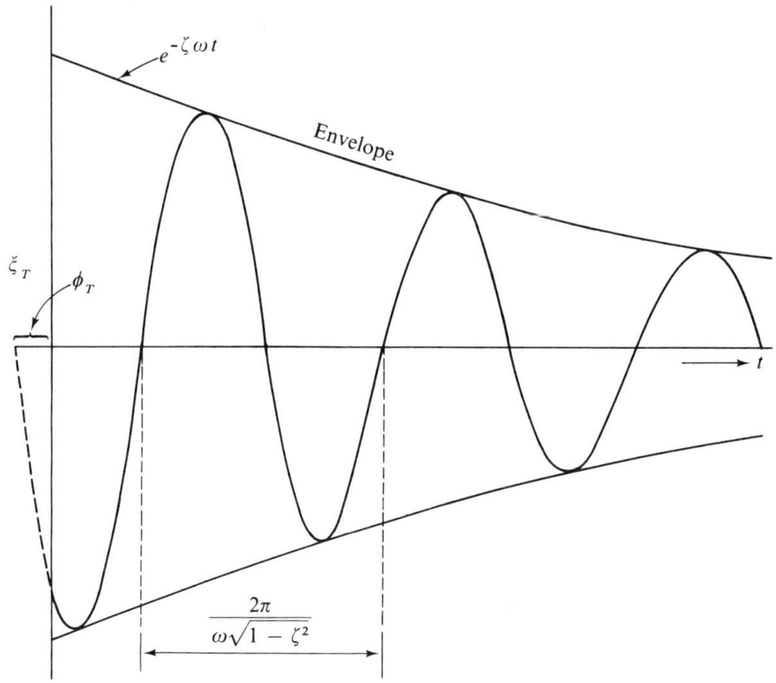

FIGURE 4.5 Second-order transient error

Example 4.9.1 A second-order system is subjected to the input function $y(t) = 2.71t^2$. If $\zeta = 0.20$ and $\omega = 10.0$, compute the relative transient error at $t = 0.80$ second.

Solution

$$\frac{\omega^2\xi_u}{K} = 2.71t^2 - 0.217t - 0.046$$

Thus

$$\xi_u(0^+) = -0.046\frac{K}{\omega^2}$$

$$\dot{\xi}_u(0^+) = -0.217\frac{K}{\omega^2}$$

from which we obtain [Equation (4.9.5)]

$$e_T \leq 0.0086 \quad \text{or} \quad 0.9 \text{ percent}$$

4.10 Higher-order Systems

Linear systems of higher than the second order can always be separated into two sums [see Equation 4.3.3]:

$$\xi = \sum_{i=1}^{\mu} A_i K \int_0^t e^{-r_i\tau} y(t-\tau)\, d\tau$$
$$+ \sum_{i=1}^{\nu} \frac{B_i K}{\omega_i\sqrt{1-\zeta_i^2}} \int_{0^+}^{\infty} e^{-\zeta_i\omega_i\tau} \sin \omega_i\sqrt{1-\zeta_i^2}\, \tau y(t-\tau)\, d\tau \qquad \textbf{(4.10.1)}$$

whenever it is possible to factor the characteristic polynomial into the form

$$\begin{matrix}(s+r_1)(s+r_2)\cdots(s+r_\mu)(s^2+2\zeta_1\omega_1 s+\omega_1^2) \\ (s^2+2\zeta_2\omega_2 s+\omega_2^2)\cdots(s^2+2\zeta_\nu\omega_\nu s+\omega_\nu^2)\end{matrix} \qquad \textbf{(4.10.2)}$$

in which

$$r_i > 0 \qquad 0 < \zeta < 1 \qquad \omega_1^2 > 0$$

and $2\nu + \mu = m$ (m = order of performance equation).

Each of the terms in the sums can be computed from the results obtained in the previous sections of this chapter. Let the partial ultimate state solutions of the system be defined as

$$\xi_{ui}^a = \sum_{n=0}^{\infty} (-1)^n \frac{Ky^{(n)}(t)}{r_i^{(n+1)}}$$

$$\xi_{ui}^b = \sum_{n=0}^{\infty} \frac{(-1)^n M_n(\zeta_i) K y^n(t)}{\omega_i^{(n+2)}} \qquad \textbf{(4.10.3)}$$

Then the complete solution has the form

$$\xi - \left[\sum_{i=1}^{\mu} A_i^{(m)} \xi_{ai}^a(t) + \sum_{i=1}^{\nu} B_i^{(m)} \zeta_{ui}^b(t)\right]$$
$$- \sum_{i=1}^{\mu} A_i^{(m)} e^{-r_i t} \xi_{ui}^a(0^+)$$
$$- \sum_{i=1}^{\nu} B_i e^{-\zeta_i\omega_i t} \left[\frac{[\xi_{ui}^b(0^+)]^2 + 2\zeta_i\omega_i\xi_{ui}^b(0^+)\dot{\xi}_{ui}^b(0^+) + \omega^2[\xi_{ui}^b(0^+)]^2}{1-\zeta_i^2}\right]^{1/2}$$
$$\sin[(\omega_i t\sqrt{1-\zeta_i}) - \phi_{Ti}] \qquad \textbf{(4.10.4)}$$

where

$$\tan \phi_{Ti} = \frac{\xi_{ui}^b(0^+)\sqrt{1-\zeta_i^2}}{\zeta_i\xi_{ui}^b(0^+) + \omega_i\dot{\xi}_{ui}^b(0^+)}$$

Equation (4.10.4) is the complete solution of the system

$$\frac{d^m\xi}{dt^m} + a_{m-1}\frac{d^{m-1}\xi}{dt^{m-1}} + \cdots + a_0\xi = Ky(t)$$

whenever the characteristic polynomial is factored into the form (4.10.2).

Example 4.10.1 A system has a characteristic polynomial which factors into

$$(s + 3.72)(s^2 + 9.00s + 225)$$

If $Ky(t) = 2.33\, t^3$, compute ξ_u from Equation (4.10).

Solution We first find A_1^3 and B_1^3 from Table 4.1:

$$A_1^3 = \frac{1}{(\zeta\omega - r_1) + \omega^2(1 - \zeta^2)}$$

$$B_1^3 = \frac{1}{(\zeta\omega - r_1)^2 + \omega^2(1 - \zeta^2)}$$

Substitution of $r_1 = 3.72$, $\omega = 15$, $\zeta = 0.3$ then gives

$$A_1^3 = 4.87 \times 10^{-3} \qquad B_1^3 = 6.98 \times 10^{-2}$$

The partial ultimate state solutions are

$$\xi_u^a(t) = 0.626t^3 - 0.505t^2 + 0.272 - 0.738$$

$$\xi_u^b(t) = 0.0914t^3 - 0.00124t^2 - 0.000276t + 0.0000182$$

and thus

$$\xi = 10^{-3}[(9.43t^3 - 2.55t^2 + 1.30t - 0.358) - 73.8e^{-3.27t} - e^{-4.50t}\, 0.0232 \sin(12.6t - \phi)]$$

$$\tan\phi = 0.0139$$

Thus $\phi = 0.80$ degree $= 0.0139$ radian.

It is evident that the second-order component does not contribute much to the transient error even though it provides much of the ultimate state dynamic error for large values of time.

4.11 A Design Limit

The reader who has followed the tortuous arguments of the preceding sections may have noticed that what has been gained in clarity and in understanding of the dynamic measurement process has been lost in notational complexity and mathematical difficulties. This was tacitly admitted in the presentation of Equation (4.10.4) wherein the sensitivity concept is totally lost, or at least buried deeply, and the solution is given in terms of the ultimate state partial solutions. There seems to be no *computational* advantage gained by dealing with sensitivity functions rather than Equation (4.10.4).

Equally difficult to justify is the use of Equation (4.10.3) to evaluate the partial solutions since there is no guarantee that the series will converge even though $y(t)$ possesses derivatives of all orders. For example, if $y(t) = (a + bt)^{-1}$, it can be shown that for any constant r_i the infinite series for the first-order partial solutions will fail to converge regardless of the values of a and b, even though it can be shown that the performance equation has a valid solution.

What has been accomplished, however, is to establish that if the solution is written in the form

$$\xi = \left\{K\left[\sum_{i=1}^{\mu}\frac{A_i^{(m)}}{r_i} + \sum_{i=1}^{\nu}\frac{B_i^{(m)}}{\omega_1^2}\right]\right\}y(t) + \epsilon(t) \qquad \textbf{(4.11.1)}$$

then the term within the braces is the ultimate state sensitivity of the system and $\epsilon(t)$ does not contain $y(t)$, but only $y^{(n)}(t)$, $n > 0$. That the mathematical results are not computable for all $y(t)$ is of no moment. It can be shown that if it is possible to write the expression for the output displacement in the form (4.11.1), the error can be computed directly without reference to (4.10.3) and (4.10.4).

Consider the general performance equation

$$\frac{d^m\xi}{dt^m} + a_{m-1}\frac{d^{m-1}\xi}{dt^{m-1}} + \cdots + a_0\xi = Ky(t)$$

and write $K = K_0a_0$. Then it is possible to write, by setting $\xi/K_0 = z$,

$$\frac{d^m z}{dt^m} + a_{m-1}\frac{d^{m-1}z}{dt^{m-1}} + \cdots + a_0 z = a_0 y(t) \qquad \textbf{(4.11.2)}$$

If now $y(t) = y_0u(t)$, the ultimate state becomes

$$z_{us} = y_0$$

or $S = 1$. What has been done here is to connect a scale factor $1/K_0$ to the displacement ξ so that the output can be read directly in the units of the input.

The solution of (4.11.2) is

$$z(t) = \int_{0^+}^{t} I(t-\tau)y(\tau)\,d\tau$$

Let $\hat{I}(t-\tau)$ be the antiderivative of $I(t-\tau)$ with respect to τ. Then integration by parts yields

$$z(t) = y(t)\hat{I}(0^+) - y(0^+)\hat{I}(t) - \int_{0^+}^{t}\hat{I}(t-\tau)\frac{dy}{d\tau}\,d\tau$$

$$\hat{I}(t) = a_0\left[\sum_{i=1}^{\mu}\frac{A_i^{(m)}e^{r_it}}{r_i} + \sum_{i=1}^{\nu}\frac{B_i^{(m)}e^{-\zeta_i\omega_it}}{\omega_i^2(1-\zeta_i^2)}\sin(\omega_i\sqrt{1-\zeta_i}\,t + \alpha_i)\right]$$

$$\sin\alpha_i = \sqrt{1-\zeta_i^2} \qquad \cos\alpha_i = \zeta_i$$

From the above it can be immediately shown that $\hat{I}(0) = S = 1$. The dynamical error is now obtained in terms of $\hat{I}(t)$.

$$|z(t) - y(t)| = |\epsilon_D| \leq |y(0^+)||\hat{I}(t)| + \left|\int_{0^+}^{t}\hat{I}(t-\tau)\frac{dy}{d\tau}\,d\tau\right| \qquad \textbf{(4.11.3)}$$

where use has been made of the well-known inequality

$$|a + b| \leq |a| + |b|$$

If $\dot{y}_M$ is the maximum value of dy/dt achieved in the interval from 0 to t, inequality (4.11.3) can now be shown to be

$$|\epsilon_D| \leq |y(0^+)||\hat{I}(t)| + |\dot{y}_M||\hat{\hat{I}}(0^+) - \hat{\hat{I}}(t)| \qquad \textbf{(4.11.4)}$$

in which $\hat{\hat{I}}(t-\tau)$ is the antiderivate of $\hat{I}(t-\tau)$ with respect to τ. By direct integration it is fairly easy to show that

$$\hat{\hat{I}}(t) = a_0 \sum_{i=1}^{\mu} \frac{A_i^{(m)} e^{-r_i t}}{r_i^2}$$
$$+ a_0 \sum_{i=1}^{\nu} \frac{B_i^{(m)} e^{-\zeta_i \omega_i t}}{\omega_i^3(1-\zeta_i^2)} \sin(\omega_i \sqrt{1-\zeta_i^2}\, t + \alpha_1)$$
$$\tan \alpha_i = \frac{2\zeta\sqrt{1-\zeta^2}}{2\zeta_i^2 - 1}$$

Inequality (4.11.4) is the *first dominating function.* With its aid it is possible to compute an upper bound to the error without detailed knowledge of the input function. All that needs to be specified is the possible maximum initial input and the greatest possible input rate. Design is accomplished by computing ϵ_D for various $\hat{I}(t)$ and $\hat{\hat{I}}(t)$ and selecting the system which meets the specification of $0 < \epsilon_D < k$ for any time $t > t_0 > 0$ in which t_0 is the relevance time (Sect. 4.8).

Much has been gained, and the computational procedure has been reduced considerably. Nevertheless, computation of (4.11.4) is not to be undertaken lightly, especially for high-order systems. It is possible to obtain a time-independent maximum error which establishes a lower bound on the r_i and $\zeta_i\omega_i$ if the system error is to be below a specified amount.

Consider an mth-order polynomial with m equal real negative roots, $s = -h_0$. Then by the binomial expansion it is possible to write

$$P_0 = (s + h_0)^m = \sum_{j=0}^{m} \binom{m}{j} h_0^j s^{m-j}$$
$$= s^m + \binom{m}{1} h_0 s^{m-1} + \binom{m}{2} h_0^2 s^{m-2} + \cdots + h^m$$

An mth-order polynomial with m real roots

$$s = -r_i \; (i = 1, 2, \ldots, m)$$

will expand into the form

$$P = s^m + (r_1 + r_2 + \cdots + r_m)s^{m-1} + (r_1r_2 + r_1r_3 + \cdots)s^{m-2}$$
$$+ \cdots + r_1r_2r_3 \cdots r_m$$

Comparison of P with P_0 will convince the reader that if $r_i > h_0$ for all i, then the individual coefficients of P are all larger than the individual coefficients of P_0. If the r_i are complex, the statement remains true, if h_0 is identified with the real part of the root.

If now h_0 is identified with the smallest real part of the roots of P, then P_0 will represent the characteristic equation of a system with a dynamic error which will vanish as a function of h_0 and vanish faster with the larger h_0. Clearly, if an error ϵ_D is specified and an h_0 found so as to satisfy the error criterion, then it will be known that any system in which the roots have real parts equal to or greater than h_0 will have a dynamic error no greater than ϵ_D.

It must be understood, however, that having the coefficients in P greater than the coefficients in P_0 is necessary, but not sufficient, to guarantee $Re(r_i) > h_0$. It is important, therefore, to work backward from the roots in the design of an instrument. The procedure is as follows:

(1) Determine h_0 to correspond to ϵ_D;

(2) Specify a damping coefficient ζ_i for any complex root pair wanted in the system (the designer has the choice of the form of the roots);

(3) The root pair can now be written in the form

$$-h_0 \pm j\frac{h_0}{\zeta_i}\sqrt{1-\zeta_i^2}$$

If m is even, there will be $m/2$ such root pairs, whereas if m is odd, there will be $(m-1)/2$ root pairs plus one real root equal to $-h_0$. The above form is obvious since any mth-order polynomial factors into either the product of $m/2$ quadratic factors or else $(m-1)/2$ quadratic factors and one first-order factor;

(4) Note that each root pair has the form

$$-K_i\left[1 \pm j\frac{1}{\zeta_i}\sqrt{1-\zeta_i^2}\right]$$

and write the polynomial in the form

$$s^m + (K_1 + K_2 + K_3 + \cdots)s^{m-1} + \cdots$$

in which $K_i > h_0$;

(5) Now juggle the system parameters so as to obtain the equation with all $K_i > h_0$, noting that changing one coefficient changes all the coefficients;

(6) Check all other specifications.

The operation is straightforward and quite expedient as long as h_0 can be found. This is done by using inequality (4.11.4) on the system

$$\left(\frac{d}{dt} + h_0\right)^m z = h_0{}^m y(t)$$

which is (4.11.1) with $P = P_0$. By direct integration it is found that

$$I(t) = \frac{h_0^m t^{m-1} e^{-h_0 t}}{m-1} \geq 0$$

$$\hat{I}(t) = e^{-h_0 t} \sum_{j=1}^{m} \frac{(h_0 t)^{m-j}}{(m-j)!} \geq 0$$

$$\hat{\hat{I}}(t) = \frac{e^{-h_0 t}}{h_0} \sum_{j=1}^{m} \frac{j(h_0 t)^{m-j}}{(m-j)!} \geq 0$$

and thus

$$\epsilon_D \leq \left[e^{-h_0 t} \sum_{j=1}^{m} \frac{(h_0 t)^{m-j}}{(m-j)!}\right] |y(0^+)| - \left[e^{-h_0 t/h_0} \sum_{j=1}^{m} \frac{j(h_0 t)^{m-j}}{(m-j)!}\right] |\dot{y}_M| + \frac{m\,|\dot{y}_M|}{h_0} \qquad \textbf{(4.11.5)}$$

The negative sign before the second term is retained since the second term is always less than or equal to $m\dot{y}_m/h_0$. Since $j > 1$, it follows that

$$\sum_{j=1}^{m} \frac{j(h_0 t)^{m-j}}{(m-j)!} > \sum_{j=1}^{m} \frac{(h_0 t)^{m-j}}{(m-j)!}$$

which, by virtue of the negative sign, permits writing

$$\epsilon_D \leq \left[e^{-h_0 t/h_0} \sum_{j=1}^{m} \frac{(h_0 t)^{m-j}}{(m-j)!}\right][h_0 |y(0)| - |\dot{y}_M|] + \frac{m}{h_0}|\dot{y}_M| \qquad \textbf{(4.11.6)}$$

This inequality is fairly easy to compute and may be used in design. A further simplification is still possible, at the expense of obtaining a somewhat larger value of h_0. It is noted that

$$\sum_{j=1}^{m} \frac{(h_0 t)^{m-j}}{(m-j)!} \leq (h_0 t)^{m-1} \sum_{j=1}^{m-1} \frac{1}{(m-j)!} = (h_0 t)^{m-1} V_m(h_0 t) \qquad \textbf{(4.11.7)}$$

with the new notation

$$V_m = \sum_{j=1}^{m} \frac{1}{(m-j)!}$$

Hence

$$\epsilon_D \leq e^{h_0 t} h_0{}^{m-2} t^{m-1} V_m \left[h_0 |y(0)| - |\dot{y}_M| + \frac{m}{h_0}|\dot{y}_m|\right] \qquad \textbf{(4.11.8)}$$

The positive function $e^{-h_0 t} t^{m-1} = f(m > 1)$ is zero at both $t = 0$ and $t = \infty$ and thus must reach a positive maximum at some time $t = t_0$. This can be found by setting the first derivative of f equal to zero, which yields

$$h_0 t_0 = (m - 1)\,(m > 1) \qquad \textbf{(4.11.9)}$$

If t_0 is set to be the relevance threshold time, solution of this function for h_0 will give a value of h_0 which will guarantee that ϵ_D is always decreasing for $t > t_0$. Obviously it is required that $h_0 > h_0$. Substitution of t_0 into (4.11.8) gives a final expression for the error:

$$\epsilon_D \leq \frac{e^{-(m-1)}(m-1)^{m-1} V_{mj}}{h_0}[h_0 |y(0)| - |\dot{y}_M|] + \frac{m}{h_0}|\dot{y}_M| \qquad \textbf{(4.11.10)}$$

Inequality (4.11.10) requires knowledge of the order of the system performance equation, the maximum rate and the maximum allowable initial input. If ϵ_D is specified, then h_0 can be found by solving the equation

$$h_0 = \frac{(m - H_m)\,|\dot{y}_M|}{\epsilon_D - H_m |y(0)|} \qquad \textbf{(4.11.11)}$$

where

$$H_m = e^{-m}(m-1)^{m-1} \sum_{j=1}^{m} \frac{1}{(m-j)!}$$

Some values of H_m are given in Table 4.4. Usually Equation (4.11.11) is sufficient to generate a valid design with a comfortable safety factor. If lower values of h_0 are needed, Equation (4.11.11) can be used as a first approximation and the result can then be refined with either Equation (4.11.6) or (4.11.5). If it were required to do even better than that, then the solution of the complete equations should be essayed. It is the opinion of the author that (4.11.11) is sufficient for most purposes.

*TABLE 4.4 Values of H_m**

m	1	2	3	4
H_m	0	0.7358	1.353	3.138
m	5	6	7	8
H_m	2.708	14.24	23.71	2194
m	9	10		
H_m	15,300	130,000		

*For $m > 10$, use formula $H_m = e^{-m}(m-1)^{m-1}$.

Example 4.11.1 A fully fedback pendulous accelerometer has the following physical characteristics (see Section 3.6 for theory and notation):

$$\begin{aligned} I &= 7.07 \times 10^{-5} \text{ slug-ft}^2 \\ C &= 3.47 \times 10^{-4} \text{ lb-ft-s} \\ m &= 1.35 \times 10^{-2} \text{ slugs} \\ \ell &= 4.17 \times 10^{-2} \text{ ft} \end{aligned}$$

The device is to be used to detect the acceleration of a small rocket vehicle during the burning time (100 seconds). It is known that the acceleration has the form

$$a(t) = \frac{(1 \pm \delta_1)10^5}{(1 \pm \delta_2)311 - 2.48(1 \pm \delta_3)t} \text{ ft/s}^2$$

in which δ_1, δ_2 and δ_3 indicate the limits of possible variations of system parameters from missile to missile. Although δ_1, δ_2 and δ_3 are not known in advance, it is known that $0 \leq \delta \leq 0.10$ in all cases tested so far. Select the system parameters so that the dynamic error does not exceed 147 fps^2 after $t_0 = 0.02$ second.

Solution The equation of motion (performance equation) was shown to be

$$\frac{d^3T}{dt^3} + \left(\frac{c + k_2}{I}\right)\frac{d^2T}{dt^2} + \left(\frac{m\ell g + k_1}{I}\right)\frac{dT}{dt} + \frac{k_3}{I}T$$
$$= k_3\frac{m\ell a}{I} + k_1 m\ell\frac{da}{dt} + k_2 m\ell\frac{d^2a}{dt^2}$$

In the ultimate state to a step input $a = a_0 u(t)$ and

$$\frac{k_3}{I}T_u = k_3\frac{m\ell}{I}a_0$$

so that $T_u/m\ell$ is directly proportional to the acceleration. Given a scale which reads directly in units of torque, we can, by dividing by $m\ell$, read directly in units of acceleration. If we set $T/m\ell = z$, then z is the actual measurement. If we also substitute the given numbers and set $k_3/I = K_3$, $k_2/I = K_2$ and $k_1/I = K_1$, we obtain the performance equation

$$\frac{d^3z}{dt^3} + (4.91 + K_2)\frac{d^2z}{dt^2} + (256 + K_1)\frac{dz}{dt} + K_3 z$$

$$= K_3 a + K_1 \frac{da}{dt} + \frac{d^2a}{dt^2}$$

We are required to find K_1, K_2 and K_3 to satisfy the requirement that the error in z be less than the specified 147 ft/s².

Immediately there arises an interesting point. The coefficients K_1 and K_2 introduce signals proportional to $\dot{a}$ and $\ddot{a}$ into the system and thus contribute only error. Why not set $K_1 = K_2 = 0$ and let it go at that? To answer this question let us consider the stability condition for the third-order system

$$(4.91 + K_2)(256 + K_1) - K_3 > 0$$

Let us also note that $K_3 = r\omega^2$ since

$$(s + r)(s^2 + 2\zeta\omega s + \omega^2) = s^3 + (r + 2\zeta\omega)s^2 + (\omega^2 + 2\zeta\omega r)s + \omega^2 r$$

is the characteristic polynomial. Since r and ω must be made very large, it is conceivable that $K_3 > (4.91)(256)$ and that the system becomes unstable. Thus either K_1 or K_2 must be retained to correct for instability. But only K_2 has any real effect. Since K_1 is added to a large number and K_2 added to a small number, we see that small changes in K_2 can produce larger increase in stability than comparable changes in K_1. Thus we could conceivably set $K_1 = 0$ and retain K_2. Then we would have $\omega(\omega + 2\zeta r) = 256$ which would limit severely our choice r and ω. Indeed we could not choose ζ, ω and r independently. Thus we are forced to retain K_1, K_2 and K_3 and hope to find values of the three constants which permit us to meet the error specification. Inequality (4.11.5) is incomplete for this problem because in the derivation it was assumed that the input to the instrument would be proportional to the input function only. In this instance, however, there exists an additional error

$$\epsilon'_D = \frac{K_1}{K_3}\int_{0^+}^{t} I(t - \tau)\frac{da}{d\tau}\,d\tau + \frac{K_2}{K_3}\int_{0^+}^{t} I(t - \tau)\frac{d^2a}{d\tau^2}\,d\tau$$

$$\epsilon'_D \leq \frac{1}{K_3}(K_1\dot{a}_M + K_2\ddot{a}_M)[\hat{I}(0) - \hat{I}(t)]$$

in which $\dot{a}_M$ and $\ddot{a}_M$ are the absolute maxima of $\dot{a}$ and $\ddot{a}$ respectively.

In the special case of equal roots,

$$1 \geq \hat{I}(t) \geq 0$$

and thus it follows that

$$\epsilon'_D \leq \frac{K_1}{K_3}\dot{a}_M + \frac{K_2}{K_3}\ddot{a}_M$$

But comparison with the performance equation shows us that

$$K_1 + 256 = 3h_0^2$$
$$K_2 + 4.91 = 3h_0$$
$$K_3 = h_0^3$$

Hence

$$\epsilon_D' \leq \left(\frac{3h_0^2 - 256}{h_0^3}\right)\dot{a}_M + \left(\frac{3h_0 - 4.91}{h_0^3}\right)\ddot{a}_M$$

The equation equivalent to (4.11.11) here is

$$[\epsilon_D - H_m a_0]h_0^3 + [H_m - m - 3]\ddot{a}_M h_0^3 - 3\ddot{a}_M h_0 + (256\dot{a}_M + 4.91\ddot{a}_M) = 0$$

Solution of this for h_0 will solve the problem. The acceleration is maximum whenever $\delta_1 = \delta_2 = \delta_3 = 0.1$ and thus

$$\max\ [a(t)] = \frac{(1.1)10^5}{(1.9)(3.1) - (2.48)(1.1)t}$$

$$a(0^+) \leq 392 \text{ fps}^2 \qquad \dot{a} \leq 6{,}128 \text{ fps}^3 \ \ddot{a} \leq 4{,}572 \text{ fps}^4$$

and also $\epsilon_D = 147$ fps^2 so that with $m = 3$ and $H_m = 1.35$, the equation becomes

$$382h_0^3 + 28{,}400h_0^2 + 13{,}700h_0 - 1{,}590{,}000 = 0$$

From the above polynomial it is found that $h_0 \approx 200$. The minimum roots are assumed to be

$$s_1 = -200$$
$$s_2 = -200 + j266$$
$$s_3 = -200 - j266$$

in which it has been decided arbitrarily (from other consideration) to set $\zeta = 0.6$ in the quadratic roots. The characteristic polynomial is now given by

$$s^3 + 600s^2 + 19.09 \times 10^4 s + 22.14 \times 10^6 = p$$

from which it follows that

$$K_1 - 189{,}834$$
$$K_2 = 595$$
$$K_3 = 2.214 \times 10^7$$

In terms of the original design parameters we have

$$k_1 = 13.42 \qquad k_2 = 0.042 \qquad k_3 = 1{,}565$$

k_2 is unrealistically small. Let us choose $k_2 = 0.20$. Then we find that $K_2 = 2{,}830$. Obviously we must apportion 2,230 units among the three roots since $s_1 + s_2 + s_3 = 4.91K_2$. This is a matter of practical knowledge of the system. Let us, for simplicity at this point, apply the whole amount to s_1; then the roots become

$$s_1 = -2{,}430$$
$$s_2 = -200 + i226$$
$$s_3 = -200 - i266$$

The improved characteristic equation becomes

$$P = s^3 + 2.835 \times 10^3 s^2 + 10.72 \times 10^5 s + 2.45 \times 10^8$$

from which it follows that

$$K_1 = 10.72 \times 10^5 \qquad k_1 = 75.8$$
$$K_2 = 2{,}830 \qquad k_2 = 0.20$$
$$K_3 = 2.45 \times 10^8 \qquad k_3 = 17{,}300$$

It remains to see if the threshold of significance and the stability of the system are both satisfied. For the threshold, $h_0 > 200$ is sufficient since $h_0 > \boldsymbol{h}_0 = 100$. Stability requires that

$$(2.835 \times 10^3)(10.72 \times 10^5) > 2.45 \times 10^8$$

which is obviously true.

Another point is worth considering: Since k_2 is so small, is it worth returning to the beginning and finding a minimum h_0 for $k_2 = 0$? We obtain a new polynomial

$$382h_0^3 + 28{,}400h_0^2 - 1.568 \times 10^6 = 0$$

for which we find $h_0 = 7.097$. Here we have a fantastic gain—we need a much cheaper and less stringent system if it were possible to satisfy the other conditions, that is, relevance threshold and stability. For the threshold we find that

$$t_0\boldsymbol{h}_0 > m - 1 = 2$$

and thus $\boldsymbol{h}_0 = 100$. Choosing, as before, $\zeta = 0.6$ we now have three minimum roots

$$s_1 = -100$$
$$s_2 = -100 + j13.03$$
$$s_3 = -100 - j13.03$$

and a characteristic polynomial

$$P = s^3 + 300s^2 + 36{,}700s + 3.13 \times 10^6$$

Hence

$$k_1 = 2.576 \qquad k_2 = 0 \qquad k_3 = 22.12$$

Unfortunately the second term of the polynomial has a coefficient of 300, which is unrealizable (since $300 \gg 4.91$). Thus, with the basic instrument we have to work with, setting $k_2 = 0$ is not possible. If it were possible to redesign the pendulum so that $C = 600$, then $k_2 = 0$ would be the proper design constant.

NOTE: Examination of typical pendulous accelerometers available commercially[8] shows that usually h_0 varies from near zero to 450 at

$\zeta = 0.60$. $h_0 = 200$ is quite typical. The range of the input is from zero to 10 g's (320 fps^2) with $h_0 = 450$ corresponding to the highest ranges. The design we obtained in this example is in the region of common practice.

PROBLEMS

4.1 A system consists of a thermocouple connected to an amplifier and the amplifier output connected to a galvanometer. The performance equations are as follows:

$$0.1\frac{dV_1}{dt} + V_1 = k_1\theta_i \qquad \text{(Thermocouple)}$$

$$\tau_1\frac{dV_2}{dt} + V_2 = k_2V_1 \qquad \text{(Amplifier)}$$

$$\frac{d^2\phi}{dt^2} + 510\frac{d\phi}{dt} + 17{,}000\phi = k_3V_2 \qquad \text{(Galvanometer)}$$

in which θ_i is the input temperature, V_1, V_2 are voltages and ϕ is the galvanometer deflection. Determine the value of τ_1 which is at the boundary between stability and instability of the system.

4.2 Does the characteristic polynomial

$$s^4 + 1.20s^3 + 6.02s^2 + 14.2s + 8.42$$

represent a stable dynamic system?

4.3 Does the characteristic polynomial

$$s^{10} + 2.12s^9 + 3.14s^8 + 7.70s^7 + 19.1s^6 + 4.12s^5 + s + 3.29$$

represent a stable dynamic system?

4.4 Select a so that the system with the following characteristic polynomial will be stable:

$$s^3 + as^2 + 11.21s + 32.0$$

4.5 A system is given by

$$\left(\frac{d}{dt} + a_0\right)\left(\frac{d^2\phi}{dt^2} + 7.29\frac{d\phi}{dt} + 275\phi\right) = ky(t)$$

Select a_0 so that the system is stable.

4.6 A first-order system is represented by $a_0 = 27.3$ and $K = 1.21$. Compute the ultimate state response to $y(t) = t^3/3$ through the use of sensitivity coefficients. Plot the relative ultimate state dynamic error.

4.7 A thermometric system is subjected to the input $\theta = 11.7 + 3.21t$. Select a_0 such that the absolute ultimate-state relative error is less than 5% for all $t \geq 0.75$ second.

4.8 Compute the relative ultimate state error of a second-order system in which $\omega = 10.2$ rad/s and $\zeta = 0.500$.

4.9 In a second-order system it is required to set $\zeta = 0.600$ and $\epsilon_D \leq 0.20$ for all $t \geq 0.5$ second. If $y(t) = 0.5t^2$, select a value of ω^2 to meet the specifications if $y_{max} = 20$.

4.10 Is there a value of $\zeta < 1$ such that $|\varepsilon_D| \leq 0.15$ for $t > 0.75$ second if $\omega^2 = 1000$ rad²/s² and $y(t) = t^2$?

4.11 A second-order system is to have its output recorded on a chart marked in 1/20-inch intervals, and has a trace 1/40-inch thick. Determine the system relevance threshold as a function of chart speed.

NOTE: *In problems 4.12 to 4.15, set* $\xi_u(0) = 1.00$ *and* $\dot{\xi}_u(0) = 0.50$.

4.12 Determine the transient dynamic error associated with the system of problem 4.6.

4.13 Determine the transient dynamic error associated with the system of problem 4.7.

4.14 Determine the transient dynamic error associated with the system of problem 4.8.

4.15 Determine the transient dynamic error associated with the system of problem 4.9.

4.16 Select ζ so that the total relative dynamic error of a second-order system subjected to the input $y(t) = 7.1t^2 + 4.0$ is less than 5% for $t \geq 1.0$ second if $\omega^2 = 730$ rad²/s².

4.17 Find the minimum amplitude of the real part of the system roots if the system is of the seventh order and the input function is $y(t) = 3.19e^{-7.74t}$ and if the acceptable error is to be no greater than $0.171y(t)$ for $t > 0.136$ second.

4.18 A pressure transducer is connected to an amplifier (first-order system), which is then connected to a galvanometer. If the pressure is given by 0.297 sin 2.20t psf, determine the system parameters such that $\epsilon_D \leq$ 0.010 psf for all $t > 0.10$ second. (This is a fairly extensive problem.)

4.19 Prove Equation (4.6.2).

NOTES

[1]In the several hundred books on controls, instrumentation and systems that the author has examined in the past twenty years, the Routh-Hurwitz criterion has always been dismissed as provable but not worth proving. It was necessary to go back to the 1905 development of E. J. Routh for a proof (Ref. 68, pp. 223–31). Even in this instance, recourse must be had to earlier works (Cauchy) to understand the proof completely.

[2]See Ref. 14, pp. 126–27. See Ref. 62 for more sophisticated criteria.

[3]The condition $a_i > 0$ is not needed since the Hurwitz determinants always yield $a_i > 0$ as a condition. It is included here as a first clue to stability.

[4]Ref. 35, p. 393, line 354.

[5]Ref. 70, Ref. 71. (See also Ref. 69.)

[6]Actually the device can be operated at several recording speeds. The author chose 5 mm/s as typical.

[7]Ref. 71, pp. 68–71.

[8]Catalogue of the Kearfott Division, General Precision, Inc., 1971.

5

THE SINUSOIDAL RESPONSE

5.1 Introduction

In those parlous days before the propagation of the digital computer made the computation of the roots of characteristic polynomials a routine matter, few attempts were made to use the direct solution of the performance equation to study the behavior of systems. With the exception of the root locus method,[1] which in practice is limited to single-parameter systems, no successful technique for design in the time plane has ever been proposed that did not founder eventually because of the difficulty of extracting the roots of a large family of polynomials. The pole-zero method, once thought to be the answer to the needs of electrical system designers, has proved to be an academic exercise, considerably less useful than the ultimate state impedance methods it aimed to replace.

There is only one complete technique for response optimization. The performance equations are written in terms of the system parameters, and families of solutions are obtained which correspond to the family of permissible design parameters. From the family of solutions a subset is selected which meets a given specification. From the subset a still smaller set is found which will satisfy another specification. The procedure is continued until all (or almost all) specifications have been met. The essence of the procedure is the systematic search through all permissible solutions for the system parameters which give the best match to an ideal solution. If the dynamical specifications are the "response time" and the "percent overshoot," graphs should be obtained of the change of response time and of overshoot with respect to the system gains.

Example 5.1.1 An amplifier is described by the performance equation

$$\frac{1}{a_0}\frac{dz}{dt} + z = y(t)$$

Determine the effect of a_0 on the characteristic response time τ if $y(t) = u(t)$

Solution Since we solved the system in the previous chapter, we can immediately write that

$$\tau = \frac{1}{a_0}$$

Let us, on the other hand, examine the system

$$\frac{d^5z}{dt^5} + 3.117\frac{d^4z}{dt^4} + 2.112\frac{d^3z}{dt^3} + a_2\frac{d^2z}{dt^2} + 7.159\frac{dz}{dt} + 3z = 3y(t)$$

and ask about the effect of a_2 on the response time. Here we cannot obtain the general solution to the problem; and either we must content ourselves with bounding a_2 by a dominating function (as discussed in the previous chapter), or else we must plot all the roots of the characteristic polynomial as functions of a_2 and then determine a_2 from the roots. Even in this simple instance the procedure becomes involved. If there were more than one parameter and more than one specification, only the modern computer would be of help.

With the general availability of digital computers or of high-speed analogue-digital hybrids, it is possible to obtain either many direct solutions to test function inputs or else the characteristic roots, so that the design of dynamical-response instruments without high-speed computers is, today, only a conservative anachronism. Most of the methods developed in the past to circumvent the lack of computing ability are obsolescent and likely to vanish from the scene. The Nyquist plot and the Hall chart will be remembered by wistful old gentlemen and by technical historians.

An exception can be made, however, for the Bode diagram (attenuation chart) since it is directly related to the concept of filtering. As rough indicators of system performance, such information as cutoff frequency, resonance-peak ratio and post-cutoff slope (roll-off) will continue to be used. Manufacturers give attenuation and phase-shift information, rather than time lags and roots of characteristic polynomials, so that the instrument designer should be able to make almost effortlessly the jump between dynamic response and ultimate-state sinusoidal response.

5.2 Ultimate-state Sinusoidal Response

One of the remarkable properties of linear ordinary differential equations in the possibility of constructing transient solutions by the superposition of ultimate-state responses to sinusoidal inputs. The essential principle is that the Fourier series of the input transforms into the Fourier series of the output. It follows, therefore, that the dynamic characteristics of the output are directly deducible from the ultimate-state response to a sinusoidal input without actually computing the dynamic response.

5.2.1 Let us perform an experiment with a recording galvanometer. Through an appropriate amplifier, an electrical signal is introduced into the galvanometer from an accurate sine-wave generator. The recorder records one volt per millimeter and the input is known to be

$$y(t) = 23 \sin \omega_0 t \text{ volts}$$

so that the double amplitude of the input signal is 46 volts (Fig. 5.1). The output trace is allowed to stabilize and its amplitude is measured.

It is soon evident that whereas for low frequencies the output double-amplitude is nearly 46 volts, for higher frequencies the output double-amplitude is less than 46 volts; indeed if $\omega_0 = 200$ Hz, the output response to a sinusoidal ± 23-volt signal is a mere 3 volts (Table 5.1). If we now compute the amplitude ratio $A = A_0/A_i$ (Amplitude *out*/Ampli-

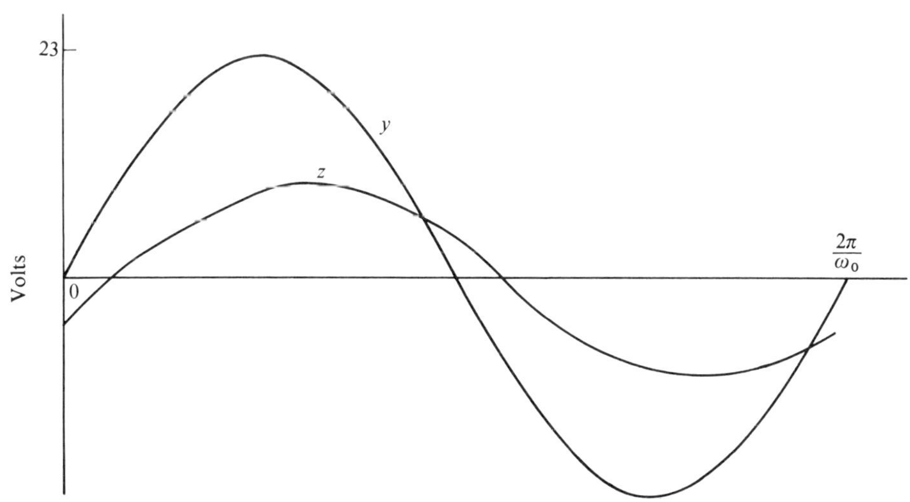

FIGURE 5.1 Attenuation and phase shift of sinusoidal signal

TABLE 5.1 Experimental response of a recording galvanometer

Freq. (Hz)	A_{out} (mm)	A_0/A_i	dB
0.0	46.0	1.000	0.000
0.5	45.5	0.989	−0.100
1.0	45.2	0.982	−0.160
2.0	44.5	0.967	−0.300
5.0	42.0	0.914	−0.782
7.0	40.0	0.869	−1.22
10.0	36.7	0.798	−1.96
15.0	34.0	0.740	−2.62
20.0	32.5	0.706	−3.02
30.0	33.0	0.717	−2.90
40.0	35.5	0.728	−2.76
50.0	34.0	0.739	−2.62
60.0	31.5	0.685	−3.29
70.0	26.0	0.566	−4.94
80.0	20.5	0.445	−7.01
90.0	16.0	0.348	−9.16
100.0	13.0	0.283	−12.0
120.0	7.5	0.163	−15.8
150.0	5.0	0.108	−19.3
200.0	3.0	0.065	−23.7

tude *in*) and plot it in the form $20 \log A_0/A_i$ vs. $\log \omega_0$, we obtain the Bode diagram for the galvanometer (Fig. 5.2). That the galvanometer system supresses the higher-frequency signals is obvious from this experiment and that this graph contains information on the dynamic characteristics of the system will become clear later.

It can be seen that the Bode chart (or attenuation chart) has an important connection with threshold matching. If a system produces signals which get smaller with higher frequencies, it is quite possible that the user will not be able to detect the high-frequency components because the component amplitude falls below the user's threshold. We say, therefore, that there is a "cutoff frequency" dictated by the threshold. It will be demonstrated later that cutoff frequencies arise more commonly from accuracy requirements than from threshold matchings.

Example 5.2.1 A first-order system, $\tau_0 = 1.25$ s, $K = 1$, is subjected to a sinusoidal input $y(t) = 3 \sin (\omega_0 t)$ mV. The output of the system is introduced into a recorder which cannot respond to an input signal smaller in amplitude than 0.01 mV. Determine a "cutoff" frequency ω_c such that for all $\omega_0 > \omega_c$ the recorder will not observe the input signal.

Solution The performance equation is

$$1.25\frac{dz}{dt} + z = 3 \sin \omega_0 t$$

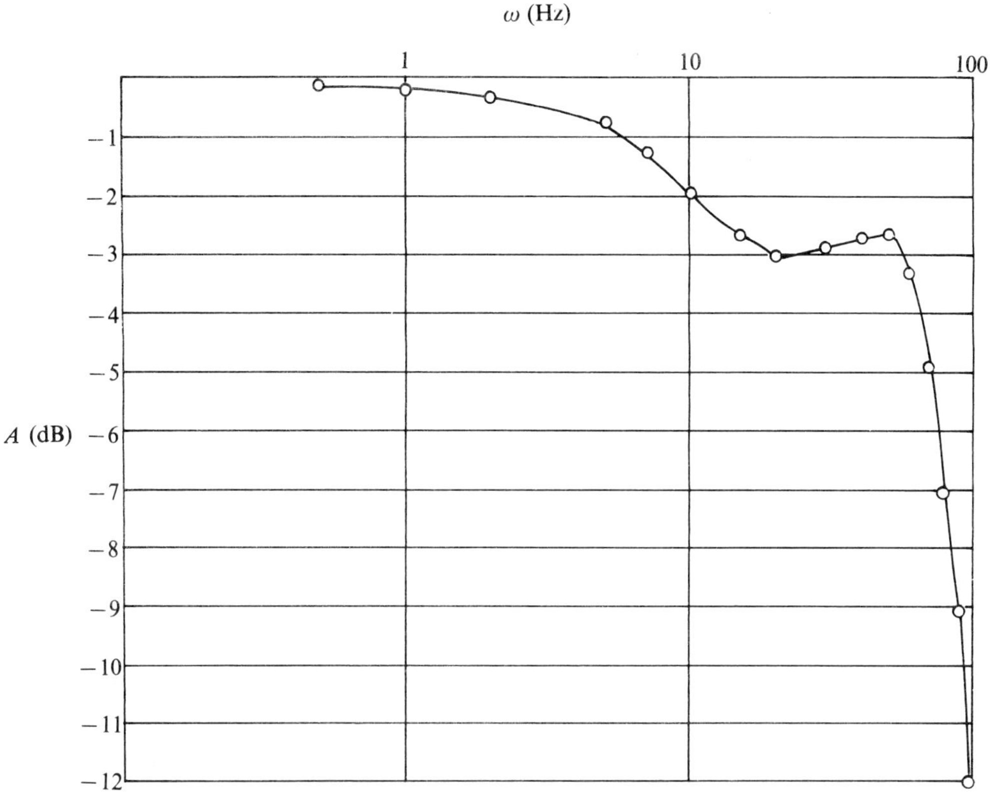

FIGURE 5.2 Experimental response of a recording galvanometer

The particular solution is

$$z_u = \frac{3}{\sqrt{1 + (1.25\omega_0)^2}} \sin(\omega_0 t + \phi)$$

$$\tan \phi = -1.25\omega_0$$

The amplitude A of z_u decreases with ω_0

$$A = \frac{3}{\sqrt{1 + (1.25\omega_0)^2}}$$

$\omega_0 = \omega_c$ is found by setting $A = 0.01$, from which we obtain

$$\omega_c = 240 \text{ rad/s} = 38.2 \text{ Hz}$$

For any input frequency $\omega_0 > 38.2$ Hz, the sinusoidal sensitivity A will be below the threshold of the recorder and will not be recorded.

5.2.2 The theory behind the dynamical design of systems by ultimate-state sinusoidal responses is best described in terms of the Fourier series. If the input and the output can both be expressed as summations of periodic functions, any difference between input and output can only be ascribed to distortion introduced by the suppression or at least by the deformation of the input sinusoids in the manner described by the Bode chart.

Any segment of input signal defined over a time interval $T/2$ can be expressed in terms of a Fourier series[2]

$$f(t) = \sum_{n=0}^{\infty} (\alpha_n \cos n\omega_0 t + \beta_n \sin n\omega_0 t) \qquad \textbf{(5.2.1)}$$

in which $\omega_0 = \pi/T$. The series represents the original function in the interval $T/2$ and is periodic with period T. Outside the original definition range, the function repeats indefinitely. The terms α_n and β_n are computed by

$$\begin{aligned}\alpha_0 &= \frac{1}{2T}\int_{-T/2}^{+T/2} f(t)\,dt \\ \alpha_n &= \frac{1}{T}\int_{-T/2}^{+T/2} f(t)\cos n\omega_0 t\,dt \\ \beta_n &= \frac{1}{T}\int_{-T/2}^{+T/2} f(t)\sin n\omega_0 t\,dt\end{aligned} \qquad \textbf{(5.2.2)}$$

Occasionally the function $f(t)$ can be expressed in the form of the Fourier integral[3]

$$f(t) = \frac{1}{\pi}\int_{-\infty}^{\infty} [\alpha(\omega_0)\cos\omega_0 t + \beta(\omega_0)\sin\omega_0 t]\,d\omega_0$$

in which

$$\alpha(\omega_0) = \int_{-\infty}^{+\infty} f(v)\sin(\omega_0 v)\,dv$$

$$\beta(\omega_0) = \int_{-\infty}^{+\infty} f(v)\cos(\omega_0 v)\,dv$$

The Fourier integral is the limiting case of a Fourier series with a very large period and infinitesimal (or continuous) harmonics, suitable for describing a function on an infinite interval whenever the integrals converge. Any piecewide continuous input function can be described by a summation of sinusoids. On the interval of definitions the Fourier series represents the function and is integrable wherever the function is integrable.

Consider the normalized performance equation

$$\sum_{i=0}^{m} a_i \frac{d^i z}{dt^i} = f(t) \qquad a_m = 1$$

in which

$$f(t) = \sum_{n=0}^{\infty} (\alpha_n \cos n\omega_0 t + \beta_n \sin n\omega_0 t)$$

Then, if all initial conditions were zero, we could solve for z in the form

$$z = \int_0^{\infty} I(\tau) \sum_{n=0}^{\infty} [\alpha_n \cos n\omega_0(t-\tau) + \beta_n \sin n\omega_0(t-\tau)]\,d\tau \qquad \textbf{(5.2.3)}$$

from which we obtain

$$z = \sum_{n=0}^{\infty} [C_n(t) \sin n\omega_0 t + D_n(t) \cos n\omega_0 t] \tag{5.2.4}$$

in which

$$\begin{aligned} C_n(t) &= \int_0^t I(\tau)[\beta_n \cos n\omega\tau - \alpha_n \sin n\omega\tau]\, d\tau \\ D_n(t) &= \int_0^t I(\tau)[\alpha_n \cos n\omega\tau + \beta_n \sin n\omega\tau]\, d\tau \end{aligned} \tag{5.2.5}$$

It is customary to define

$$\begin{aligned} \tan \psi_n &= \frac{C_n(t)}{D_n(t)} \\ \underline{A}'_n(t) &= [D_n^2(t) + C_n^2(t)]^{1/2} \end{aligned} \tag{5.2.6}$$

so that the solution has the form

$$z(t) = \sum_{n=0}^{\infty} \underline{A}'_n(t) \sin [n\omega t + \psi_n(t)] \tag{5.2.7}$$

In (5.2.7) we have the normal form of the sinusoidal output.

It is customary to write the amplitude of the outputs as a product of the amplitude of the inputs and a dimensionless sensitivity ratio. Let

$$\underline{a}_n = [\alpha_n^2 + \beta_n^2]^{1/2}$$

and

$$\underline{A}'_n(t) = \underline{A}_n(t)\underline{a}_n$$

Then

$$z(t) = \sum_{n=0}^{\infty} \underline{a}_n \underline{A}_n(t) \sin (n\omega t + \psi_n) \tag{5.2.8}$$

in which $\underline{A}_n(t)$ is the "dimensionless sinusoidal sensitivity." Of great value is the ultimate state of $\underline{A}_n(t)$ and to this we give a special symbol

$$A_n = \lim_{t\to\infty} \underline{A}_n(t)$$

The ultimate state of $\psi_n(t)$ is also quite important and is denoted by

$$\phi_n = \lim_{t\to\infty} \psi(t)$$

Example 5.2.2 A periodic function

$$f(t) = \begin{cases} \dfrac{\pi}{4} t & 0 < t < \dfrac{\pi}{2} \\ \dfrac{\pi}{4}(\pi - t) & \dfrac{\pi}{2} < t < \pi \end{cases}$$

is represented by the Fourier sine series

$$f(t) = \sum_{n=0}^{\infty} \frac{(-1)^n}{(2n+1)!} \sin (2n+1)t$$

Determine $A_n(t)$ and $\phi_n(t)$ for the first-order system $K = 1$, $a_0 = 1$.

Solution In this instance

$$\frac{dz}{dt} + z = f(t)$$

and thus $I(t) = e^{-t}$ and

$$z = \sum_{n=0}^{\infty} \sin(2n+1)t\left[\frac{(-1)^n}{(2n+1)!}\int_0^t e^{-\tau}\cos(2n-1)\tau\, d\tau\right]$$
$$+ \sum_{n=0}^{\infty} \cos(2n+1)t\left[\frac{(-1)^n}{(2n+1)!}\int_0^t e^{-t}\sin(2n+1)\tau\, d\tau\right]$$

The two coefficients can be integrated to yield

$$\tan\psi_n = \frac{\sin\phi_n + e^{-t}\sin[(2n \pm 1)t + \phi_n]}{\cos\phi_n - e^{-t}\cos[(2n+1)t + \phi_n]}$$

$$\underline{A}'_n(t) = \frac{1}{(2n+1)!\,[(2n+1)^2+1]^{1/2}}[1 + e^{-2t} - 2e^{-t}\cos(2n+1)t]^{1/2}$$

where $\tan\phi_n = -(2n+1)$.

The dimensionless sensitivity ratio is

$$\underline{A}_n = \frac{(-1)^n[1 + e^{-2t} + 2e^{-t}\cos(2n+1)t]^{1/2}}{(2n+1)!\,[(2n+1)^2+1]^{1/2}}$$

In the ultimate state $\tan\psi_n - \tan\phi_n$ and

$$A_n = \frac{(-1)^n}{(2n+1)!\,[(2n+1)^2+1]^{1/2}}$$

The ultimate state will be a distorted wave form only roughly similar to the input function.

In Example 5.2.3 it would have been just as easy to make the period different from 2π, say, some arbitrary interval $0 < t < T$. Then in the time period $0 < t < T$, the Fourier series representation would be indentical to the solution of the ordinary differential equation

$$\frac{dz}{dt} + z = \frac{\pi}{2}t$$

This solution is

$$z = \frac{\pi}{2}(e^{-t} + t - 1)$$

If $T = 10$ s, then at the time $T/2 = 5$ s the true solution and the Fourier solution will begin to diverge. At that instant $e^{-T/2} = 4 \times 10^{-5}$ and z will differ from its ultimate state solution by only 0.001%. The Fourier series represents the complete dynamical solution, since what happens after $t = T$ s is unimportant.

The Fourier series solution will eventually reach an ultimate state so that each period will not have e^{-t} in it. Yet, since at $t = T$ the input is switched to a down ramp, the solution must represent the dynamical solution of the problem

$$\frac{dz}{dt} + z = \frac{\pi}{4}(a_1 - t)$$

with the initial conditions $z = (a_1 - 1)\pi/2$.

At time $t > (3/2)T$, the Fourier series will represent the solution to

$$\frac{dz}{dt} + z = \frac{\pi}{4}\left(t_1 - \frac{3}{2}T\right)$$

Obviously the Fourier series obtainable by summing the ultimate-state sinusoidal solution will represent a dynamical solution to some input and some initial condition.[4] We can state the principle, therefore, that knowledge of the ultimate state response of a system to all pure sinusoidal inputs is equivalent to knowledge of some dynamical results.

5.3 Step Response

Knowledge of the amplitude and phase-shift variations with input frequency can be used to obtain an estimate of the dynamical behavior of the system. Let the Bode plot and the phase plot be on hand. Then the ultimate state response to any component of a Fourier series is known, and thus the ultimate state response to any periodic function representable by a Fourier series is also known. Consider, for example, the square wave

$$f(t) = \begin{cases} M & 0 < t < T/2 \\ -M & T/2 < t < T \end{cases}$$

which can be represented as a function periodic in T (Fig. 5.3) by the series[5]

$$f(t) = \frac{4M}{\pi} \sum_{n=0}^{\infty} \frac{1}{(2n+1)} \sin\,(2n+1)\frac{2\pi}{T}t$$

Each harmonic, when processed through the instrument, will be modified by the dimensionless sinusoidal sensitivity $\underline{A}_n(t)$ and phase shift $\psi_n(t)$ so that the response is given by

$$z(t) = \frac{4M}{\pi} \sum_{n=0}^{\infty} \frac{\underline{A}_n(\omega_n, t)}{(2n+1)} \sin\left[(2n+1)\frac{2\pi}{T}t + \psi_n(t)\right]$$

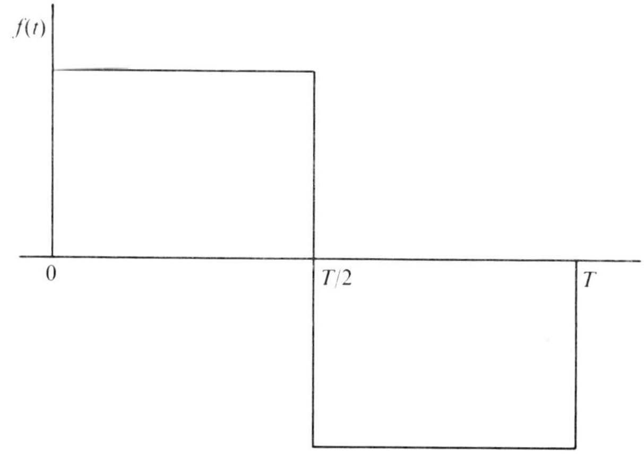

FIGURE 5.3 Square wave of period T

If a sufficient amount of time is allowed to pass, $z(t)$ will become a series of the form

$$z(t) = \frac{4M}{\pi} \sum_{n=0}^{\infty} \frac{A_n(\omega_n)}{2n+1} \sin\left[(2n+1)\frac{2\pi}{T}t + \phi_n\right]$$

in which A_n and ϕ_n are to be found in the Bode and phase plots for $\omega_n = (2n+1)2\pi t/T$. If the period T is sufficiently great, the response $z(t)$ will look as in Figure 5.4. Obviously for sufficiently large values of t and T, $z(t)$ will represent the dynamical response of the system to a step input when the initial conditions are

$$z(t) = \pm M \frac{d^n z}{dt^n} = 0 \quad \text{at } t = mT \qquad n > 0 \qquad m = 1, 2, 3, \ldots$$

Given the Bode diagram and the phase shift graph, a step output response can be evaluated.

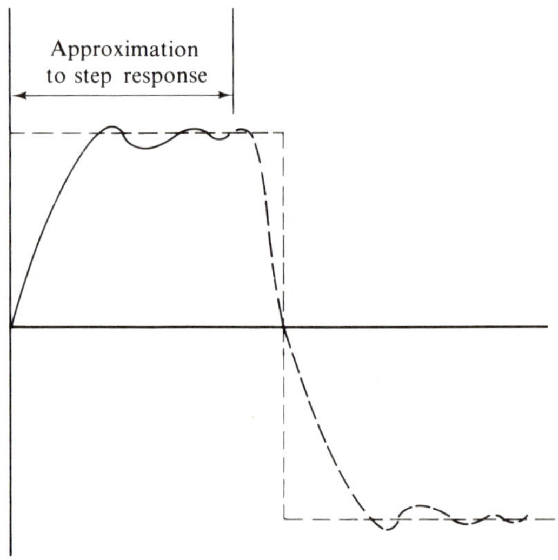

FIGURE 5.4 Approximation of unit step response by Fourier series

If the output is now shifted through the distance M and if $M = 1/2$, the result will be the response to a unit step input at times $\omega_1 t = 0$, $\omega_3 t = 2\pi$, etc.

$$z(t) = \frac{1}{2} + \frac{2}{\pi} \sum_{n=0}^{\infty} \frac{A_{2n+1}(\omega_{2n+1})}{2n+1} \sin(\omega_{2n+1}t + \phi_{2n+1})$$

in which

$$\phi_{2n+1} = (2n+1)\frac{2\pi}{T}$$

The selection of T is a rather touchy point. A reasonably good practice is to select

$$A_{dB_{11}} = 20 \log A_{11} = -12 \text{ dB}$$

and let $\omega_1 = \omega_{11}/11$ in which ω_{11} is the frequency which corresponds to $A_{dB_{11}}$.

Example 5.3.1 A second-order system is given by $a_0 = 64$, $a_1 = 8$, $K = \omega^2 = a_0$. Determine the unit step response by direct integration and by the approximate method.

Solution The performance equation is

$$\ddot{z} + 8\dot{z} + 64z = 64u(t)$$

Since the characteristic polynomial has the form

$$(s + 4)^2 + 6.93^2 = 0$$

the solution can be written as

$$z(t) = 1 + 1.15e^{-4t} \sin (6.93t - 2.10)$$

This result is easily computed and is plotted by the solid line in Figure 5.5.

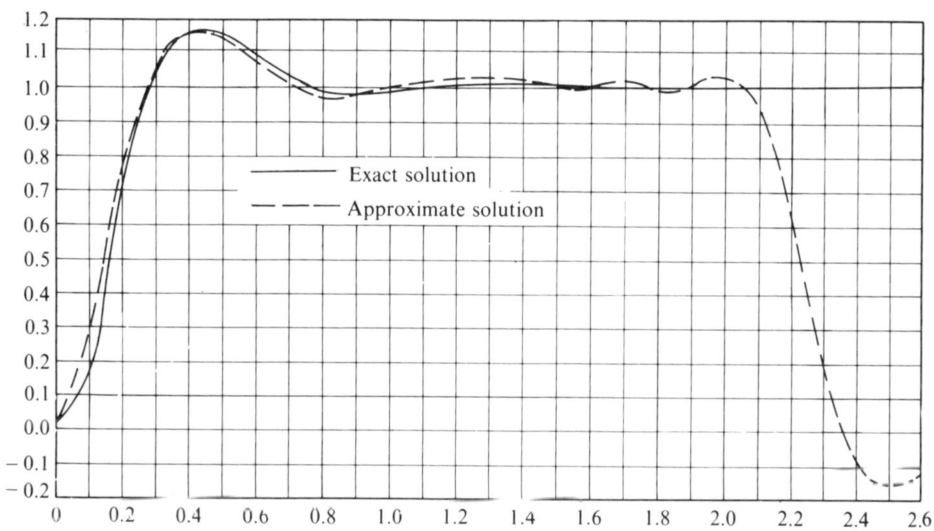

FIGURE 5.5 Comparison of solutions of Example 5.3.1

To compute the approximated solution from the sinusoidal ultimate state sensitivity and the sinusoidal ultimate state phase shift, it is first necessary to obtain the Bode chart and the phase shift graph. Usually the necessary charts are obtained from experimental data. In this instance, however, it is quite easy to obtain the graphs theoretically. (See Section 5.4 for the details.) The amplitude ratio is

$$A = \frac{64}{[(64 - \omega_0^2)^2 + 64\omega_0^2]^{1/2}}$$

$$\tan \phi = \frac{-8\omega_0}{64 - \omega_0^2}$$

The next task is to determine T, and hence ω_{2n+1}. It is necessary to take T large enough to give accurate answers with reasonable computation. A common practice is to choose the 11th harmonic to correspond to the frequency at which the amplitude ratio (ultimate state sinusoidal sensitivity) is -12 dB. A number of other combinations are possible, for example, the 17th harmonic could be equivalent to the frequency at which $A = -19$ dB. The point is to make the last harmonic quite small as compared to the fundamental and to have sufficient terms to permit a good approximation. Since the 11th harmonic is the sixth term, and since $A(11\omega_1) = -12$ dB is less than 2% of $A(\omega_1)$, we have an approximation which is reasonable. In our case we wish to set $\omega_{11} = 11\omega_1$ in such a fashion that $A_{11} = -12$ dB, or $A_{11} = 0.251$. Thus we obtain

$$0.251 = \frac{64}{[(64 - \omega_{11}^2)^2 + 64\omega_{11}^2]^{1/2}}$$

which yields $\omega_1 = 1.53$ rad/s. We prepare a table:

n	Harmonic	ω_{2n+1}	A_{2n+1}	$A_{2n+1}/(2n+1)$	ϕ_{2n+1}
0	1	1.53	1.02	1.02	−0.1958
1	3	4.59	1.13	0.377	−0.7021
2	5	7.65	1.04	0.208	−1.488
3	7	10.71	0.644	0.092	−2.104
4	9	13.77	0.384	0.042	−2.421
5	11	16.83	0.251	0.023	−2.591

We shall therefore evaluate the sum

$$\begin{aligned} z(t) = 0.50 + 0.6360[&1.02 \sin(1.53t - 0.1958) \\ &+ 0.377 \sin(4.59t - 0.7021) + 0.208 \sin(7.65t - 1.438) \\ &+ 0.092 \sin(10.71t - 2.104) + 0.042 \sin(13.77t - 2.421) \\ &+ 0.023 \sin(16.85t - 2.591)] \end{aligned}$$

The computation is rather laborious but completely straighforward and can be accomplished with a digital computer. Alternately a "quick and dirty" graphical procedure can be applied by making sinusoidal templates for graphical computation.[6] The results are plotted as a dotted line in Figure 5.5. The close match between the actual solution and the approximate solution is striking.

Example 5.3.2 Obtain the coefficient S_1 in terms of the Bode diagram and phase-shift chart.

Solution We know that

$$S_1 = \int_0^t I(\tau)(t - \tau)\,d\tau - t\int_0^t I(\tau)\,d\tau$$

The first integral represents the response of the system to the unit ramp input $y(t) = t$ and the second integral is S_0. We now obtain the Fourier series which represents the function

$$y(t) = t$$

This is given by the sine series[7]

$$y_1(t) = 2 \sum_{n=1}^{\infty} \frac{(-1)^{n+1}}{\omega_n} \sin \omega_n t$$

in which

$$\omega_n = \frac{2n\pi}{T}$$

When processed through the system, the input generates the response

$$z_1(t) = 2 \sum_{n=1}^{\infty} \frac{(-1)^{n+1} A_n}{\omega_n} \sin (\omega_n t + \phi_n)$$

If we now note that

$$\frac{1}{2}[1 + (-1)^{n+1}] = \begin{cases} 0 & n = 0, 2, \ldots \\ 1 & n = 1, 3, \ldots \end{cases}$$

then we can combine z_1 with the results of Example 5.3.1 to obtain

$$S_1 = \frac{1}{2}\left(1 - \frac{t}{T}\right) + \sum_{n=1}^{\infty} \frac{\{[(-1)^{n+1} - 1][1 - (t/T)] + 1\}}{\omega_n} A_n \sin (\omega_n t + \phi_n)$$

in which

$$\omega_n = \frac{2n\pi}{T}$$

As before we choose ω_{11} as the frequency for which $A_{11} = -12$ dB.

The construction of the other sensitivity coefficients is now quite simple if it is noted that the Fourier series representation of t^m is

$$t^m = \sum_{n=1}^{\infty} b_n \sin \omega_n t$$

in which

$$\omega_n = \frac{2n\pi}{T}$$

and

$$b_n = \frac{(1)^{n+1} 4}{\omega_n^{m+1}} \sum_{p=0}^{q} \frac{(n\pi)^{m-2p}}{(m - 2p)!}$$

where

$$q = \begin{cases} \dfrac{m-1}{2} & \text{for } m \text{ odd} \\[2ex] \dfrac{m-2}{2} & \text{for } m \text{ even} \qquad m \neq 0 \end{cases}$$

It is clear, at this point, that since all the sensitivity coefficients are computable by application of Fourier series methods, it is also possible to compute the response of any linear system to any piecewise linear input by the same methods. Since Fourier series techniques lead uniquely to general system dynamics, it becomes obvious that knowledge of system response to sinusoidal inputs also can give information with respect to dynamical response.

Experience tells us how a certain collection of sinusoids combine in a dynamic solution and which parameters are relevant. We then observe the parameters themselves and carry out our design strictly with the sinusoidal information. The technique is known as *frequency plane analysis.*[8]

The values of A_n and ϕ_n can often be obtained experimentally much more easily than step response. In many physical systems it is possible to oscillate the input and record the output fairly easily, whereas it is difficult to apply a step input or a ramp input or other test functions successfully. In such a case it is better to obtain the phase shift and sinusoidal sensitivity by direct measurement and then to construct the test function response from it.

5.4 The General Sinusoidal Ultimate State

If the performance equation of a linear system is known, the Bode and phase plots can be obtained routinely. Since a sinusoid can be either a sine or a cosine, let us use a general input of the form

$$y(t) = C(\cos \omega t + j \sin \omega t) = Ce^{j\omega t}$$

The response $z(t)$ will be a complex number, the real part of which will be the solution to the input $C \cos \omega_0 t$, and the imaginary part, the solution to the input $C \sin \omega_0 t$. If the transform of the impulsive response connecting a specific input y to a specific output z is $G(s)$, then the performance relationship will have the form

$$\mathcal{L}(z) = \frac{CG(s)}{s - j\omega_0} \qquad \textbf{(5.4.1)}$$

in which $G(s) = \mathcal{L}[I(t)]$. By partial fractions we can write

$$\mathcal{L}(z) = \frac{CG(j\omega_0)}{s - j\omega_0} + R(s)$$

in which $R(s)$ is the remainder of the transform due to poles other than the one at $s = +j\omega_0$. $R(s)$ represents the transform of the transient part so that in a stable system

$$\lim_{t\to\infty} \mathcal{L}^{-1}R(s) = 0$$

and thus

$$\lim_{t\to\infty} z(t) = z_{us} = CG(j\omega_0)e^{j\omega_0 t} \qquad \textbf{(5.4.2)}$$

Now $G(j\omega_0) = R_e[G(j\omega_0)] + jIm[G(j\omega_0)]$ and thus

$$G(j\omega_0) = G(j\omega_0)e^{j\phi(\omega_0)} \quad \text{where } \tan \phi(\omega_0) = \frac{Im(G)}{R_e(G)} \qquad \textbf{(5.4.3)}$$

If Equations (5.4.2) and (5.4.3) are combined, the ultimate state solution becomes

$$z_{us}(t) = C|G(j\omega_0)|e^{j(\omega_0 t+\phi)}$$
$$= C|G(j\omega_0)|[\cos(\omega_0 t + \phi) + j\sin(\omega_0 t + \phi)]$$

The real part of z_{us} is the response to a cosine input, and the imaginary part, the response to a sine input. In either case

$$A(\omega_0) = |G(j\omega_0)|$$

and

$$\tan\phi = \frac{ImG(j\omega_0)}{R_eG(j\omega_0)}$$

are the dimensionless sinusoidal sensitivity and the phase shift, respectively.

Example 5.4.1 A first-order linear system $a_0 = 1/\tau$, $K = 1/\tau$, is driven by the input function $\alpha \sin \omega_0 t$. Compute the amplitude and phase angle of the ultimate state response.

Solution

$$\frac{dz}{dt} + \frac{1}{\tau}z = \frac{1}{\tau}f(t)$$

$$\mathfrak{L}z = \frac{1}{\tau}\frac{\mathfrak{L}f(t)}{s + 1/\tau}$$

Clearly $G = 1/(\tau s + 1)$ and thus

$$G(j\omega_0) = \frac{1}{\tau j\omega_0 + 1} = \frac{1 - j\tau\omega_0}{1 + (\tau\omega_0)^2}$$

Finally

$$\tan\phi = -\tau\omega_0$$

$$A = \frac{1}{\sqrt{1 + (\tau\omega_0)^2}}$$

A and ϕ are plotted in Figure 5.6.

Example 5.4.2 Obtain the phase angle and the dimensionless sinusoidal sensitivity of a second-order system $a_0 = \omega^2$, $a_1 = 2\zeta\omega$, $K = \omega^2$.

Solution

$$G = \frac{\omega^2}{s^2 + 2\zeta\omega s + \omega^2}$$

Thus

$$A = \frac{\omega^2}{[(\omega^2 - \omega_0^2)^2 + (2\zeta\omega\omega_0)^2]^{1/2}}$$

$$\tan\phi = \frac{-2\zeta\omega\omega}{\omega^2 - \omega_0^2}$$

A and ϕ are plotted in Figures 5.7a, 5.7b.

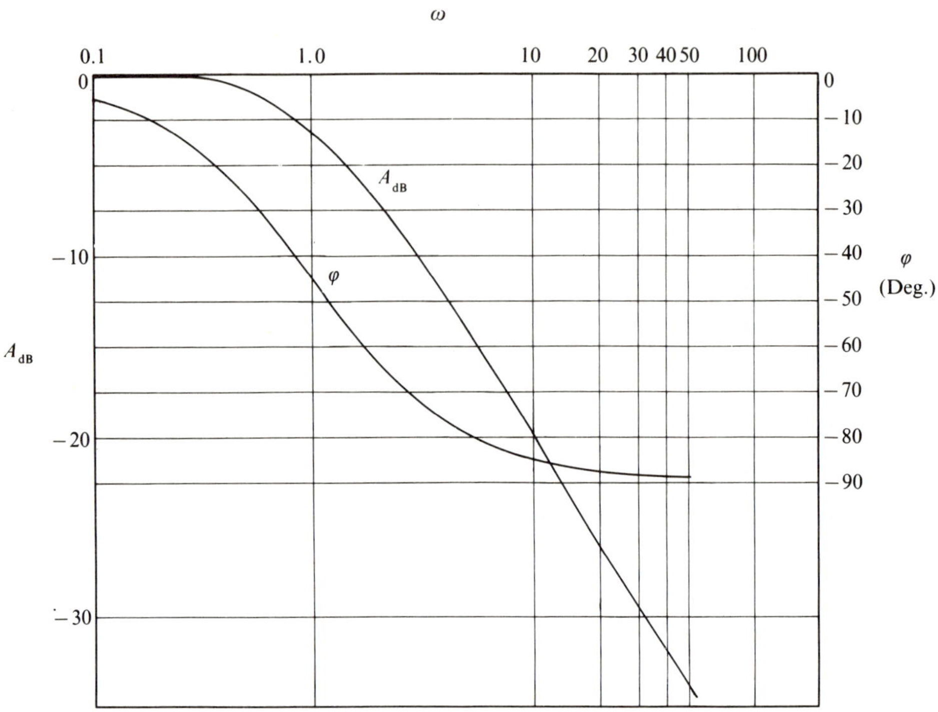

FIGURE 5.6 Phase shift vs. dimensionless frequency for second-order system

The amplitude and phase angle of the first- and second-order ultimate-state sinusoidal response are functions of both the system parameters and the input frequency ω_0. In both instances the amplitude decreases as the driving frequency increases, so that eventually it is possible to find a quantity ω_c such that for all $\omega_0 > \omega_c$ the output amplitude ratio (sensitivity) falls below a prespecified minimum A_c, the cutoff amplitude ratio. For example, A_c could either be an amplitude which is the threshold of the user or else embody the largest amplitude error tolerated (or accepted) by the user.

Ideally an instrument should respond with $A = 1, \phi = 0$ for all input frequencies. Then all the harmonics of an input would be undistorted by the measurement system. That such an ideal situation is unachievable should be clear by now. In the design process, one must decide in advance the percentage of allowable error and the highest harmonic which must be held within such an error—all lower harmonics to be less distorted. For the designer it is a question of how much distortion to allow, that is, to find the highest value of ω_c for which $A < 1 \pm \epsilon$ where ϵ is a preselected error specification.

For example, in a first-order system,

$$A = \frac{1}{\sqrt{1 + (\omega_0 \tau)^2}} = 1 - \epsilon$$

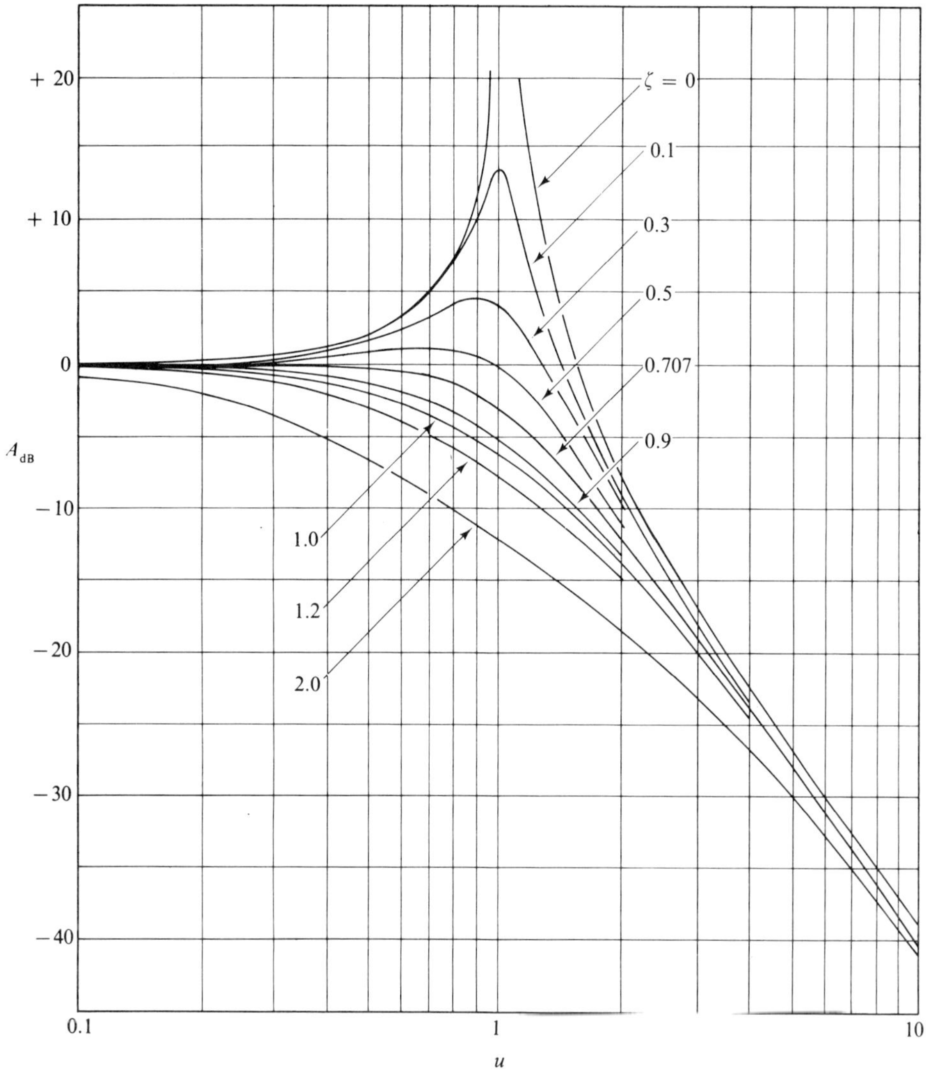

FIGURE 5.7a Amplitude vs. frequency for second-order system

from which we can prepare an allowable error table:

% error	0.1	1.0	5.0	10.0	15.0	25.0
$\omega_0\tau$	0.045	0.141	0.330	0.484	0.620	0.880

If it were required to have an instrument of the first order respond with no more than 0.1% amplitude reduction at an input frequency of 1,000 Hz $\approx$ 6,280 rad/s we would need an instrument with response time $\tau = 0.045/6{,}280 = 7.17 \times 10^{-5}$ seconds.

In commerical practice it is customary to state that an instrument

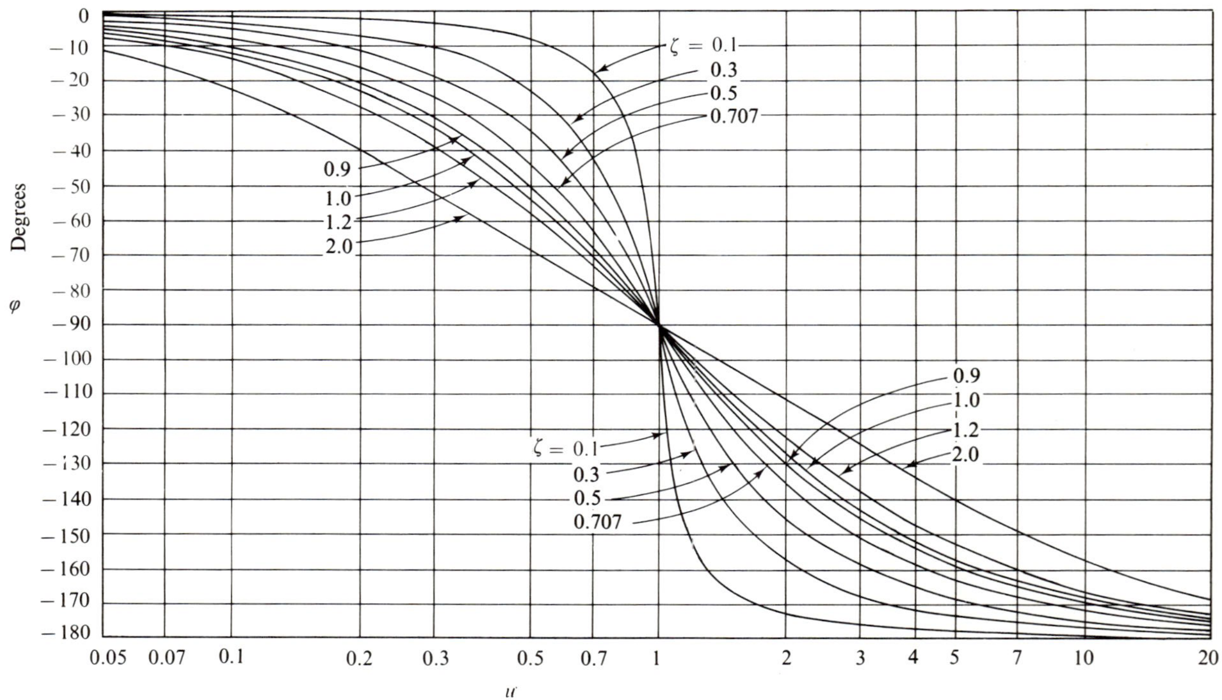
ζ = 0.1
0.3
0.5
0.707
0.9
1.0
1.2
2.0
Degrees
φ
0
−10
−20
−30
−40
−50
−60
−70
−80
−90
−100
−110
−120
−130
−140
−150
−160
−170
−180
0.05
0.07
0.1
0.2
0.3
0.5
0.7
1
2
3
4
5
7
10
20
u

FIGURE 5.7b

"has a flat response curve within blank percent up to blank Hz." "Flat response" implies a nearly horizontal line of the curve of A vs. ω_0. As a matter of convenience the information is presented in semilog form with the frequency plotted in dimensionless form along the logarithmic ordinate decibels:

$$A_{\mathrm{dB}} = 20 \log A$$

Although the logarithmic presentation and the decibel are not particularly useful to us (indeed the decibel is a hangover from simpler days and the logarithmic presentation suppresses the relevant portion of the sensitivity curve), we shall continue to use them since they are so firmly embedded in practice that they are a brand mark of the system designer.

This type of representation, which was called the *Bode plot* or *attenuation chart* in Section 5.2, is usually presented in dimensionless form, in which A is the ratio of the output amplitude to the input amplitude and the frequency is presented as the ratio or the input frequency to some frequency-like parameter.

For example, the Bode plot of the first-order system is obtained by graphing (Fig. 5.6)

$$A = \frac{1}{[1 + u^2]^{1/2}}$$

in which $u = \tau\omega$, and the second-order plot is obtained by graphing (Fig. 5.7)

$$A = \frac{1}{[(1 - u^2)^2 + (2\zeta u)^2]^{1/2}}$$

where $u = \omega_0/\omega$. These general Bode plots for first- and second-order systems can be applied to specific instruments by adjusting the scale to a given τ or a specific ω and ζ.

5.5 The Second-order Bode Plot and Amplitude Distortion

The attenuation of pure sinusoidal signals implies that signals composed of an infinite collection of sinusoids will be changed in transmission through the system. It is possible, therefore, to compute the distortion characteristics of an instrument if the linear performance equations are known.

5.5.1 Examination of Figure 5.7a shows that for $\zeta < 0.707 = \sqrt{2}/2$, the amplitude ratio for lower frequencies is near zero dB (that is, $A \approx 1$), increasing slowly for some range of u until at some point the curve separates rapidly from the axis and climbs to the resonance peak at some dimensionless frequency $u_R = \omega_{0_R}/\omega$. For $u > u_R$, the amplitude ratio decreases rapidly, approaching $-\infty$ dB (that is, $A \longrightarrow 0$) as u approaches positive infinity. It can be seen that if a system is insufficiently damped ($\zeta \ll 1$), it is quite likely that some of the higher harmonics will be overemphasized. If the first harmonic is on the flat portion of the curve and the third harmonic is near the resonance peak, then the third harmonic will swamp the first.

Example 5.5.1 A second-order system, $\zeta = 0.30$, $\omega = 10$ rad/s, $K = \omega^2$, is subjected to an input signal

$$y(t) = \sum_{n=1}^{11} \frac{1}{n^2} \sin(n^2 0.2t)$$

The threshold of the user system is $z = 0.007$. Determine which harmonics are distorted more than 10% and which are completely suppressed insofar as the user is concerned (neglect phase shift).

Solution We plot the Bode diagram of the system by scaling the $\zeta = 0.30$ line in Figure 5.6 by multiplying u by 10 rad/s. We then prepare a table as follows:

n	Harmonic	ω_0	A	A/n^2
1	1	0.2	1.00	1.000
2	4	0.8	1.00	0.250
3	9	1.8	1.02	0.113
4	16	3.2	1.06	0.066
5	25	5.0	1.29	0.051
6	36	7.2	1.72	0.046
7	49	9.8	1.65	0.033
8	64	12.8	1.00	0.016
9	81	16.2	0.46	0.006
10	100	20.0	0.30	0.003
11	121	24.2	0.20	0.001

All harmonics past the 16th (4th term) are distorted more than 10% and all harmonics past the 64th, i.e., the last three terms, are completely suppressed.

Some care must be exercised here in interpretation of the results. Although the 25th harmonic is distorted 29%, the actual amplitude of the harmonic (0.04) is only 4% of the first harmonic so that the 25th harmonic introduces only an error of 1.16% in the response. The 36th harmonic, which is the most highly distorted, introduces an error of 1.80% even though distorted 72%. If all the possible maximum errors are added, it is found that the maximum possible signal distortion is less than 3%. Within the limits of our specification this signal is undistorted.

Example 5.5.2 What must the value of ω be if a second-order system with $\zeta = 1$, $K = \omega^2$ is to reproduce the input

$$f(t) = \sum_{n=1}^{11} \frac{1}{n^2} \sin(n^2 0.2t)$$

with each harmonic distorted less than 10% (neglect phase shift)?

Solution From Example 5.5.1 we see that the frequency of the highest-order term is 24.2 rad/s. Since the amplitude ratio curve for $\zeta = 1$ is monotonically decreasing, it suffices to choose ω so that $A = 0.9$ at the 11th term. From the dimensionless Bode chart we obtain $A = 0.9$ at $\omega_0/\omega = 0.3$. Thus

$$\omega = \frac{24.2}{0.3} = 80 \text{ rad/s}$$

5.5.2 It is clear from Figure 5.7a that $\zeta = 0.707$ is a partition parameter between the region in which $A = G(j\omega)$ shows a resonance peak ($\zeta < 0.707$) and the region in which $G(j\omega)$ shows no resonance peak ($\zeta > 0.707$). This can be demonstrated by computing the value of $u = u_R$ at which a resonance peak exists. We write

$$\frac{d}{du}|A(u)| = 0$$

and obtain

$$u_R = [1 - 2\zeta^2]^{1/2}$$

and hence $u_R = 0$ at $\zeta = \sqrt{2}/2$ and is imaginary for $\zeta > \sqrt{2}/2 = 0.707$. The quantity $\zeta = 0.707$ is the lowest damping ratio for which it is not possible to increase the importance of higher harmonics.

It is quite easy to deceive oneself into giving more importance to the number $\zeta = 0.707$ than it deserves. Many systems engineers (including some textbook writers) categorically state that $\zeta = 0.707$ represents an optimum damping ratio which should be achieved to guarantee the largest band width. This conclusion seems reasonable from a casual examination of Figure 5.7, from which it can be seen that the $\zeta = 0.7$ curve remains closer to the $A = 0$ dB line than any other line in the figure. A little reflection, however, shows that for a given allowable error, a value of $\zeta < 0.707$ is not always the optimum. Let us plot the figure with u vs. ζ and with $A = 1.01$ and $A = 0.99$ as parameters (Fig. 5.8). In the figure, the $\zeta = 0.707$ line intersects the $A = 0.99$ line at the cutoff frequency ratio $u_c = 0.376$, whereas the $\zeta = 0.655$ line intersects at a cutoff frequency ratio of $u_c = 0.469$.

For all allowable errors, the ζ vs. u line for constant $A = 1 + \epsilon$ reaches a peak and then ζ decreases for increasing u. The $\zeta = \zeta_0$ line which is tangent to the peak will intersect the $A = 1 - \epsilon$ line to establish a cutoff frequency ratio. It is not difficult to show, by differentiating the expression for the amplitude ratio, that the cutoff frequency (maximum frequency determined by the allowable error) is

$$u_c = \left\{\left[1 - \frac{1}{(1+\epsilon)^2}\right]^{1/2} + \left[\frac{1}{(1-\epsilon)^2} - \frac{1}{(1+\epsilon)^2}\right]\right\}^{1/2}$$

The two expressions can be computed and compared with the cutoff frequency ($u = u_{0.707}$) obtained at $\zeta = 0.707$ (see Table 5.2).

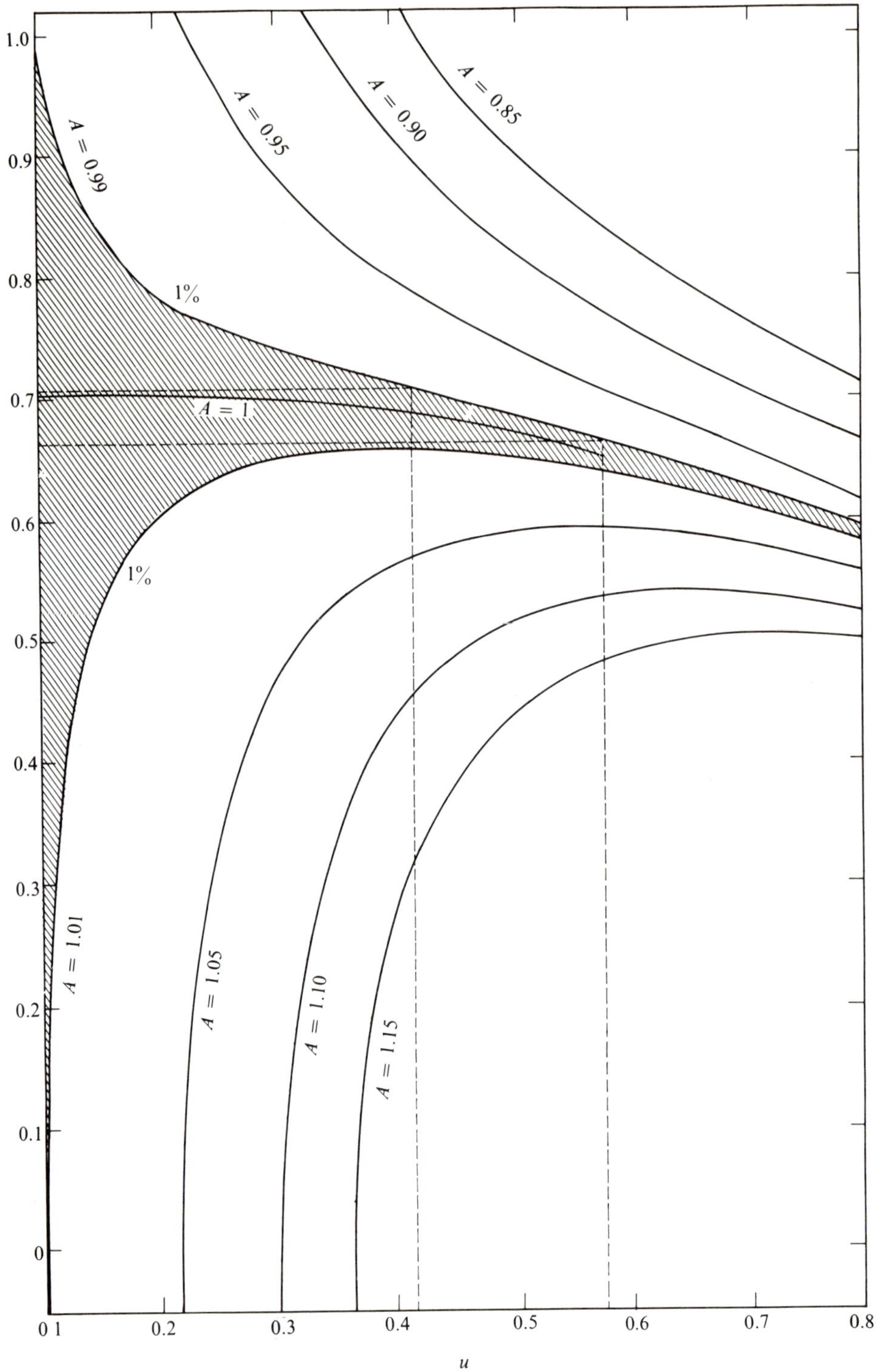

FIGURE 5.8 Optimum damping for a permissible error range

TABLE 5.2

ϵ	ζ	u_c	$u_{0.707}$
0	0.707	0	0
0.005	0.671	0.389	0.316
0.010	0.655	0.469	0.376
0.050	0.588	0.767	0.572
0.100	0.539	0.902	0.699
0.150	0.503	1.010	0.787
0.200	0.454	1.141	0.866

For design purposes the relevant region is represented by $0 < \zeta \leq 1$, since, in practice, very large damping requires the selection of a large value of ω to guarantee a sufficiently large band width. This can be seen from Figure 5.9 which is an expansion of the relevant region. Let us, for

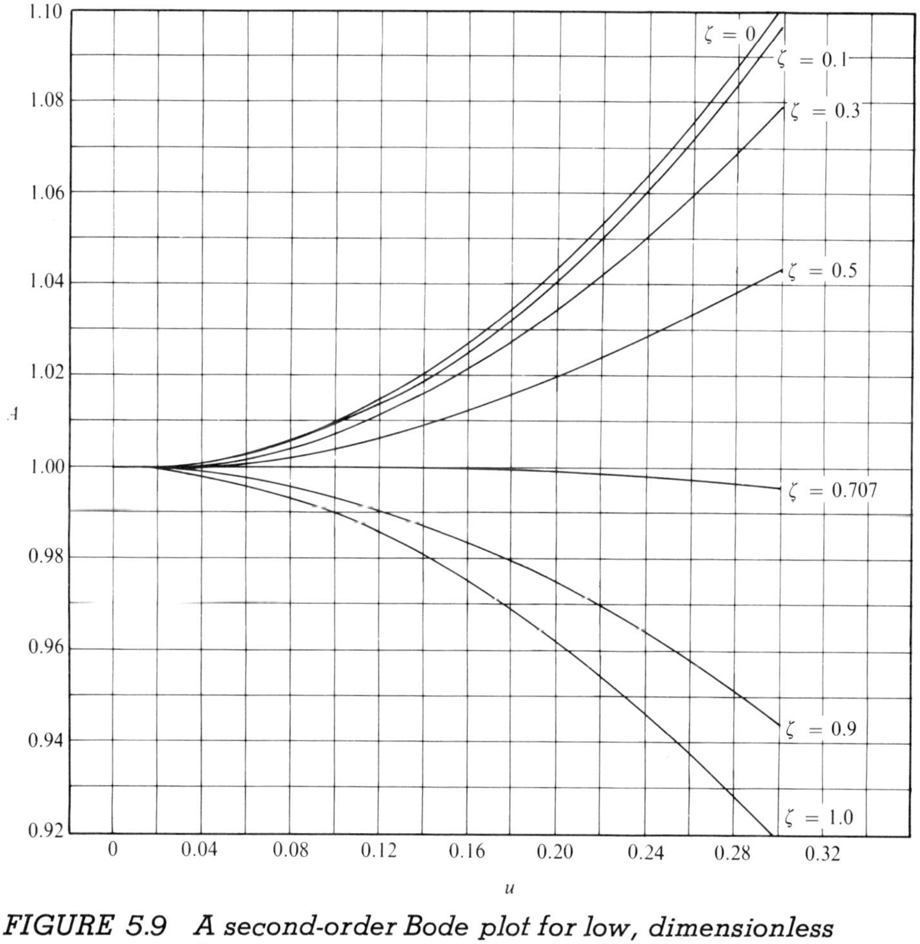

FIGURE 5.9 A second-order Bode plot for low, dimensionless frequencies

example, require a system to have a band width of 100 rad/s at the 10%-error level. From Figures 5.8 and 5.9 it is easy to obtain a set of values of ω for which an input with $\omega_0 = 100$ rad/s will generate a response with an amplitude no more than 10% in error.

ζ	0.2	0.3	0.4	0.5	0.6	0.7	1.00	5.0
u	0.33	0.34	0.38	0.48	0.91	0.71	0.33	0.05
ω	333	294	264	208	111	141	303	2000

Obviously the use of a damping coefficient greater than 1 is wasteful of ω, and hence of system amplification. In a linear spring-mass system, for example, increase of ω implies either an increase of spring constant or else a decrease of system inertia. Since the ultimate state sensitivity is $S = K/\omega^2$, it follows that an increase of ω implies a reduction of ultimate state amplitude unless balanced by an increase of the gain K.

It has been demonstrated that specification of response implies an equivalent specification of cutoff frequency and phase shift. Unfortunately it is not possible to make an easy transfer from one system to another, and the simplicity of frequency-plane analysis cannot be utilized. The full equivalence is noticed if the amplitude ratio and phase shift as obtained theoretically were used to obtain specification equivalences. The result will reduce to the solution of the characteristic equation of the system.

5.6 The Second-order Phase Shift and Associated Distortion

The distortion introduced by the phase shift is not easily described since its effect is to leave the maximum value of the component harmonics unchanged but to shift the components in time. The result can be remarkable. Consider, for example, two square waves (Fig. 5.10), each composed of an infinite number of harmonics. If the two waves are 1/2 cycle out of phase, their sum is zero. If, on the other hand, they are out of phase by 1/2 cycle plus some amount τ_0 (that is, each harmonic of the lower signal is shifted through τ_0), the resulting wave will consist of two rectangular pulses per cycle of width τ_0 and height $2M$.

The effects of phase shift, though noticeable, are not often as dramatic as those shown in Figure 5.10. Two sinusoids of amplitude M, displaced by 10 degrees from each other, can be shown to add to

$$1.995M \sin (\theta + 6^\circ)$$

which is a reasonably undistorted signal. Interestingly enough the phase shift, which leaves the individual harmonics undisturbed, actually affects the amplitudes of the sum.

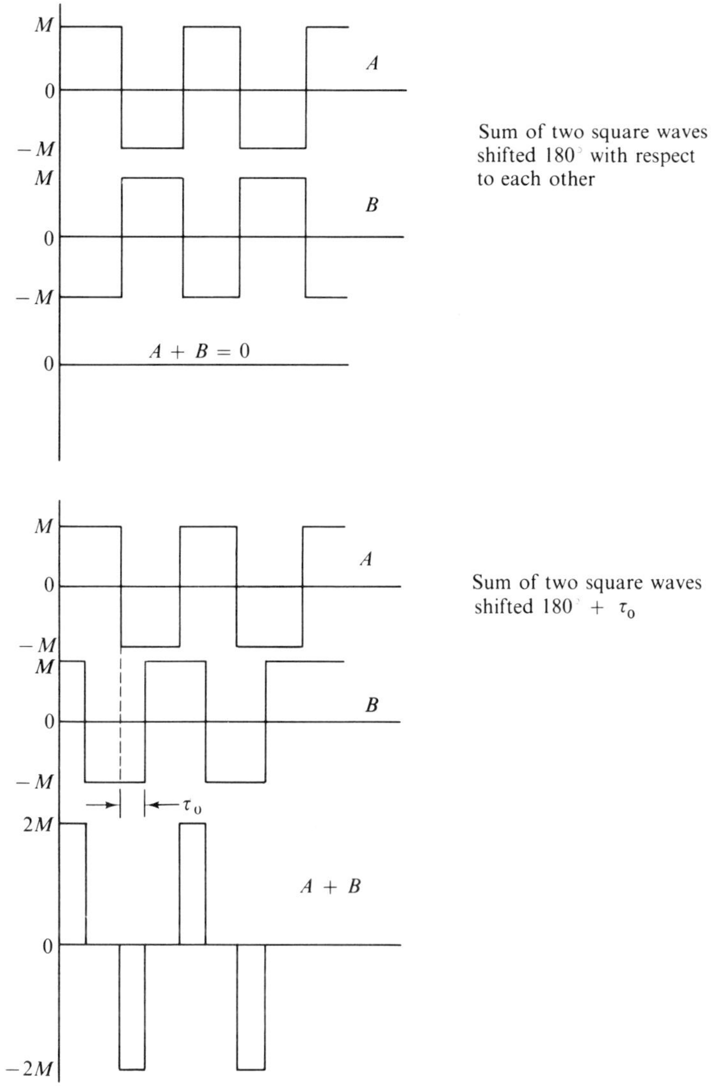

FIGURE 5.10 Effect of phase shift on superimposed square waves

Typical of phase shift effects is the shift angle in a second-order linear system,

$$\tan\phi = \frac{-2\zeta\omega\omega_0}{\omega^2 - \omega_0} = \frac{-2\zeta u}{1 - u^2}$$

It is seen that for $u^2 < 1$, ϕ lies between zero and $-90°$; and for $u^2 > 1$, ϕ lies between $-90°$ and $-180°$ with $\phi = -90°$ at $u^2 = 1$ (Fig. 5.7b).

The same parameters which optimize the amplitude ratio A also optimize the phase angle. We prefer $\tan\phi \approx 0$. There is, however, no possibility of achieving such a result. The best we can do is to specify some minimum phase shift and then try to achieve it. For example, the 10% amplitude error requirement carries with it a phase shift penalty.

ζ	0.1	0.2	0.4	0.707	1.00	5.00
u	0.30	0.30	0.38	0.707	0.13	0.05
ϕ	$-4°$	$-8°$	$-25°$	$-70°$	$-15°$	$-27°$

The so-called optimum ($\zeta = 0.707$) is, as a matter of fact, no optimum at all since at the 10% amplitude error level it carries a 70° phase shift penalty! If we require ϕ to be less than 4°, $\zeta = 0.707$ would impose a requirement of $u = 0.1$, whereas at $\zeta = 0.1$, u would equal 0.30 (see Figure 5.7). Hence the "optimum" is more wasteful of gain than the non-optimum $\zeta = 0.1$. Since amplifier gain correlates with cost and moderately low damping is usually no more expensive than high damping,

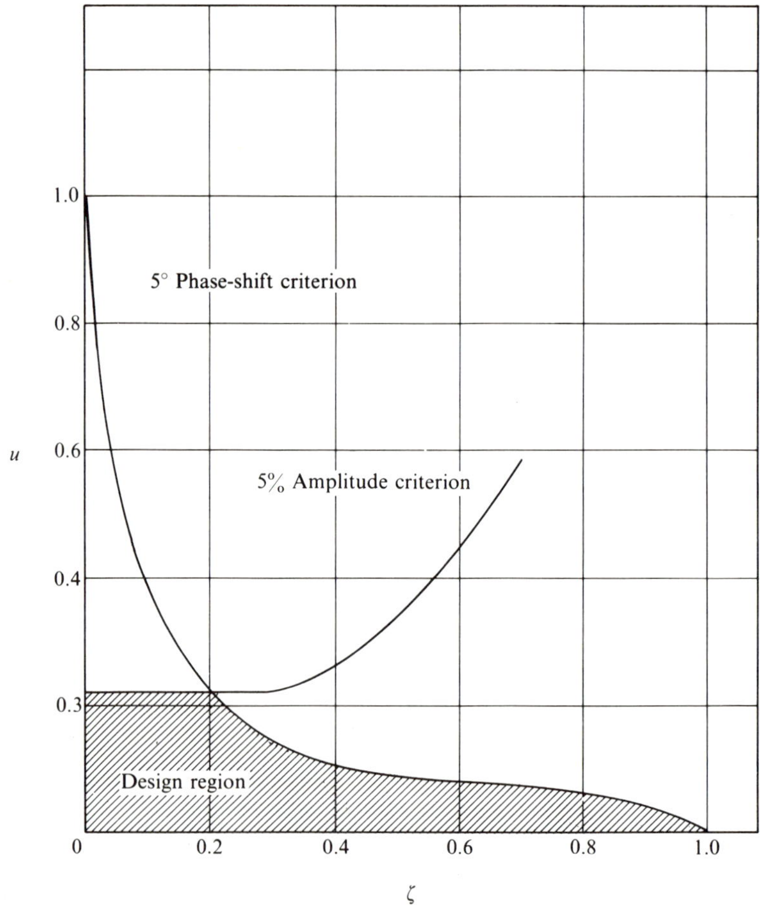

FIGURE 5.11 Selection of a design region for Example 5.6.1

the low-damping system is really the best design answer under the condition specified.

Example 5.6.1 Find a second-order system for which each harmonic is shifted no more than 5° and distorted no more than 5%.

Solution We set

$$\frac{2\zeta u}{1-u^2} = \tan 5° = 0.0875$$

from which we obtain

ζ	0	0.05	0.3	0.5	0.8	1.00
u	1	0.58	0.14	0.09	0.06	0.00

We now write

$$\frac{1}{[(1-u^2)^2 - 2\zeta u]^{1/2}} = \begin{cases} 1.05 & \text{if } \zeta < 0.707 \\ 0.95 & \text{if } \zeta > 0.707 \end{cases}$$

and obtain

ζ	0	.1	.3	.5	.7	1.00
u	0.22	0.22	0.22	0.34	0.58	0.24

If the two lines are plotted on the same curve (Fig. 5.11) it is evident that the design criterion, or widest band width consistent with 5° and 5%, is at $u = 0.22$ and $\zeta = 0.24$, a considerable distance from the "optimum" $\zeta = 0.707$.

5.7 Distortion

The distortion of signals by instruments with an insufficiently high band width is quite common in practice. Figure 5.12 is the result of a Fourier analysis of a time-vs.-pressure curve in the pulmonary artery.[9] The reconstructed tracing is obtained by the suppression of harmonics higher than the sixth. The matching of the curves is fairly good although from the point of view of analysis of the system, the important flattening of the top of the tracing is lost. It can be shown that the 14th harmonic is needed to make the error of the peak of the same order as the error of the rising and descending limbs.

In measurement arrays it is the recording device which is usually to blame for distortion. Transducers and amplifiers can be found with a sufficiently fast response for most applications, whereas recorders are usually frequency limited. Whenever the input to the recorder oscillates with a frequency greater than 10,000 Hz, a permanent recording can only be made on high-speed magnetic tape or on the screen of a retentive phosphor oscilloscope.

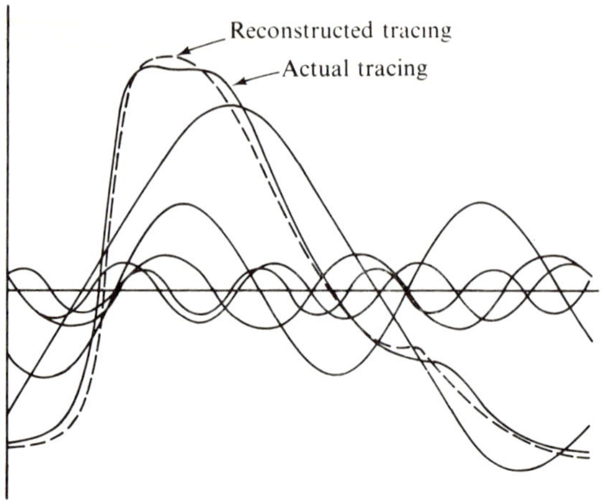

FIGURE 5.12 Fourier analysis of a pressure pulse in a pulmonary artery

Laboratory type pen-and-ink recorders suppress even the low-frequency components. A good example is shown in Figure 5.13, a photograph of the output traces of two inputs of identical amplitude sinusoidal input signals which differed only by 2.7 Hz in frequency. The higher-frequency signal generated an output amplitude decrease of 68%.

Let it be desired to control the distortion. Then it shall be required to control the error $\epsilon = y - z$ or equivalently (Section 1.3) to obtain the minimum of the variance computed from ϵ. Since ϵ varies with time, we must select K values of ϵ_k at various times and compute their variance.

$$\sigma_K^2 = \frac{1}{K} \sum_{k=1}^{K} \epsilon_k^2$$

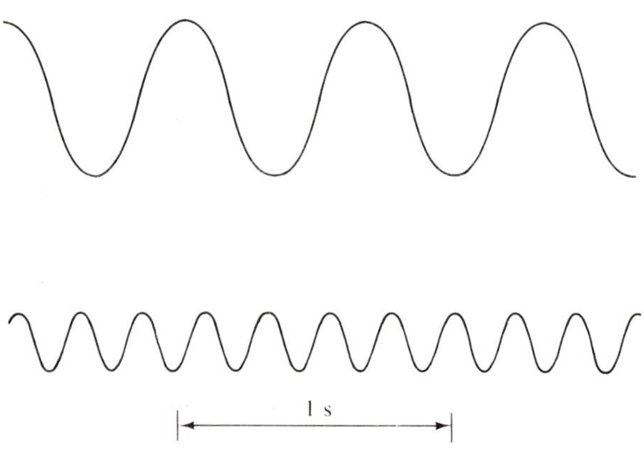

FIGURE 5.13 Attenuation of low-frequency signals by a recorder

A limiting procedure is now possible. We obtain ϵ_k^2 by considering the function

$$\epsilon_k^2 = \epsilon^2(t_k) = \int_0^\infty \epsilon^2(t)\delta(t - t_k)\,dt$$

Let the t_k be separated by an interval Δt. Then

$$\sigma_K^2 = \sum_{k=1}^{K} \frac{\Delta t}{K\Delta t} \int_0^\infty \epsilon^2 \delta(t - t_k)\,dt$$

The quantity $K\Delta t$ is the total time elapsed from which it follows that as K goes to infinity, the summation becomes

$$\sigma^2 = \lim_{T_0 \to \infty} \frac{1}{T_0} \int_0^{T_0} \epsilon^2(\tau)\,d\tau$$

If both y and z are represented by Fourier sine series

$$y = \sum_{n=1}^{\infty} a_n \sin \omega_n t$$

$$z = \sum_{n=1}^{a} a_n A_n \sin(\omega_n t + \phi_n)$$

valid in the interval $T_0 = T/2$, and if we set

$$\epsilon^2 = (y - z)^2 = y^2 - 2yz + z^2$$

it follows that

$$\frac{2}{T} \int_0^{T/2} \epsilon^2\,dt = \frac{2}{T} \sum_{n=1}^{\infty} a_n^2 \epsilon_n^2$$

where $\epsilon_n^2 = 1 - 2A_n \cos \phi_n + A_n^2$.

The above result represents the sum of the squared differences between the input and the output. To make the result dimensionless we divide by the root mean square input (compare with definition of "dimensionless error" in chapter 1):

$$\bar{y} = \frac{2}{T} \int_0^{T/2} y^2\,dt = \frac{2}{T} \sum_{n=1}^{\infty} a_n^2$$

$$\sigma^2 = \frac{\sum_{n=1}^{\infty} a_n^2 (1 - 2A_n \cos \phi_n + A_n^2)}{\sum_{n=1}^{\infty} a_n^2}$$

Example 5.7.1 A second-order system ($\zeta = 0.5$) is subjected to a signal given by the following Fourier series representation:

Harmonic	1	3	5	7	9
Frequency Ratio	0.10	0.30	0.50	0.70	0.90
Input Amplitude	1.00	0.70	0.30	0.10	0.50

Compute the output vs. input error.

Solution From the Bode diagram (Fig. 5.7a) and the phase diagram (Fig. 5.7b) for a second-order system we find

K	1	3	5	7	9
ϕ_n	$-5°$	$-17°$	$-34°$	$-54°$	$-77°$
A_n	1	1.02	1.10	1.13	1.08
ϵ_n^2	0.008	0.070	0.276	0.801	1.59
a_n^2	1	0.49	0.09	0.01	0.002
$(a_n\delta_n)^2$	0.008	0.034	0.025	0.008	0.003

and thus

$$\sigma^2 = \frac{0.078}{1.59} = 0.049 \qquad \sigma = 0.221$$

The system creates a mean-square error of 22.1%. Thus, even though the most distorted component is "up" only 13%, the phase shift makes the total error greater than 20%.

5.8 Filtering

There are times during which the suppression of the higher harmonics of an input signal is a desirable characteristic of an instrument. Consider, for example, the signal shown in Figure 5.14—a single, low-frequency sinusoid marred by noise. If the expected output sinusoid is sketched (dotted line), it is clear that the noise has a much higher frequency than the signal. If instead of a large band-width recording instrument a less expensive slower instrument had been used, the resulting trace would have been cleaner.

Example 5.8.1 A recording system, $\zeta = 0.707$, $\omega = 10$ Hz, is used to record a signal $\omega_0 = 1$ Hz. If the output threshold occurs at a sensitivity ratio of 0.1,

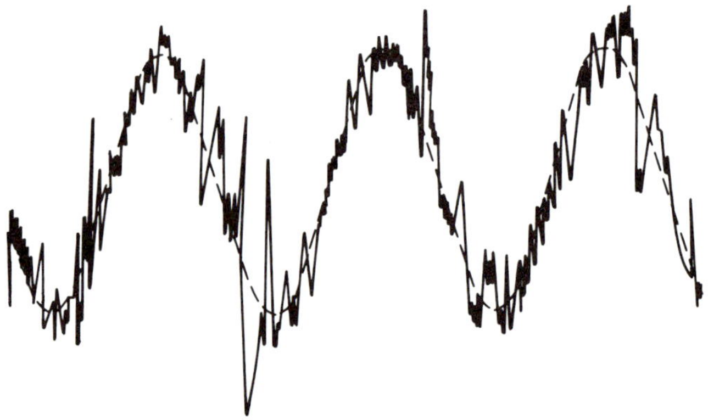

FIGURE 5.14 Noisy, low-frequency output

determine the frequency of the lowest harmonic which is totally suppressed.

Solution $u = 1/10 = 0.1$ and the amplitude ratio is $20 \log 0.1 = -20$ dB. From Figure 5.7a we find that $u_c = 3$ and thus $\omega_c = 3 \times 10 = 30$ Hz. In this instance, a noise of 30-Hz frequency is no longer recorded.

The rejection of higher harmonics is the most common concept in filtering. It is possible, however, to design filters which will reject low-frequency noise (high-pass filters) or else select only a narrow band of signals and reject the rest (band pass filters). For example, a filter which suppresses all signals below 10 Hz is a high-pass filter, whereas a filter which suppresses all inputs with frequency above 20 Hz is a low-pass filter. The high-pass filter followed by the low-pass filter passes only frequencies between 10 Hz and 20 Hz and hence forms a band pass filter.

Filters are not perfect devices—in fact, the filtering procedure occurs only if the required signal and the unwanted signal operate on two different regions of the Bode plot. Basically it is required that the input signal be on the flat part of the curve and the noise far out on the sloping part. Ideally one should set the noise level below the threshold of the user: the implication is that for high values of ω small change of input frequency can generate a large reduction in amplitude ratio (Fig. 5.15).

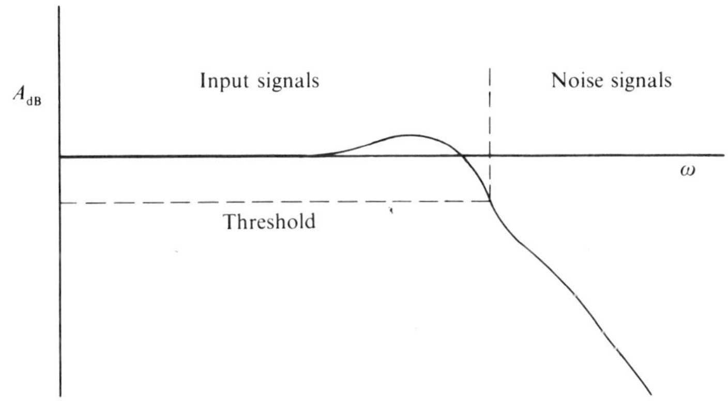

FIGURE 5.15 Separation of signal and noise with a low-pass filter

The transform of any impulsive response can be written (see chapter 4) in the form

$$\frac{1}{(s+a_1)(s+a_2)\ldots(s+a_n)(s^2+2\zeta_1\omega_1 s+\omega_1^2)\ldots(s^2+2\zeta_m\omega_m S+\omega_m^2)}$$

so that the ultimate state amplitude becomes

$$A = \frac{K}{\prod_{i=1}^{n}[(1/\tau_i)^2+\omega_0^2]^{1/2}\prod_{i=1}^{m}[(\omega_0^2-\omega_i^2)^2+(2\zeta\omega_0\omega_i)^2]^{1/2}}$$

If ω_0 is made very large, then

$$\lim_{\omega_0 \to \infty} A = \frac{K}{\omega_0^{2m+n}}$$

In decibel notation

$$A(\omega)_{\text{dB}} = 20 \log K - (2m + n)20 \log \omega$$

On semilog paper the slope will be $-20(2m + n)$ so that high-order systems suppress high harmonics more effectively than low-order systems. Unfortunately, high-order systems invariably reduce the band width.

Consider, for example, a cascade (series sequence) of first-order amplifiers. If 10% attenuation defines the band width we can write

$$A_c = \frac{1}{(1 + u_c^2)^{n/2}} = 0.90$$

where n is the number of amplifiers. Then

$$u_c = (0.234)^{2/n}$$

and hence

n	1	2	3	4	5	10
u_c	0.484	0.330	0.234	0.229	0.204	0.101

It is obvious that increasing the number of amplifiers not only increases the roll-off slope but also decreases the cutoff frequency.

PROBLEMS

5.1 A second-order instrument, $\zeta = 0.3$, $\omega = 200$ Hz, is connected to a system which responds only to signals greater than 0.1 volt. If the input signal is

$$f(t) = \sum_{n=1}^{10} \frac{4}{n^2 + 2} \sin n\omega_0 t$$

determine the highest value of ω_0 for which no component discernible to the user (i.e., components for which $a_n > 0.1$) is completely suppressed.

5.2 Obtain the ultimate-state sinusoidal dimensionless sensitivity and phase shift of the linear ordinary third-order system

$$\sum_{i=0}^{3} a_i \frac{d^i y}{dt^i} = Kf(t)$$

5.3 The system discussed in problem 5.2 can be represented by

$$\left(\frac{d}{dt} + \frac{1}{\tau}\right)\left(\frac{d^2}{dt^2} + 2\zeta\omega\frac{d}{dt} + \omega^2\right)z = \frac{\omega^2}{\tau}K_0 f(t)$$

in which $K = K_0\omega^2/\tau$. If $u_1 = \omega_0\tau$ and $u_2 = \omega_0/\omega$, obtain a curve $u_1 = f(u_2)$ for which A is distorted no more than 10% if $\zeta = 0.5$.

5.4 Repeat problem 5.3 to obtain a curve for which $\tan \phi = 0.10$. Is there a solution for which A is distorted 10% and $\tan \phi = 0.107$?

5.5 Consider the fourth-order system

$$\left(\frac{d^2}{dt^2} + 2\zeta_2\omega_2\frac{d}{dt} + \omega_2^2\right)\left(\frac{d^2}{dt^2} + 2\zeta_1\omega_1\frac{d}{dt} + \omega_1^2\right)z = (\omega_1\omega_2)^2 y(t)$$

The input signal has a frequency range of 0–150 Hz and the noise a frequency range 1,000–∞ Hz. Select ζ_1, ζ_2, ω_1, ω_2 in such a fashion that the input signal amplitude is reduced no more than 1% and shifted no more than 2° and such that the noise is reduced at least 70% in amplitude.

5.6 Plot the third-order Bode diagram and phase shift diagram with $\tau =$ 0.01 second, $\zeta = 0.45$ and $\omega = 350$ rad/s. Compute the unit step response under the assumption that the frequency of the 11th harmonic corresponds to an A down 12 dB from the zero line.

5.7 The input to a third-order system is

$$f(t) = \sum_{n=0}^{\infty} \frac{(-1)^n}{(2n+1)!} \sin(2n+1)t$$

Compute the dynamical sinusoidal amplitude $A'_n(t)$.

5.8 What are the filtering characteristics of the system

$$\frac{dz}{dt} + 0.52 = 4\frac{d^2y(t)}{dt^2}$$

Can one use this system to process a 500-Hz input signal while suppressing a 60-Hz line-noise?

5.9 A third-order system is given by $a_0 = 10 = K$, $a_1 = 3$, $a_2 = 3$, $a_3 = 1$. An input is represented by the following harmonics:

n	1	2	3	4	5	6
ω_n	10	20	30	40	50	60
a_n	3.0	1.1	0.53	0.11	0.04	0.01

in which ω_n is given in radian/second. Compute the dimensionless variance between the input and the output.

5.10 Obtain the Bode and phase diagrams of the system described in problem 5.9 and compute the unit step response of the system.

NOTES

[1]Ref. 71, 76. See also any recent textbook on control system design. For amusement see also Ref. 47, p. 371.

[2]There exist mathematically pathological cases of functions which cannot be expanded in a Fourier series on an interval. Nature does not produce such functions, and the physical scientist need not be concerned with them.

[3]Ref. 24, p. 311 et. seq.

[4]Ref. 73, pp. 144–50.

[5]Ref. 24, pp. 238–39.
[6]Many methods are known. Some of them even work. See Ref. 71, p. 145. Ref. 73, Ref. 74, pp. 332–65.
[7]Ref. 35, p. 145.
[8]Ref. 29, pp. 311–63.
[9]Ref. 75, p. 630.

6

ELEMENTS OF STATISTICS

6.1 The Expected Value

Statistical procedures provide the best methods for the abstraction of the expected signal from a noisy output. With the advent of the space program statistical estimation has become an important part of the engineering armamentarium. The result has been an extraordinary enhancement of measurement practice. Radar measurements, long-distance communications and optical observations have all been improved by at least one order of magnitude through statistical analysis. The effective engineer should, therefore, understand the basic principles of statistical inference.

6.1.1 The aim of the instrument designer is to provide an output which is the best approximation to the information-bearing signal. He has to answer two questions:

(a) Given an input which is noise laden, what is the most likely representation of the noise-free signal?

(b) Is there an instrument which will modify the noisy signal so that the output has a high probability of presenting a noise-free signal with a maximum of information, or with a minimum of information suppression?

The first question has been answered in the theory of classical data reduction of noisy signals,[1] what may be called *manual ex-post-facto filtering*. A set of n data is recorded, each datum being the response z_{nk} to the nth measurand y_n and laden with the error ϵ_{nk}, the kth error corresponding to the nth measurand. By statistical analysis the best approximation to z_n is found and an estimate of the "risk" taken in accepting the estimate as the expected value is found. Such a mode of thinking is rather foreign to the engineer. The designer faced with a set of measurements is inclined to take the computed estimate as the correct value regardless of the risk. If a number is available and it is the only number, what choice is there but to take it and go on?[2]

From an action point of view, there is really no choice. The expected value is not known; otherwise we need not measure it. The assumptions which underlie statistical inference are so poorly anchored in reality that we have no choice but to go on with our eyes closed. This is not to imply that we can never place any confidence in our measurements and our analysis; after all, if we use the numbers and the system works we feel that even if the computation did not yield *the* right measurements the result is acceptable. Be it as it may, we do have a technique for extracting a representative measurement from noisy data even though we cannot be sure of the truth of our inference.

The answer to the second question is nothing more than the mechanization of the answer to the first one. If the data are recorded in a permanent manner, it is possible to design a device which will automatically extract the representative measurements or estimate of the expected value. A digital computer is usually sufficient.

Much more difficult to solve is the "on-line filtering problem." Here we are faced with the removal of noise from instantaneous data measurements. No truly satisfactory answer exists to the problem since the filtering must be carried out with only a single signal and hence with no knowledge of the future statistical character of the noise. All that can be done is to extremize the operation of the instrument response with respect to some assumed noise pattern and then hope for the best.

6.1.2 Let us examine a typical noisy experiment. It is hoped to determine the motion of the diaphragm of a pressure transducer presumably describable by an ordinary linear second-order differential equation with constant coefficients. The diaphragm is excited by a unit step input and the output recorded as in Figure 6.1. It can be shown[3] that if the output were truly the response to the second-order system, the system damping could be computed by

$$\frac{\ln R_{rp}}{[\pi^2(p-r)^2 + (\ln R_{rp})^2]^{1/2}}$$

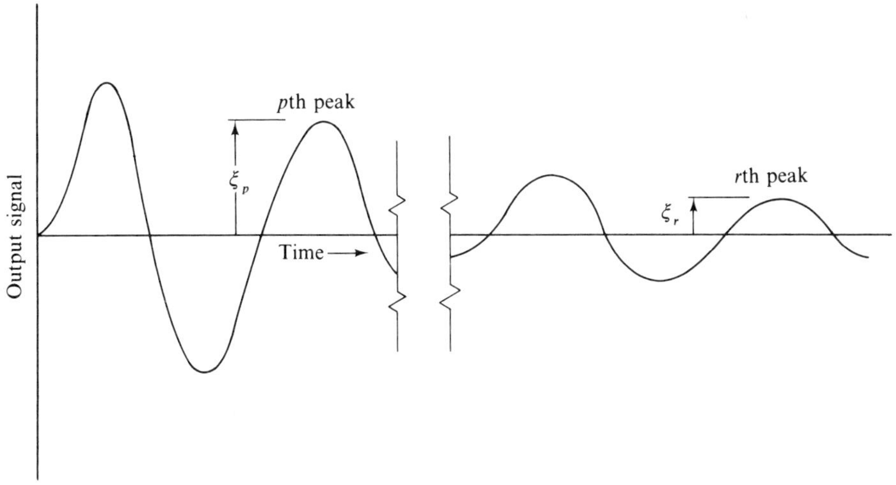

FIGURE 6.1 Output signal of a pressure transducer excited by a unit step

where

$$R_{rp} = \frac{\xi_r}{\xi_p}$$

in which ξ_r is the height of the rth peak of the output, and ξ_p is the height of the pth peak of the output. Usually, however, the computed ratio R_{rp} varies from peak to peak as a result of errors, systemic or random. In addition, the operator who reads the graph introduces small personal measurement errors (errors of observation) which affect the readings. Some operators are less random-error prone (more precise) than others but introduce small systemic errors. If a sufficient number of operators are used to read the graph, it is hoped that the systemic errors will cancel out. What "cancel out" means will become evident later. As an experiment, the author induced seven graduate students to measure the height of fourteen subsequent peaks on an actual tracing and to compute the peak ratios of all adjacent peaks, one cycle apart. The results are presented in Table 6.1. It can be seen that the ratios vary (have a range) from 0.880 to 1.000, and that great variation exists both in the various peak measurements and in the measurement of an individual peak ratio by seven different operators.

Consideration of the variety of numbers in the table gives us the idea that there is a number (0.9XX) which is the "true value," and a collection of errors which depend on the time, method of measurement and identity of the operator. We must decide what digits to put in the (0.0XX) spots. To do this we must make the following "reasonable" assumptions.

TABLE 6.1

Peak Pair	Operator Number 1	2	3	4	5	6	7
1–3	0.901	0.885	0.895	0.920	0.900	0.888	0.880
2–4	0.929	0.965	0.970	0.903	0.937	0.886	0.908
3–5	0.927	0.925	0.915	0.930	0.910	0.938	0.955
4–6	0.962	0.955	0.940	0.912	0.962	0.936	0.918
5–7	0.902	0.920	0.935	0.943	0.940	0.916	0.933
6–8	0.920	0.920	0.905	0.941	0.920	0.931	0.930
7–9	0.979	0.980	0.970	0.921	0.980	0.981	0.950
8–10	0.979	0.980	0.965	0.980	0.978	0.981	0.968
9–11	0.956	0.980	0.965	1.000	0.955	0.963	0.968
10–12	0.933	0.925	0.930	0.958	0.910	0.915	0.968
11–13	0.977	0.955	0.965	0.913	0.964	0.971	0.940
12–14	0.929	0.935	0.960	0.978	0.927	0.940	0.978
z	0.941	0.944	0.943	0.942	0.940	0.937	0.941
σ^2	7.47×10^{-4}	8.25×10^{-4}	6.61×10^{-4}	8.91×10^{-4}	6.92×10^{-4}	10.1×10^{-4}	7.80×10^{-4}
σ	2.73×10^{-2}	2.87×10^{-2}	2.57×10^{-2}	2.98×10^{-2}	2.63×10^{-2}	3.18×10^{-2}	2.79×10^{-2}
σ_ϵ^2	8.41×10^{-4}	9.24×10^{-4}	7.40×10^{-4}	9.98×10^{-4}	7.84×10^{-4}	11.5×10^{-4}	8.76×10^{-4}

Assumption 1 THE EXPECTED VALUE LIES BETWEEN THE TWO EXTREME LIMITS OF MEASUREMENT.

In our case we are assuming that the expected value lies between 0.880 and 1.000. If there exists a systemic error we shall assume that the sum of the expected value and the systemic error are bounded by 0.880 and 1.000.

Assumption 2 THE ERROR IN A SINGLE MEASUREMENT IS INDEPENDENT OF THE ERROR IN ANY OTHER MEASUREMENT.

Thus it is assumed that the causes of error are "random" and that the history of error has no bearing upon future errors.

Assumption 3 GIVEN THE EXPECTED VALUE AND THE MEASUREMENT z_i, $i = 1, 2, \ldots, n$, THEN FOR $n \longrightarrow \infty$ THE ERROR WILL BE APPROXIMATELY DIVIDED INTO $n/2$ POSITIVE ERRORS AND $n/2$ NEGATIVE ERRORS.

This symmetry property of the errors is not invariably found in measurement data. Nevertheless, the distribution of random measurement errors is so often symmetrical that the experienced engineer expects little else (occasionally incorrectly).

Assumption 4 THERE ARE MORE POINTS NEAR THE EXPECTED VALUE THAN AWAY FROM IT. IN OTHER WORDS A LARGER NUMBER OF THE $|\epsilon_i|$ ARE SMALL THAN LARGE.

Here we have an expression of the centrality of the data. An operator who measures the height of a curve with a scale is more likely to miss by one millimeter than by one centimeter or a meter. He will obtain a set of misses of many sizes but since he is trying to be accurate, he will always try to make his error as small as possible.

The assumptions we have made lead us to assume that if a measure of the difference or *distance* between points and expected value were found, the expected value would be that point which minimizes the measure of the distance. In terms of our assumptions we would obtain a number which would make $\bar{z}$ the "most central" value, with a minimum error and with the greatest symmetry.

This intuitive approach can be given a slightly more formal shape by the use of an "n dimensional hyperspace." Suppose, for example, that two measurements z_1 and z_2 have been made. They can be plotted as a single point on the (z_1, z_2) plane. Since these points are measurements of the same value of z, a perfect measurement would place (z_1, z_2) on the line $z_1 = z_2 = \bar{z}$ (Fig. 6.2). Now we can use our centrality assumptions to select the point on the line which is the "best" guess of the value of $\bar{z}$. If we draw a line of length R from (z_1, z_2) to the locus of $\bar{z}$ we select as the most likely value of $\bar{z}$ that point on the line which is nearest to (z_1, z_2), that is, that value of $\bar{z}$ which minimizes

$$R^2 = (z_1 - \bar{z})^2 + (z_2 - \bar{z})^2 = \epsilon_1^2 + \epsilon_2^2$$

Suppose, now, that n measurements had been made. Then the n-tuple

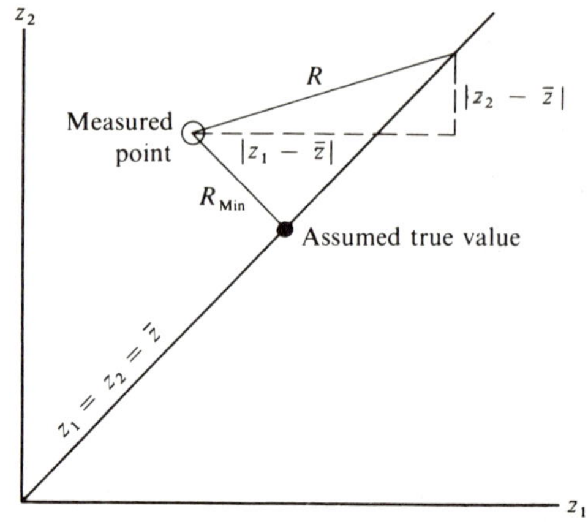

FIGURE 6.2 A two-point least squares approximation

vector $(z_1, z_2, \ldots, z_n)$ could be plotted (in imagination at least) on a hypercoordinate system with n mutually perpendicular coordinates. On this hyperspace the locus of expected values would be the line $z_1 = z_2 = \ldots = z_n = \bar{z}$, and the square of the distance from point to line would be

$$R^2 = \sum_{i=1}^{n} (z_i - \bar{z})^2 = \sum_{i=1}^{n} \epsilon_i^2$$

Our aim is to minimize R^2, that is, to minimize the sum of the errors squared. This procedure is called the *method of least squares* in the literature. It is our aim to find the value of $\bar{z}$ which causes R^2 to become a minimum:

$$\frac{dR^2}{d\bar{z}} = \sum_{i=1}^{n} \frac{d}{d\bar{z}}(z_i - \bar{z})^2 = -2 \sum_{i=1}^{n} (z_i - \bar{z}) = 0$$

That the extreme is a minimum is obvious. The equation transforms into

$$\sum_{i=1}^{n} \bar{z} = \sum_{i=1}^{n} z_i$$

and hence

$$\bar{z} = \frac{1}{n} \sum_{i=1}^{n} z_i \tag{6.1.1}$$

This expected result, that the most likely value is the arithmetic average, shall now serve as our basic theorem.

6.1.3 The method of least squares is a general principle of great power in the analysis of data. Let ϵ_i be an error; then by our principles the set of errors should be symmetrical about $\epsilon = \bar{\epsilon}$ in which $\bar{\epsilon}$ is the systemic error. Thus by the previous argument, the most probable value for $\bar{\epsilon}$ is found by minimizing

$$R^2 = \sum_{i=1}^{n} (\epsilon_i - \bar{\epsilon})^2 \tag{6.1.2}$$

Let ϵ_i depend on a set of parameters $a_j, j = 1, 2, \ldots, k$. Then the principle of least squares can be stated in general form.

Definition 6.1.1 IF IN A SYSTEM A SET OF MEASUREMENT ERRORS IS COMPUTABLE IN TERMS OF CERTAIN UNKNOWN QUANTITIES $a_j, j = 1, 2, \ldots, n$, THEN THE BEST VALUE OF THE a_j IS OBTAINED BY MINIMIZING

$$R^2 = \sum_{i=1}^{n} (\epsilon_i - \bar{\epsilon})^2$$

WITH RESPECT TO ALL THE a_j.

One of the most common applications of the principle is the fitting of algebraic curves to a set of points.

Example 6.1.1 It is known that the system properties obey the relationship $y = az + b$. A set of n measurement pairs (z_i, y_i) is obtained in the laboratory. If it is known that the measurements are free of systemic errors, determine the values of a and b which best fit the data in the absence of systemic error.

Solution If a and b were known it would be possible to compute a value y_i' corresponding to a specific z_i. Thus $y_i - y_i' = \epsilon_i$ would be a measure of the error. Under this assumption the least squares expression becomes

$$R^2 = \sum_{i=1}^{n} (y_i - ax_i - b)^2$$

Extremization is accomplished by setting

$$\frac{\partial R^2}{\partial a} = \frac{\partial R^2}{\partial b} = 0$$

from which it follows that

$$\sum_{i=1}^{n} y_i z_i - a \sum_{i=1}^{n} z_i^2 - b \sum_{i=1}^{n} z_i = 0$$

$$\sum_{i=1}^{n} y_i - a \sum_{i=1}^{n} z_i - nb = 0$$

Let us define $R_{yz} = 1/n\,(\sum yz)$; then the two equations can be rewritten, after division by n, in the form

$$R_{yz} - aR_{zz} - b\bar{z} = 0$$

$$\bar{y} - a\bar{z} - b = 0$$

and thus

$$a = \frac{\begin{vmatrix} R_{yz} & \bar{z} \\ \bar{y} & 1 \end{vmatrix}}{\begin{vmatrix} R_{zz} & \bar{z} \\ \bar{z} & 1 \end{vmatrix}}$$

$$b = \frac{\begin{vmatrix} R_{zz} & R_{yz} \\ \bar{z} & \bar{y} \end{vmatrix}}{R_{zz} - \bar{z}\bar{z}}$$

Extensions to higher-order curves are straightforward.[4]

6.1.4 It is now time to return to the numbers in Table 6.1. The mean for each of the seven operators has been computed and the degree of uncertainty has become much smaller. The expected value ranges from 0.937 to 0.944, a range of 0.007. To two significant figures, therefore, $\bar{z} = 0.94$. Is it possible to estimate that last digit?

Blithely we could average the seven $\bar{z}_j$ and accept the result as an estimate of the expected value. However, we should realize the implication of such an average. In particular, we assume either that there is no systemic error or else that the various systemic errors of the several operators are random as defined in the assumptions. That is, if an operator is taken at random from the population, his own particular systemic error in graph reading is equal to the error (though opposite in sign) to that of another operator in the same population. Then the average of the individual averages is the best estimate of the expected value. In our example it is 0.941 (Table 6.2).

TABLE 6.2

Operator	z_j	ϵ_j	ϵ_j^2
1	0.941	0.000	0.00×10^{-6}
2	0.944	0.003	9.00×10^{-6}
3	0.943	0.002	4.00×10^{-6}
4	0.942	0.001	1.00×10^{-6}
5	0.940	0.001	0.00×10^{-6}
6	0.937	0.004	16.00×10^{-6}
7	0.941	0.000	0.00×10^{-6}
	$\bar{z} = 0.941$		$\sigma_E^2 = 4.43 \times 10^{-6}$
			$\sigma_E = 2.10 \times 10^{-3}$

6.2 Precision and the Variance

The precision of an instrument remains unknown to us no matter what we do. It is possible, however, to measure the input a number of times and to abstract from the collected measurements a statistic which can be theoretically connected with the precision. The standard deviation can be shown to be a reproducible predictor of measurement precision.

6.2.1 In chapter 1, a relationship was obtained between the error and its components in the form

$$\sigma_B^2 = \sigma_C^2 + \sigma_D^2 + R_{CD}$$

in which

$$\sigma_B^2 = \frac{1}{m} \sum_{k=1}^{m} i_{Bk}^2$$

$$\sigma_C^2 = \frac{1}{m} \sum_{k=1}^{m} i_{Ck}^2$$

$$\sigma_D^2 = \frac{1}{m}\sum_{k=1}^{m} i_{Dk}^2$$

$$R_{CD} = \frac{1}{m}\sum_{k=1}^{m} i_{Ck} i_{Dk}$$

and i represents the appropriate dimensionless error as indicated by the subscript. In the absence of systemic error ($i_{Dk} = 0$) the measurement is *certain* but *imprecise* and hence the inaccuracy is given by

$$\sigma_B^2 = \sigma_C^2 = \frac{1}{m}\sum_{k=1}^{n} i_{Ck}^2$$

$$= \frac{1}{m}\sum_{k=1}^{n} (z_{nk} - s_n)^2 \frac{1}{y_n^2}$$

in which z_{nk} is the kth measurement of the nth measurand, and s_n is the expected output. It is now possible to assume that $s_n = \bar{z}_n$ and that σ_B^2 is the sum of the squared deviations divided by the true value squared of the nth measurement. The quantity

$$\sigma^2 = y_n^2 \sigma_B^2 = \frac{1}{m}\sum_{k=1}^{m} (z_{nk} - \bar{z}_n)^2 \qquad \textbf{(6.2.1)}$$

contains all the elements of the measurement error. The quantity on the right is a measure of the dispersion of the errors of individual measurements, and hence of the inaccuracy of the individual measurements. Given a specific measurement with expected output $\bar{z}_n$, σ^2 is a measure of the imprecision of the measurement. If a number of instruments are used to measure the same y_n, the instrument which consistently yields a set of measurements with the lowest value of σ^2 will be the most precise instrument. The operators discussed in Section 6.1.1 exhibit different values of the variances σ^2 and thus may be ranked in the order (see Table 6.1) 3, 5, 1, 7, 2, 4, 6 from most precise to least precise.

6.2.2 The seven means $\bar{z}_j$ in Table 6.1 are seven separate estimates of the expected value, and the seven σ_j^2 are estimates of the imprecision of individual measurements by each operator. The seven means vary from the overall mean. Thus it is possible to compute a variance for the sets of means (Table 6.2). This variance, $\sigma_E^2 = 4.43 \times 10^{-6}$, is much smaller than the variance of individual measurements. The imprecision of the mean is computed from Equation (6.2.1) if we let z_{nk} represent the individual means and $\bar{z}_n$ the mean of the means.

$$\sigma_{\text{mean}} = \left[\frac{1}{\bar{z}^2 m^2}\sum_{k=1}^{m} (\bar{z}_j - \bar{z})^2\right]^{1/2}$$

$$= \frac{1}{0.941}\sigma_E = 0.00221$$

The precision of the mean is (to three digits) $1 - \sigma_{\text{mean}} = 0.998$.

The question now arises, "Is it possible to obtain an estimate of σ_{mean} with only one set of measurements?" Let m sets of n measurements be made and let the overall mean be $\bar{z}$. Then the variance of the individual measurements is written as

$$\sigma^2 = \frac{1}{nm} \sum_{j=1}^{m} \sum_{i=1}^{n} (z_{ij} - \bar{z})^2$$

in which $\bar{z}_{ij}$ is the ith measurement in the jth set. In addition,

$$\sigma_E^2 = \frac{1}{m} \sum_{j=1}^{m} (\bar{z}_j - \bar{z})^2$$

where $\bar{z}_j$ is the mean of the jth set. But

$$\bar{z}_j = \frac{1}{n} \sum_{j=1}^{m} (\bar{z}_j - \bar{z})^2$$

and hence

$$\sigma_E^2 = \frac{1}{mn^2} \sum_{j=1}^{m} \left[\sum_{i=1}^{n} (z_{ij} - \bar{z}) \right]^2$$

The quantity in the square bracket can be written as

$$\left[\sum_{i=1}^{n} (z_{ij} - \bar{z}) \right]^2 = \left[\sum_{i=1}^{n} (\epsilon_{ij}) \right]^2 = \epsilon_{1j}^2 + \epsilon_{2j}^2 + \cdots + \epsilon_{nj}^2$$
$$+ \cdots + 2\epsilon_{1j}\epsilon_{2j} + 2\epsilon_{1j}\epsilon_{3j} + \cdots + 2\epsilon_{n-1j}\epsilon_{nj} + \cdots$$

By the symmetry property, the sum of the terms of the form $\epsilon_{Aj}\epsilon_{Bj}$ for $A \neq B$ will tend to zero for a sufficiently large sample and thus

$$\sigma_E^2 = \frac{1}{mn} \sum_{j=1}^{m} \sum_{i=1}^{n} \epsilon_{ij}^2 = \frac{\sigma^2}{n} \quad \textbf{(6.2.2)}$$

If the variance of the population σ^2 is known, then the variance of a subset of n elements is estimated by σ^2/n.

We seem to be no nearer to our goal, since we have no knowledge of σ^2. Thus we must try to obtain an estimate of σ^2 from the σ_j^2 of each set of data. As before, we write

$$\sigma^2 = \frac{1}{\sum_{j=1}^{m} n_j} \sum_{j=1}^{m} \sum_{i=1}^{n_j} (z_{ij} - \bar{z})^2$$

in which we have used m sets each with n_j elements. Letting

$$\sum_{j=1}^{m} n_j = N \quad \text{and} \quad \epsilon_j = \bar{z}_j - \bar{z} \qquad \epsilon_{ij} = z_{ij} - \bar{z}_j$$

we obtain

$$N\sigma^2 = \sum_{j=1}^{m} \sum_{i=1}^{n_j} (\epsilon_j + \epsilon_{ij})^2$$
$$= \sum_{j=1}^{m} \sum_{i=1}^{n_j} \epsilon_j^2 + \sum_{j=1}^{m} \sum_{i=1}^{n_j} \epsilon_{ij}^2 + 2 \sum_{j=1}^{m} \sum_{j=1}^{n_j} \epsilon_{ij}\epsilon_j$$

For a sufficiently large N and with a set of data which follows the symmetry assumption, the last summation approaches zero and thus in the limit

$$N\sigma^2 = \sum_{j=1}^{m} m_j\epsilon_j^2 + \sum_{j=1}^{m} n_j\epsilon_j^2 \quad \textbf{(6.2.3)}$$

Let us consider the case in which $n_j = n$ and $N = nm$. Equation (6.2.3) becomes

$$m\sigma^2 = m\sigma_E^2 + \sum_{j=1}^{m} \sigma_j^2$$

If we had only one set of measurements, $m = 1$, then

$$\sigma^2 = \sigma_E^2 + \sigma_j^2$$

whereupon, remembering that $\sigma_E^2 = \sigma^2/n$ we obtain as an estimate of σ^2

$$\sigma^2 = \frac{n\sigma_j^2}{n-1} = \frac{\sum \epsilon_i^2}{n-1} \tag{6.2.4}$$

Hence

$$\sigma_{\text{mean}} = \left[\frac{\sigma_j^2}{n-1}\right]^{1/2} \tag{6.2.5}$$

Thus we have a measurement theorem to the effect that if the variance of a sample is known, the variance of the mean is estimated by dividing the variance of the sample by $\sqrt{n-1}$.

Example 6.2.1 Estimate the precision of the means of the data in Table 6.1 for each operator.

Solution Since $n - 1 = 11$, the precision of the mean as estimated from each set of data is as follows:

Operator	σ_j^2	$\bar{z}_j$	σ_e^2	σ_{mean}^2	C
1	7.47×10^{-4}	0.941	6.79×10^{-5}	7.67×10^{-5}	0.9912
2	8.25×10^{-4}	0.944	7.50×10^{-5}	8.42×10^{-5}	0.9908
3	6.61×10^{-4}	0.943	6.01×10^{-5}	6.76×10^{-5}	0.9918
4	8.91×10^{-4}	0.942	8.10×10^{-5}	9.13×10^{-5}	0.9904
5	6.92×10^{-4}	0.940	6.29×10^{-5}	7.12×10^{-5}	0.9916
6	10.1×10^{-4}	0.937	9.18×10^{-5}	10.5×10^{-5}	0.9898
7	7.80×10^{-4}	0.941	7.09×10^{-5}	8.01×10^{-5}	0.9910

There is some variation of the last two digits but not enough to affect the second digit. Thus from the measurements we can state that the estimate of the precision of the mean lies between 99.0 and 99.1.

Example 6.2.2 If the 84 tests in Table 6.1 were considered a single set of data, compute an estimate of the precision of the mean $\bar{z} = 0.941$.

Solution The variance of all the measurements is 8.02×10^{-4} so that $\sigma_E^2 = 8.02 \times 10^{-4}/83 = 9.66 \times 10^{-6}$ and thus

$$C = 1 - \frac{\sigma_E}{0.941} = 0.997$$

It will be demonstrated later that if the computation of the mean were repeated with 84 more trials, the new mean would have a 273-to-1 chance of lying inside the interval 0.941 ± 0.009.

The results of the computations in Example 6.2.1 demonstrate the meaning of the word "estimate." Statistical analyses never give a definite risk-free answer. Statistics are quantifications of experience in the sense that if statistic a has been extracted from n trials or tests, then a can be extracted from $N > n$ trials or tests to a better approximation. There is, however, a finite chance that no matter how large the sample, the property a will not be extracted. Consider, for example, the mean of the 84 data points in Table 6.1. If the process of measurement is repeated, we may believe that we have a very good chance of obtaining 0.941 as a mean, yet we really have 1 chance in 4.30 that $\bar{z}$ will exceed 0.942, 1 chance in 41.96 that x will exceed 0.943, and 1 chance in 738.8 that x will exceed 0.943. Thus whenever a range is selected for measurement error, a chance is taken that a single test or set of tests may fall outside the range. The question to be answered is, therefore, how big a chance is the experimenter willing to take? Consider again the numbers in Table 6.1. What is the probability of the occurrence of a measurement as large as 0.9466? By a method to be described later it can be shown that such a number can occur once in 7.8×10^{11} trials.[5] The author would call this certainty—but from a statistical-theoretical point of view there is still a finite chance of occurrence.

Practical considerations, however, tell us that such a finite chance of occurrence is pure nonsense. An operator reading a graph with the aid a scale may miss one quarter of a millimeter but is not likely to miss by one meter. For practical purposes, therefore, certain extreme measurements can be assumed never to occur. Thus the experimenter is required to decide in advance what chance he is willing to take, that is, what he will call *certainty*. He is "certain" that the sun will rise tomorrow for, after all, the sun has risen millions of times before and has never failed to rise. Based on human experience, therefore, the sun shall certainly come up tomorrow. On the other hand an observer living in Seattle, Washington, is quite sure (but not certain) that there will be at least one cloud in the sky tomorrow. If he prophesies "clouds" he would be right most of the time and fail perhaps once in ten thousand days. Thus he can claim that he is certain that at least one cloud will appear since he calls one chance in 10,000 "certainty."

In the instrument field such hesitation as to the meaning of "certainty" does not exist. All instruments exhibit an amount of random error as a result of internal noise and external interference with the measurand. The best instrument is that which approximates the expected value most closely, that is, the instrument whose measurements exhibit the smallest σ^2. Given a measurand and a specified noise component, the best instrument is that which presents an output with the smallest variance.

6.3 The Concept of Distribution

Collections of measurements possess statistics which tend to characterize the collection. The collections can also be described by certain algebraic and geometrical relationships which describe the interrelationships of the data and serve as predictors of the patterns into which the statistics fall. Such general descriptions of data are called *distributions*, and permit the designer to predict the properties of future experiments.

6.3.1 Let the test of a property yield a set of values z_i, $i = 1, 2, \ldots, k$ and let each of the values occur n_i times. Then it follows immediately from the definition of the mean that

$$\bar{z} = \frac{1}{n} \sum_{i=1}^{k} n_i z_i$$

in which $n = \sum_{i=1}^{k} n_i$ is the total number of individual tests. It follows that the k possible values of z_i do not contribute equally to the mean, but contribute according to a proportion of weight factor $p_i = n_i/n$ or according to the *frequency* of occurrence of the individual event out of a total number of tests. The *weighted mean* is defined as

$$\bar{z} = \sum_{i=1}^{k} p_i z_i \tag{6.3.1}$$

Suppose now that the number of tests be made very large and that a limit exists, $\lim n_i/n = p_i$; then p_i will become the "probability" of the occurrence of z_i and is called $p(z_i)$.

Definition 6.3.1 IF IN A POPULATION n A CERTAIN EVENT z_i OCCURS n_i TIMES, THEN IN A SET OF m TESTS THE EVENT WILL PROBABLY OCCUR A NUMBER OF TIMES $m_i = p(z_i)\, m = (n_i/n)m$ IN WHICH $p(z_i)$ IS THE PROBABILITY OF THE OCCURRENCE OF z_i.

It is postulated that if in an exhaustive collection a certain event occurs as a specific fraction of the totality of events, then a small subset of the collection is likely to have the same event in the same ratio to the totality of events in the same subset. Thus if the occurrence of events in a population is known, the probability of the occurrence of the events in a subset is also known.

In practice it is difficult to measure the whole population—there is no end to possible tests. Often it is even impossible to ascertain in a reasonable number of tests that n_i/n approaches a limit. Normally it is *assumed* that the weights obtained in a sufficiently large sample (whatever that may mean) are approximations to the probability of occurrence of a set of values z_i for which i is finite. The quantity p_i is called the *probability distribution function of* z_i (PDF).

Example 6.3.1 An instrument measures the passage per unit time of certain particles which can attain any of ten speeds. After some time the results are tabulated as follows in a frequency chart:

Speed (cm/s)	No. of Particles
2150	1017
2500	1320
2720	2300
2900	2610
2950	5060
3025	2530
3125	2500
3540	1551
3910	1001
4122	810

Estimate the probability distribution from the sample and obtain the mean velocity of a particle.

Solution The work can be laid out in tabular form:

	S_i = Speed	n_i	p_i	$p_i S_i$
	2150	1017	0.0491	105.56
	2500	1320	0.0634	158.50
	2720	2300	0.1111	302.19
	2900	2610	0.1261	365.69
	2950	5060	0.2444	720.98
	3025	2530	0.1222	369.65
	3125	2500	0.1208	378.71
	2540	1551	0.0749	265.15
	2910	1001	0.0484	189.24
	4122	810	0.0391	161.17
Total		20699	0.995	3016.84

The mean is 3,017. It is not a particularly meaningful quantity since it is not the result of an error-bearing input but rather of expected states. The distribution p_i, on the other hand, is quite important. If the value of the p_i is correct we can use the table to predict the number of particles of a given speed which will be found in a set of particles selected in such a way that the conditions operative in the test are reproduced exactly.

6.3.2 More typical of what happens in measurement is the continuous probability distribution. If the output of the instrument has the form $z_i = \bar{z} + \epsilon_i$ in which ϵ_i is a number (positive or negative) which lies between two limits and which can take *any* of the values between the limits, it is not likely that in a finite number of tests any single error will be repeated too many times. Given an infinite collection of points the likelihood that any particular number will arise in a test is essentially zero. Thus it is theoretically impossible to obtain a frequency chart connected to a single value of z_i. Consider, for example, a test in which the error can be obtained to two significant figures. Even though it is known that ϵ_i can take *any value* in the interval, in practice it can take only one hundred values, $(.00)c$, $(.01)c$, . . . , $(.98)c$, $(.99)c$, in which c is some constant. To obtain a distribution, however, we need many times 100 tests. If the numerically smallest probability in the distribution is p_m, and if it were wanted to obtain a correct distribution, then the event connected with p_m should occur at least once. Hence the number of tests should be

$$n \geq \frac{1}{p_m}$$

If, for example, $p_m = 0.001$, n should be greater than 1,000. Since there is no guarantee that the event will occur even once in a 1,000 trials, it is likely that several thousand tests would have to be made.

To avoid the difficulty, and in keeping with the concept of a continuous probability distribution function, the tests are classified in intervals. The resulting distribution will then be interpreted to mean that there exists a probability p_i that the error will lie in a given interval.

Example 6.3.2 Obtain a frequency-distribution chart and an approximate probability distribution for the data in Table 6.1, in ten equal intervals.

Solution The range of the variable is $1.000 - 0.880 = 0.120$. Thus the intervals have the range 0.012. Let us place the upper bound of each interval in the next interval so as to exclude common points. The half-open intervals contain elements as indicated in the following table.

Interval	n_i	p_i
0.880–0.892$^-$	4	0.0476
0.892–0.904$^-$	5	0.0595
0.904–0.916$^-$	9	0.1071
0.916–0.928$^-$	11	0.1310
0.928–0.940$^-$	13	0.1548
0.940–0.952$^-$	7	0.0833
0.952–0.964$^-$	13	0.1548
0.964–0.976$^-$	8	0.0952
0.976–0.988$^-$	13	0.1548
0.988–1.000$^-$	1	0.0015
	$\sum p_i =$	0.9996

As should be expected, $\sum p_i = 1.000$ within the limits prescribed by the accepted round-off error.

The class interval representation of measurement data leads us to the definition of another useful concept, the probability density function. We first draw a bar histogram of the data in the table (Fig. 6.3) in which

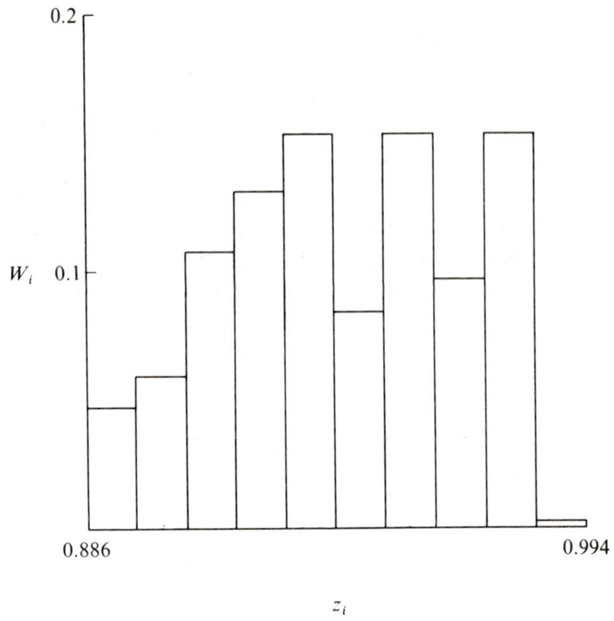

FIGURE 6.3 Bar histogram for Example 6.3.2

the height of a single box is $p_i/0.120$ and the base is 0.0120, or, in general,

$$w_i = \frac{p_i}{\Delta z_i} \qquad w_i \,\Delta z_i = p_i$$

Obviously the area under the individual bars is the probability of the occurrence of events in the interval so that we can write

$$p\left(z_i - \frac{\Delta z}{2} < z < z_i + \frac{\Delta z}{2}\right) = w_i \,\Delta z_i$$

If a and b were two values of z, then it is obvious that

$$p(a < z < b) = \int_a^b w(z)\,dz \qquad \textbf{(6.3.2)}$$

that is, the probability of an event occurring in the interval (a, b) is expressed as the integral of a *probability density function* $w(z)$ integrated in the interval. Obviously the mean over the interval has the form

$$\bar{z} = \sum_{i=1}^{n} z_i w_i \,\Delta z_i$$

which in the limit becomes

$$\bar{z} = \int_a^b z w(z)\,dz \qquad \textbf{(6.3.3)}$$

Example 6.3.3 Compute the average from the frequency table of Example 6.3.1.

Solution For the value of z_i in each interval we take the average value $(z_a + z_b)/2$. The computation then becomes

i	z_i	$w_i = p_i/0.012$	$z_i w_i$
1	0.886	0.0476	0.04217
2	0.898	0.0595	0.05343
3	0.910	0.1071	0.09746
4	0.922	0.1311	0.12087
5	0.934	0.1548	0.14458
6	0.946	0.0833	0.07880
7	0.958	0.1548	0.14830
8	0.970	0.0952	0.09234
9	0.982	0.1548	0.15201
10	0.994	0.0015	0.00149
		$\bar{z} =$	0.931

This differs from the correct value in the second digit as would be expected if we consider the "rough net" employed, that is, if we realize that the intervals were very large and hence that the approximate integration is quite poor.

If the weight of a discrete distribution of n elements is given as $w = p_i(y)\delta(z - z_i)$ in which $\delta(z - z_i)$ is the dirac delta function (see Appendix B), the equation for the mean becomes

$$\bar{z} = \int_{-\infty}^{+\infty} zw(z)\,dz = \int_{-\infty}^{+\infty} zp_i\,\delta(z - z_i)\,dz = \sum_{i=1}^{n} z_i p_i(z_i)$$

With the delta function convention the integral definition of the mean is extended to cover the case of discrete distribution.

6.3.3 The variance of the continuous distribution is easily computed by the integral

$$\sigma^2 = \int_{-\infty}^{+\infty} w(z)(z - \bar{z})^2\,dz \tag{6.3.4}$$

which reduces to a summation on i whenever $w(z) = p_i(z)\delta(z - z_i)$.

The quantity σ^2 is often called the *second central moment* or *moment of inertia* of the distribution $w(z)$, by analogy to a similar equation in elementary dynamics. In statistics the function

$$\alpha_n = \int_{-\infty}^{+\infty} w(z)(z - \bar{z})^n\,dz \tag{6.3.5}$$

is called the *nth central moment of the distribution*, the word "central" implying that the "center of rotation" is the mean. The zero-order moment is the sum of the distribution over the interval; hence $\alpha_0 = 1$. $\alpha_1 = 0$ as can be demonstrated by direct expansion.

$$\alpha_1 = \int_{-\infty}^{+\infty} w(z)(z - \bar{z})\,dz = \int_{-\infty}^{+\infty} w(z)z\,dz - z\int_{-\infty}^{+\infty} w(z)\,dz$$

The first integral on the right is $\bar{z}$ and the second integral on the right is

$\alpha_0 = 1$; hence, $\alpha_1 = 0$. Other moments can be computed in the same fashion.

6.4 Two Distributions

In the practice of instrument design two distributions are universally used as test functions, the uniform distribution and the normal distribution. (Note: White noise is often discussed in the literature. The "whiteness" or nonwhiteness of a signal, however, is not characterized by distributions of the signal amplitude. The Fourier analysis of a recorded noise segment is said to be "white" if the sinusoidal components present at any one time have nearly equal amplitude and if the appearance of a particular component, i.e., of a particular frequency, is random. Discussion of white noise is deferred to a later portion of this chapter.)

6.4.1 In measurement system design it is usually sufficient to determine some statistics at the input and compare the output statistics of various designs. For such an experiment the uniform distribution is quite valuable. A set of numbers is said to be uniformly distributed if each number is as likely to be drawn during an experiment as any other number in the region of definition. This implies that $w(z) = c$, a constant. Since the area under the $w(z)$ line is a rectangle, it is not possible to have an infinite range distribution. Over the whole number system there must be an interval in which the probability is constant and the remainder of the number system must have a probability of zero. The zero moment must be

$$\int_a^b c\,dz = c(b - a) = 1$$

Thus the uniform distribution in the interval $a \leq z \leq b$ must be given by the probability density function

$$w(z) = \frac{1}{b - a} = \frac{1}{L} = c$$

where L is the length of the interval. The mean value of the variable z is

$$\bar{z} = \int_a^b cz\,dz = \frac{c}{2}(b^2 - a^2) = \frac{b + a}{2}$$

that is, $\bar{z}$ is the midpoint value between the limits. The variance is computed as easily

$$\sigma^2 = \int_a^b \frac{(z - \bar{z})^2}{b - a}\,dz = \frac{(b - a)^2}{12}$$

6.4.2 The normal distribution has a probability density function of the form

$$w(z) = \frac{1}{\sigma\sqrt{2\pi}} \exp\left[\frac{(z - \bar{z})^2}{2\sigma^2}\right]$$

in which, as can be shown from direct integration,[6] together with the theorem

$$\int_{-\infty}^{+\infty} e^{-\xi^2}\, d\xi = \sqrt{\pi}$$

$\bar{z}$ and σ^2 are the mean value and the variance of the distribution respectively.[7]

Most modern statistical theory of error is based on the normal distribution—with the exception of the distribution-free techniques which have gained some popularity in recent industrial use.[8] The popularity of the normal distribution depends on three concepts:

(1) Many error phenomena in nature, and many measurements are normally distributed.

(2) Many continuous distributions can be transformed into the normal distribution by appropriate choice of variables.

(3) The central limit theorem.

Items one and two are self-explanatory. Item three is a remarkable theorem of great utility and even greater theoretical import. It has been called the "most important theorem in statistics"[9] and "one of the most remarkable theorems in the whole of mathematics."[10]

Theorem 6.4.1 WHATEVER THE DISTRIBUTION OF THE RANDOM VARIABLE OF A POPULATION WITH EXPECTED VALUE $\bar{z}$ AND VARIANCE σ_s^2, THE DISTRIBUTION OF A COLLECTION OF SAMPLE MEANS APPROACHES A NORMAL DISTRIBUTION WITH $\sigma^2 = \sigma_s^2/n$ IN WHICH n IS THE SIZE OF THE SAMPLE.

The general proof of this theorem is beyond the scope of this book; indeed it is beyond the scope of all except the most advanced books on statistics.[10] The important point is that whenever a mean value is obtained from a set of tests, it is likely to be drawn from a normal distribution.

Perhaps more important is the likelihood that much of the noise found in instruments is likely to be normally distributed. Any deviation from a mean can be conceived of as being generated by a large number of small, randomly distributed, independent causes and hence the actual deviation is of the nature of a mean of the many inputs. If this assumption be true, then the deviations from the expected value should, by the central limit theorem, be normally distributed with mean value zero.

The integral

$$F = \int_a^b w(z)\, dz = \int_a^b \frac{1}{\sigma\sqrt{2\pi}} \exp \frac{(z - \bar{z})^2}{2\sigma^2}\, dz \qquad \textbf{(6.4.1)}$$

represents the probability that number z, which is drawn from a normally distributed population, lies between the limits

$$a \leq z \leq b$$

The quantity F is well tabulated[11] so that the probabilities can be obtained easily. An interesting variation of this integral is obtained by substituting $t = (z - \bar{z})/\sigma$ into the integral so that t represents multiples of σ and hence the integral represents the probability of drawing a deviation inside an interval $\pm t\sigma$ about the mean. From the tables of the probability integral,[11] we obtain

t	Prob.
1	0.6826
2	0.9546
3	0.9974
4	0.999366

In a normally distributed system we have a 68.26% probability of obtaining a deviation within the interval $z \pm \sigma$, or 31.73% of obtaining a deviation outside the interval. We have less than 0.000634 chance, or one chance in 1,577, of finding an absolute deviation greater than 4σ. The connection between relative precision and the variance is now obvious. We established that

$$C = 1 - \frac{\sigma}{\bar{z}} = 1 - \sigma_c$$

Thus C represents a dimensionless ratio of the lower limit below which z has only a $(1 - 0.6826)/2 = 0.1587$ probability of occurrence. Figure 6.4 indicates this relationship clearly. 68.26% of the normal curve lies between $1 - \sigma_c$ and $1 + \sigma_c$. The shaded area to the left of C is, therefore, 15.87% of the total area under the curve.

From Figure 6.4 it can be seen, therefore, that the smaller is σ, i.e.,

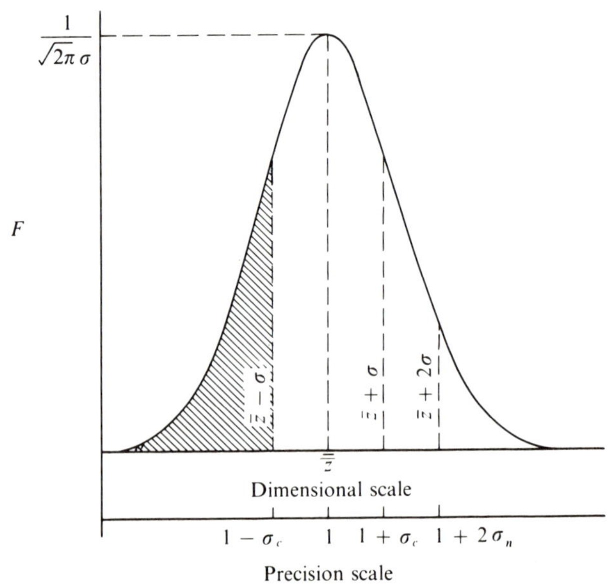

FIGURE 6.4 Standard deviation and precision

the nearer is C to 1, the more precise is the instrument. In the limit, as $\sigma_c \longrightarrow 0$ or $C \longrightarrow 1$ the "spread" of the measurements is made to vanish. (F becomes an impulse.)

Example 6.4.1 An instrument which measures two constant measurands is subjected to a normally distributed noise. The expected values are $\bar{z}_1 = 10$, $\bar{z}_2 = 7$. What must the value of σ_2/σ_1 be to make the local precision of the two measurements equal?

Solution We require

$$C_1 = C_2 = 1 - \sigma_{c1} = 1 - \sigma_{c2}$$

Hence

$$\sigma_{c1} = \sigma_{c2} = \frac{\sigma_1}{\bar{z}_2} = \frac{\sigma_2}{\bar{z}_2}$$

and thus

$$\frac{\sigma_2}{\sigma_1} = \frac{\bar{z}_2}{\bar{z}_1} = 0.7$$

The answer in the Example 6.4.1 indicates that precision is a local characteristic which implies lower precision at lower inputs. From a measurement point of view it must be remembered that C $\longrightarrow 0$ as $\bar{z}$ goes to zero. It is best, therefore, to measure at the high end of the scale if the most precise measurements are expected. Whenever the bulk of the measurements are indicated on the lower half of a scale, it is good practice to change to an instrument with a lower full scale range.

6.5 Ensemble Statistics

Up to now discussion has centered on the interpretation of statistics obtainable from a set of measurements of a single time-fixed measurand. Whenever the input signal is a time-varying function which is not reproduced periodically, it becomes almost impossible to repeat the measurands enough times to compute statistics. In general the instrument will yield an output $z(t)$ which is the sum of the wanted output signal plus an error. Let it be desired to extract the expected value of the output signal at some time $t = t_0$ with an instrument which disturbs the signal randomly. We desire, therefore, to extract the noise introduced into the signal chain by the instrument. An obvious technique is to deliver the input signal to a large number of identical instruments and the output of each instrument recorded for a period of time (Fig. 6.5). It is assumed that all instruments draw the noise from the same probability distribution. Since the internally generated noise is random, each instrument will probably disturb the signal by an amount which is different from the disturbance introduced by any other instrument in the system, even though all disturbances are drawn from the same distribution. From the recorded set of events mean, variance and probability

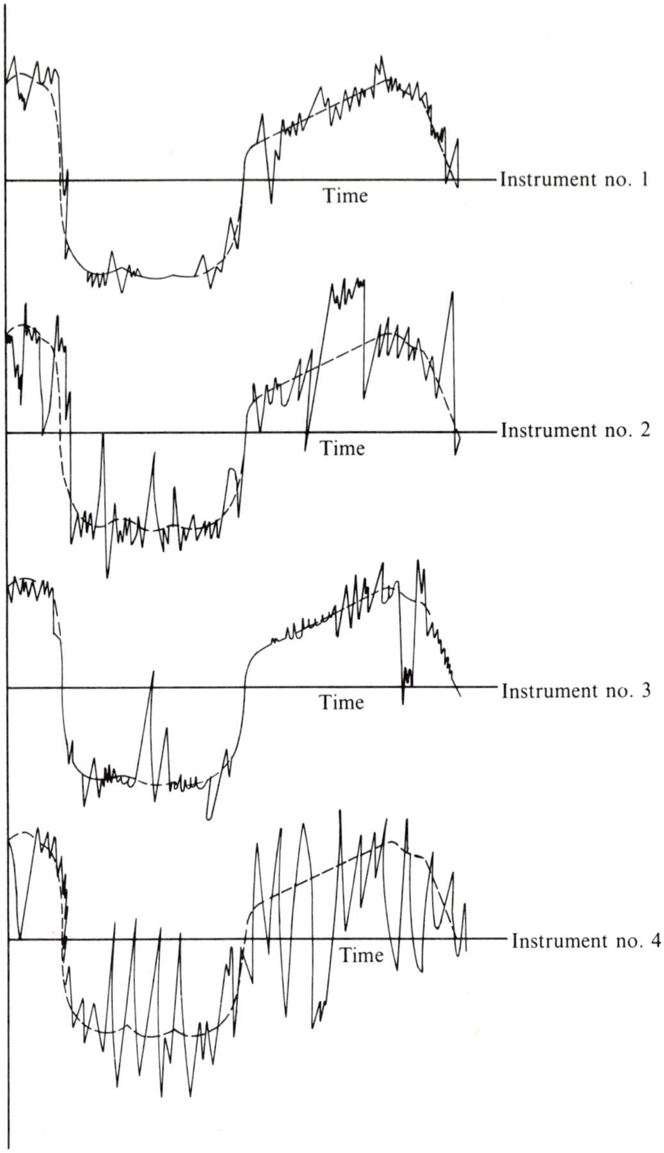

FIGURE 6.5 Ensemble recording of same input by four instruments. Dotted line is expected signal.

distributions can be obtained at any time $t = t_0$. The mean should be an approximation to the expected value and has the form

$$\bar{z}(t_0) = \int_{-\infty}^{+\infty} w(t_0, z)z(t_0)\,dz \qquad \textbf{(6.5.1)}$$

The quantity $\bar{z}(t_0)$ is the "best estimate" of the value of the expected output. Known as the "ensemble average," $\bar{z}(t_0)$ is rather difficult to obtain. The required large number of identical instruments is usually unattainable, the cost of measurement is too high and there is no guarantee that the instruments will be identical and will be used in an identical fashion.

A more subtle problem is that posed by an ideal instrument used to measure an inherently noisy signal. Here the ensemble average as presented above is meaningless. If the signal were known, the statistical properties of the noise could be computed and used to extract the *statistical properties* of future unknown signals under the assumption that the noise distribution is time invariant. The process is called *statistical filtering.*

The technique of statistical filtering and of computation of the ensemble average become identical if the signal is statistically constant, that is, if the signal has time-invariant statistics. One trick used in current design[12] is to design the filter to pass a statistically constant input in the presence of a noise with a constant probability density distribution. The "stationary" signal permits the designer to make a notable simplification. If the input is statistically constant with a superposed noise of non-time-varying statistics, or more precisely, if the input is a random variable with constant statistical properties, then the properties can be determined from the response of a single system. This is embodied in the "ergodic principle."

Principle 6.5.1 n OBSERVATIONS OF RANDOMLY CHOSEN TIMES $t_1, t_2, \ldots, t_n$ OF A STATIONARY SIGNAL WILL YIELD THE SAME STATISTICS AS A SET OF OBSERVATIONS AT A SINGLE TIME MADE ON n IDENTICAL SYSTEMS.

We observed that the ensemble mean was given by

$$\bar{z} = \int_{-\infty}^{+\infty} zw(z, t)\, dz$$

It is possible to define a time average

$$\bar{z}_T = \lim_{T\to\infty} \frac{1}{2T} \int_{-T}^{+T} z(t)\, dt \tag{6.5.2}$$

If the process is stationary, that is, with time-invariant statistics, then

$$\bar{z} = \bar{z}_T$$

It is quite clear that all the ensemble moments of any order will equal the equivalent time statistic of the same order in stationary systems.

6.6 Correlation Functions

Consider a process in which observations are made at time $t = t_1$ and time $t = t_2$. If the measurements are made over an ensemble we obtain a set of pairs $[z_i(t_1), z_i(t_2)]$. From the frequency of occurrence of elements of the set, a probability density function can be obtained for the occurrence of a particular pair, or better yet, for the computation of the probability that the first member of the pair will fall between the limits of an interval $(z_1, z_1 + \Delta z_1)$ at time t_1, and the second number of the pair fall in the interval $(z_2, z_2 + \Delta z_2)$ at a time $t = t_2$. The resulting "joint-probability density function" $w_1(z_1, z_2; t_1, t_2)$ will yield a probability in the rectangular interval $(\Delta z_1, \Delta z_2)$ of the form

$$P(z_1, z_2; t_1, t_2) = w_1 \, \Delta z_1 \, \Delta z_2$$

If we now write

$$R_{zz}(t_1, t_2) = \int_{-\infty}^{+\infty} \int_{-\infty}^{+\infty} (z_1 - \bar{z}_1)(z_2 - \bar{z}_2) w_1 \, dz_1 \, dz_2 \qquad \textbf{(6.6.1)}$$

we call $R(t_1, t_2)$ the auto correlation between $z(t_1)$ and $z(t_2)$. A simple extension is to correlate two random functions y and z to obtain the cross correlation function

$$R_{yz}(t_1, t_2) = \int_{-\infty}^{+\infty} \int_{-\infty}^{+\infty} [y(t_1) - \bar{y}(t_1)][z(t_1) - \bar{z}(t_1)] w_1 \, dy \, dz$$

To understand the meaning of the correlation functions let us examine a simple example for the case in which w_1 is a discrete probability distribution function, $w_1 = p_i \delta(y - y_i)\delta(z - z_i)$. The integral then becomes

$$R_{yz} = \sum_{i=1}^{m} p_i (y_i - \bar{y}_i)(z_i - \bar{z})$$

But $p_i = n_i/n$ and thus

$$R_{yz} = \frac{1}{n} \sum_{i=1}^{n} (y_i - \bar{y})(z_i - \bar{z}) n_i \qquad \textbf{(6.6.2)}$$

is the cross correlation coefficient for a discrete system.

Example 6.6.1 An experiment is conducted in which a stimulus (input) of magnitude x produces a response (output) y. The experiment is carried out under three conditions so that each x_i corresponds to three reactions y_{ai}, y_{bi}, y_{ci}. Plot three figures (x, y_a); (x, y_b); (x, y_c) and compute R_{xy} in each case. What can we conclude?

x	y_a	y_b	y_c
1.0	2.00	2.20	2.19
1.1	2.20	2.24	2.15
1.2	2.40	2.08	3.52
1.3	2.60	2.93	3.74
1.4	2.80	2.40	3.20
1.5	3.00	2.23	2.12
1.6	3.20	3.02	2.94
1.7	3.40	2.88	3.80
1.8	3.60	3.42	3.00
1.9	3.80	2.64	2.29
2.0	4.00	3.40	3.39

Solution The respective means are $x = 1.50$, $y_a = 3.00$, $y_b = 2.68$ and $y_c = 2.94$. If the data are plotted (Fig. 6.6a, b, c), it becomes apparent that $y_a = 2x$, y_b is a noisy function of x and y_c appears to be unrelated to x.

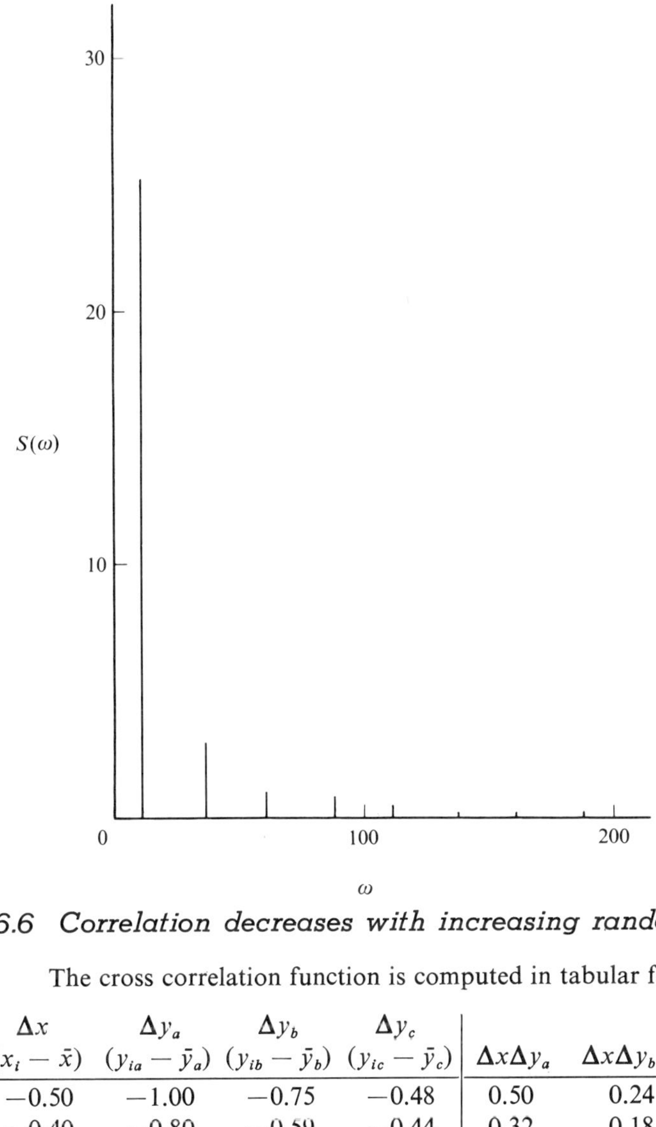

FIGURE 6.6 Correlation decreases with increasing randomness.

The cross correlation function is computed in tabular form:

x	Δx $(x_i - \bar{x})$	Δy_a $(y_{ia} - \bar{y}_a)$	Δy_b $(y_{ib} - \bar{y}_b)$	Δy_c $(y_{ic} - \bar{y}_c)$	$\Delta x \Delta y_a$	$\Delta x \Delta y_b$	$\Delta x \Delta y_c$
1.0	−0.50	−1.00	−0.75	−0.48	0.50	0.24	0.37
1.1	−0.40	−0.80	−0.59	−0.44	0.32	0.18	0.24
1.2	−0.30	−0.60	0.58	−0.60	0.18	0.18	−0.17
1.3	−0.20	−0.40	0.80	0.25	0.08	−0.05	−0.16
1.4	−0.10	−0.20	0.26	−0.28	0.02	0.03	−0.03
1.5	0.00	0.00	−0.82	−0.45	0.00	0.00	0.00
1.6	0.10	0.20	0.00	0.34	0.02	0.03	0.00
1.7	0.20	0.40	0.86	0.20	0.08	0.04	0.20
1.8	0.30	0.60	−0.06	0.44	0.18	0.22	−0.02
1.9	0.40	0.80	−0.65	−0.08	0.32	−0.03	−0.26
2.0	0.50	1.00	0.45	0.72	0.50	0.36	0.22
				SUMS	2.20	1.20	0.39

Hence $R_{xy_a} = 0.20$, $R_{xy_b} = 0.11$, $R_{xy_c} = 0.035$.

It appears that the set of data with the least noise, that is, with the best functional relationship, has the highest value of R_{xy}. The correlation functions essentially measure how closely the two variables are related by a linear expression in the form $y = ax + b$. If x and y are unconnected, the correlation function is zero.

R_{xy} has the dimensions of xy and is not easily adaptable to comparing systems with various inputs. If, however, R_{xy} were divided by $(R_{xx} R_{yy})^{1/2} = \sigma_x \sigma_y$, then a general dimensionless coefficient would be obtained which is free of scale.

$$\rho_{xy} = \frac{\sum_{i=1}^{n} (x_i - \bar{x})(y_i - \bar{y})}{\left[\sum_{i=1}^{n} (x_i - \bar{x})^2 \sum_{i=1}^{n} (y_i - \bar{y})^2\right]^{1/2}} \tag{6.6.3}$$

$|\rho_{xy}|$ is bounded by 0 and 1 with the maximum linear correlation at 1 and minimum at zero. With the results of Example 6.6.1 we compute $\rho_{xy_a} = 1.00$, $\rho_{xy_b} = 0.784$, $\rho_{xy_c} = 0.25$. The degree of correlation is quite evident from the relative magnitudes of these numbers.

6.7 Spectral Density

If a system response is stationary, the auto correlation can be related to the concept of ultimate state sinusoidal response (Chap. 5) in an elegant, physically significant fashion. In this instance the time average replaces the ensemble average and thus

$$R_{ff}(\tau) = \lim_{T\to\infty} \frac{1}{2T} \int_{-T}^{+T} f(t)f(t+\tau)\,d\tau \tag{6.7.1}$$

in which $t_1 = t$ and $t_2 = t + \tau$. Let $f(t) = 0$ for $t < 0$ and for $t \geq T$. Then

$$R_{ff}(\tau) = \frac{1}{2T} \int_{-\infty}^{+\infty} u(t)u(t+\tau)f(t)f(t+\tau)\,dt$$

Let us define the spectral density $S(\omega)$ as the Fourier transform of $R(\tau)$

$$S_{ff}(\omega) = \int_{-\infty}^{+\infty} R(\tau)e^{-j\omega\tau}\,d\tau \tag{6.7.2}$$

$$= \frac{1}{2T} \int_{-\infty}^{+\infty} \int_{-\infty}^{+\infty} f(t)f(t+\tau)e^{-j\omega\tau}\,dt\,d\tau$$

$$= \frac{1}{2T} \left[\int_{-\infty}^{+\infty} f(t)e^{j\omega\tau}\,dt\right]\left[\int_{-\infty}^{+\infty} f(\mu)e^{-j\omega\mu}\,d\mu\right] \tag{6.7.3}$$

in which the substitution $t + \tau = \mu$ has been made. The spectral density has now been shown to be transformable to the product of two integrals, each easily recognizable. Since the Laplace transform of $f(t)$ is

$$F(s) = \int_{0^+}^{+} f(t)e^{-st}\,dt$$

it can be demonstrated (as in chapter 5) that if $f(t)$ were the impulsive response of a system, and if the system were subjected to a sinusoidal input, the ultimate state would take the form

$$\left[\int_{0^+}^{\infty} f(t)e^{-j\omega t}\,dt\right]e^{j\omega t} = G(j\omega)e^{j\omega t}$$

But in (6.7.3) $f(t) = 0$ for $t \leq 0$ and thus the integrals can be recognized in terms of $G(jw)$. Hence

$$S_{ff} = \frac{1}{2T}G(-j\omega)G(j\omega) = \frac{1}{2T}A^2(\omega) \qquad \textbf{(6.7.4)}$$

so that under certain restricted circumstances a spectral density graph bears a relationship to the square of the amplitude in the Bode diagram of the function $f(t)$.

Of course, the many assumptions made in the derivation of (6.7.4) make the relationship, as given, somewhat useless. It suffices to say that Equation (6.7.2) represents the spectral density and that the spectral density bears a relationship to the amplitude vs. frequency distribution, or to the Fourier analysis of the function $f(t)$. A clearer picture can be obtained by computing the spectral density of some nonstatistical signals.

Example 6.7.1 Determine the spectral density of the time variable deterministic function

$$f(t) = \sum_{n=1}^{\infty} a_n \sin \omega_n^t$$

Solution We note that $\bar{f}(t) = 0$ since

$$\frac{1}{2T}\int_{-T}^{+T} \sin \omega t\,dt = \frac{1}{T}\frac{\cos \omega t}{\omega}\Bigg|_{-T}^{+T} = 0$$

Thus the signal is "stationary." (Properly we should also establish that the higher moments are also constant.) Let us now compute the "correlation function" by (6.7.1). Under the integral sign will appear integrals of the type

$$\frac{1}{2T}\int_{-T}^{+T} a_i a_j \sin \omega_i\tau \sin \omega_j(t + \tau)\,d\tau = \phi_{ij}$$

Whenever $\omega_i \neq \omega_j$ the integrals can be integrated only in terms of the form (as can be shown by direct integration)

$$\frac{\sin(\omega_i + \omega_j)T}{T} \qquad \frac{\cos(\omega_j - \omega_i)T}{T}$$

which vanish as $T \longrightarrow \infty$. Thus $\phi_{ij} = 0$ for $i \neq j$. Whenever $\omega_i = \omega_j$, and thus $a_i = a_j$, the integral becomes $(a_i/2) \cos \omega_i\tau$ plus terms which vanish as T becomes large. Thus

$$R(\tau) = \sum_{n=1}^{\infty} \frac{a_n^2}{2} \cos \omega_n\tau$$

The spectral density is

$$S(\omega) = \sum_{n=1}^{\infty} \frac{a_n^2}{2} \int_{-\infty}^{+\infty} \cos \omega_n \tau e^{+j\omega\tau} \, d\tau$$

The sin $\omega\tau$ term does not contribute, the integrand being an odd function. The cos $\omega\tau$ portion integrates into a dirac delta function (Appendix B), and thus[13]

$$S(\omega) = \sum_{n=1}^{\infty} \frac{a_n^2}{4} \delta(\omega - \omega_n)$$

The spectral density of a sum of sinusoids consists of a set of values located at $\omega = \omega_n$, each of amplitude $a_n^2/4$. For example, the spectral density of a square wave, which is represented by the Fourier series

$$f(t) = \sum_{n=1}^{\infty} \frac{A}{2n+1} \sin (2n+1)\frac{2\pi}{T}t$$

is given by

$$S(\omega) = \sum_{n=0}^{\infty} \frac{A^2}{4(2n+1)^2} \delta\left[\omega - (2n+1)\frac{2\pi}{T}\right]$$

For the case in which $T = 0.5$ second and $A = 20$, the spectral density is plotted in Figure 6.7. For all practical purposes the distribution involves only the first harmonic; the others are nearly negligible.

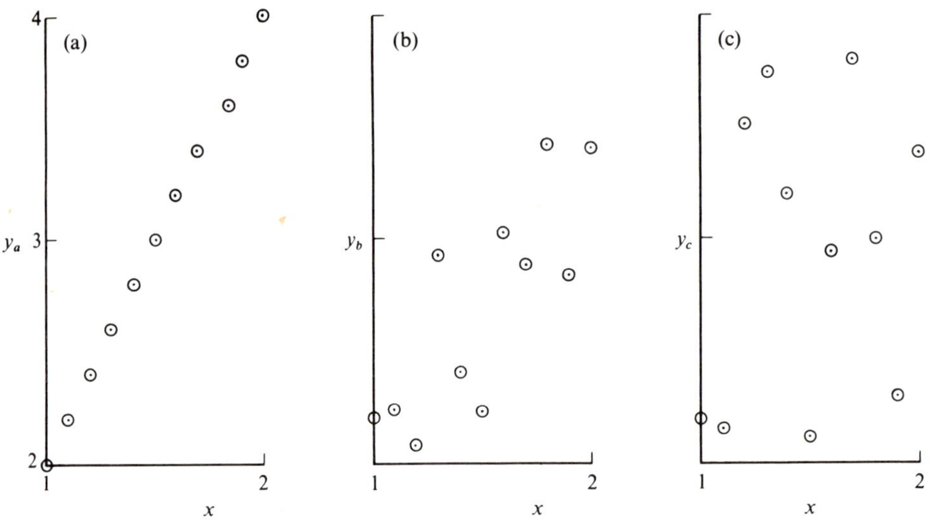

FIGURE 6.7 Spectral density for a sum of sinusoids

Example 6.7.2 The auto correlation function of a random signal is $R(\tau) = K\delta(\tau)$. Determine the spectral density.

Solution

$$S(\omega) = \int_{-\infty}^{+\infty} K\delta(\tau) \, d\tau = K$$

Thus a system with an impulsive correlation function has a constant spectral density, that is, possesses spectral components of all frequencies in equal proportion. In analogy to light, such noise is said to be "white."

White noise is a most useful concept in the design of optimum instruments. In the first place, white noise represents certain types of signals (such as static) reasonably well. Secondly, white noise is an extreme input since it contains signals of all frequencies with equal amplitude. If an instrument can handle white noise it can do well with other more restricted spectra. Finally white noise has a very simple mathematical representation.

PROBLEMS

6.1 It is known that the output to an instrument is a constant. The output, however, is noisy. One hundred measurements are made. Assuming no systemic error, estimate the input and determine the accuracy of the instrument.

1.217	1.207	1.185	1.121	1.298
1.207	1.152	1.102	1.185	1.258
1.222	1.254	1.235	1.153	1.141
1.224	1.167	1.270	1.248	1.168
1.158	1.171	1.254	1.107	1.233
1.237	1.202	1.276	1.160	1.128
1.122	1.119	1.110	1.153	1.179
1.299	1.229	1.264	1.127	1.257
1.104	1.200	1.210	1.108	1.196
1.110	1.205	1.295	1.186	1.226
1.115	1.108	1.216	1.214	1.195
1.229	1.200	1.216	1.270	1.269
1.137	1.200	1.182	1.160	1.135
1.191	1.131	1.127	1.100	1.258
1.190	1.226	1.143	1.139	1.117
1.193	1.192	1.170	1.298	1.157
1.149	1.108	1.184	1.187	1.144
1.202	1.182	1.103	1.221	1.218
1.199	1.154	1.161	1.254	1.290
1.242	1.108	1.232	1.254	1.137

6.2 Two noisy instruments record the same constant input with the following results:

Instrument A		Instrument B	
9.000	11.121	9.086	11.211
9.500	9.569	9.028	9.383
9.900	9.863	9.569	9.805
10.122	10.021	9.789	10.115
10.241	10.433	10.211	10.517
10.279	9.679	10.322	10.412
11.000	10.005	10.329	9.489

Determine the expected value of the output for each instrument. Estimate the expected value of the measurement on the basis of all the measurements. Which instrument is more accurate? How precise is the mean?

6.3 Develop a least squares formula for a system describable by the equation

$$y = ax^2 + bx + c$$

6.4 A system yields two outputs which, it is suspected, can be fitted to the line $ax + b = y$.

(a) Fit the best straight line to the data.

(b) Compute the correlation coefficient. Is the straight line assumption reasonable?

y	2.00	5.97	5.21	8.62	10.03	12.81	16.33	15.71
x	4.81	4.87	5.21	5.07	5.26	5.57	5.49	5.22

y	16.42	18.40	20.40	20.80	22.00	22.80
x	5.51	5.68	5.62	5.73	5.72	5.89

6.5 Obtain a distribution in ten equal intervals of the errors for the data in problem 6.1. Compute the mean, the variance and the third moment from the distribution chart.

6.6 Obtain the spectral density of the periodic function shown in Figure 6.8.

6.7 The system $\ddot{x} + 2.97\dot{x} + 5x = f(t)$ is subjected to the input

$$f(t) = \frac{1}{2} - \frac{4}{\pi^2} \sum_{n=1}^{\infty} \frac{1}{(2n+1)^2} \cos \frac{(2n+1)\pi b}{T}$$

Obtain the spectral density of $f(t)$ and of the ultimate state output $\lim_{t \to \infty} x(t)$.

6.8 The first two columns of data in problem 6.1 may be considered signals at time t and time $t + \tau$ respectively, each point having been obtained 0.1 second after the previous one. Estimate $R(\tau)$.

NOTES

[1]Ref. 77, 78, 79, 84, 90.

[2]There is something to be said for the "better a wrong answer than no answer at all" school of engineering. The important thing is that a start should be made in the preliminary design and that some reasonable numbers be used even though decent experimentation is not available. There is, however, much to be said for the fussbudget school of engineering, if the fussbudgeting extends throughout the whole design operation. The only reprehensible measurement technique is that which mixes precise and imprecise methods, the "string, chalkmark and micrometer" school of engineering measurement.

[3]Ref. 1, pp. 271–81; Ref. 69, 70.

[4]Ref. 77, pp. 67–72.

[5]See Table A, p. 230, Ref. 78; or p. 250, Ref. 35.

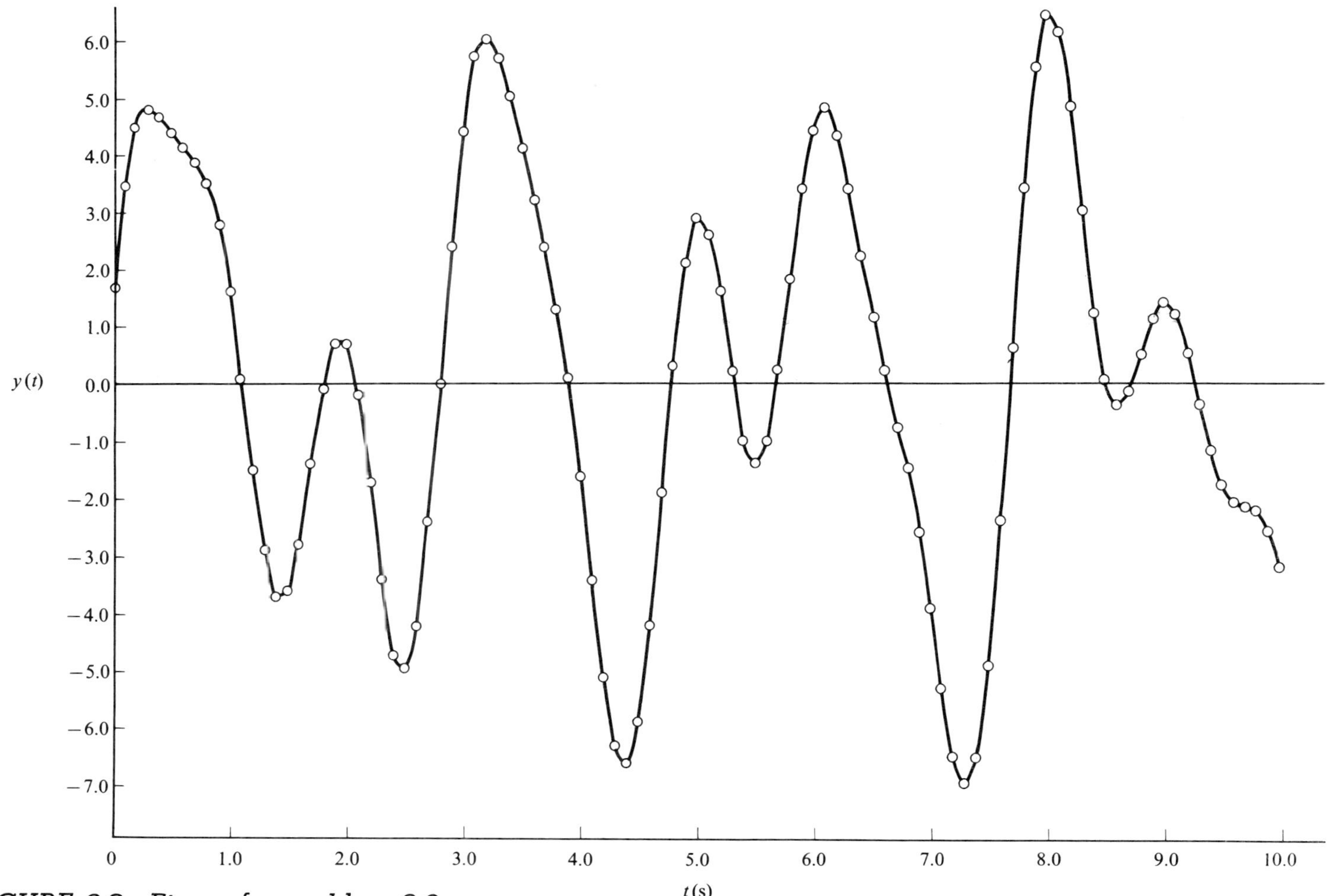

FIGURE 6.8 Figure for problem 6.6

[6]"Directly" does not mean "easily." See, for example, Ref. 81, pp. 767–76.

[7]Ref. 18, pp. 99–100, 208–12.

[8]Ref. 79, Chapt. 16.

[9]Ref. 79, p. 136.

[10]Ref. 18, pp. 213–20.

[11]Ref. 35, p. 250. Most modern computers are programmed to give normally distributed statistics on demand.

[12]Ref. 60.

[13]Direct integration of the cos $\omega_n\tau$ cos ω_τ terms will not yield the dirac delta in an obvious fashion. It must first be proved that as $\tau \longrightarrow \infty$ the integral tends to vanish except when $\omega_n = \omega$. It is best to use transform methods to obtain the same result. See, for example, Ref. 60, pp. 28–32.

7

PROPAGATION OF ERROR

7.1 Concepts

All elements in a signal chain tend to enhance, suppress or modify the input error in the same fashion as they tend to enhance, suppress or modify the input signals. We say that error is *propagated* by the element to the next element in a modified form. Proper design procedure requires us to see what the propagation (or transmission) characteristics of a system are and to determine the effects of error at the output. In general, it appears that the errors superposed on a variable propagate in the same manner as the variable and that suppression of a propagated error implies partial suppression of a useful signal.

7.1.1 The simplest procedure for the estimation of output error is to replace the error-bearing quantity in the performance equation with a nominal value added to an expected error and then to solve the resulting expression for the output error. It is customary to assume that the error is small and that higher-order terms can be neglected.

Example 7.1.1 A fluid flow meter obeys the equation (see chapter 8 for a more intensive study of this problem)

$$Q = C(p - p_A)^{1/2}$$

where

$$Q = \text{flow } (m^3/\text{s})$$
$$p = \text{pressure } (N/m^2)$$
$$p_A = \text{pressure to the atmosphere } (N/m^2)$$
$$C = \text{calibration constant } (m^7 K^{-1})^{1/2}$$

Let there be the following errors in the system:

$$C = C_o + \epsilon_o$$
$$p - p_A = \Delta p + \epsilon_p$$

in which C_o and Δp are the nominal design values of the quantities, and ϵ_c and ϵ_p are the errors introduced into the system. Find the error ϵ_Q, where $Q = Q_o + \epsilon_Q$ and Q_o is the nominal expected value of Q.

Solution

$$Q = (C_o + \epsilon_c)(\Delta p + \epsilon_p)^{1/2}$$

If we assume that ϵ_c and ϵ_p are small numbers, we can write, by expansion of the second term into a bionominal series,

$$Q = (C_o + \epsilon_c)\left(\Delta p^{1/2} + \frac{\epsilon_p}{2\Delta_p^{1/2}} + \cdots\right)$$

$$= C_o\,\Delta_p{}^{1/2} + \frac{C_o}{2\Delta_p^{1/2}}\epsilon_p + \Delta_p^{1/2}\epsilon_c + \cdots$$

The higher-order terms contain $(\epsilon_p\, \epsilon_c)$, ϵ_p^2, ϵ_c^2, etc., and may be neglected if ϵ_p and ϵ_c are sufficiently small. The expected value of the output is $Q_o = C_o\sqrt{\Delta_p}$, and thus

$$\frac{\epsilon_Q}{Q_o} = \frac{Q - C_o\,\Delta_p^{1/2}}{C_o\,\Delta_p^{1/2}} = \frac{\epsilon_p}{2\Delta_p} + \frac{\epsilon_c}{C}$$

This manipulation indicates that an error in the quantity C is propagated directly as an error in flow measurement, whereas an error in pressure measurement is halved in transmission. In design and use of such an instrument it becomes clear that care in the determination of C is more fruitful than improvement in the measurement of Δp.

7.1.2 The propagation of error in an element can be very difficult to estimate. Consider, for example, a system with the nominal performance equation

$$\frac{d^2z}{dt^2} + a_1\frac{dz}{dt} + a_2 z = S_o y(t)$$

but with a_1 and a_2 in error. If the changes in a_1 and a_2 are reasonably constant in time, it is reasonably easy to solve for the value of z and to compare it to the nominal expected value. If, on the other hand, a_1 and a_2 change with time (drift) or randomly with time (noise) the solution becomes quite difficult and in most instances impossible. Consider, for example, a simple series RC circuit. [See Section 8.2, Figure 8.1 and Equation (8.2.1)]. The performance equation is

$$RC\frac{dq}{dt} + q = CE$$

and let $E = E_o\,u(t)$ where E_o is a constant. In the ultimate state $q_u = CE_o$, and thus in this situation an error in R does not affect the ultimate state charge. Suppose, however, that R drifts, say, that $R = R_o + At$. Can the same conclusion he made? Examination of the equation leaves us with no information. In this instance, however, the equation can be solved and it can be shown that there is no ultimate state effect but that the ultimate state is delayed. The error in R propagates as an error in the time constant τ so that instead of $\tau_o = R_oC$ we obtain (Section 8.2)

$$\tau = -\frac{R_o}{A}(0.3679^{-AC} - 1) \qquad \textbf{(7.1.1)}$$

which approaches R_oC as $A \longrightarrow 0$.

Now that we have obtained τ for a steady input we could plot τ/R_oC as a function of AC but would not be able to conclude much about the system. If the input were not a constant there would be no reason to assume that Equation (7.1.1) applies to the new input. For lack of better information we can use the estimate of Equation (7.1.1) but cannot guarantee the results.

7.2 Propagation of Error in Systems with Algebraic Performance Equations and Deterministic Errors

Algebraic performance equations lend themselves readily to the determination of system error. In addition, if it were possible to solve the system differential equations, the theory discussed below would be directly applicable to most performance equations.

7.2.1 Let a system be represented by a performance equation with no derivatives or integrals:

$$z = z(y_1, y_2, \ldots, y_n, a_1, a_2, \ldots, a_m) \qquad \textbf{(7.2.1)}$$

where the y_i are the several inputs to the system and the a_j are system parameters which could bear error. Let $y_i = y_i^o + \epsilon_{y_i}$ and $a_j = a_j^o + \epsilon_{a_j}$ and expand (7.2.1) into a Taylor series. Then the output error ϵ_z can be written as

$$\epsilon_z = z - z(y_i^o, a_j^o) = \frac{\partial z_1}{\partial y_1}\epsilon_{y_1} + \frac{\partial z_2}{\partial_{y_2}}\epsilon_{y_2} + \cdots$$
$$+ \frac{\partial z}{\partial a_1}\epsilon_{a_1} + \frac{\partial z}{\partial a_2}\epsilon_{a_2} + \cdots$$

in which the derivatives are evaluated with $\epsilon_{y_i} = \epsilon_{a_j} = 0$. If the errors are small so that the higher-order terms are negligible, we obtain

$$\epsilon_z = \sum_{i=1}^{n} \frac{\partial z_i}{\partial y_i} \epsilon_{y_i} + \sum_{j=1}^{m} \frac{\partial z}{\partial a_j} \epsilon_{a_j} \tag{7.2.2}$$

The quantities ϵ can be either positive or negative and so can the derivatives. The total error, therefore, can achieve a maximum if the errors and the derivatives all achieve their positive values simultaneously. Thus the maximum design estimate can be found by

$$|\epsilon_z| = \sum_{i=1}^{n} \left|\frac{\partial z}{\partial y_i}\right| |\epsilon_{y_i}| + \sum_{i=1}^{m} \left|\frac{\partial z}{\partial a_j}\right| |\epsilon_{a_j}| \tag{7.2.3}$$

If the sign of the errors is known and is stable, it is possible that Equation (7.2.3) can give outsize estimates, so that (7.2.2) could give the best design estimate. However, when the sign of the error depends on the measurements conditions, Equation (7.2.3) gives an estimate of the maximum achievable error and of the inaccuracy of the instrument.

Example 7.2.1 In the design of the galvanometer discussed in Section 3.5, it is wished to use a resistance R_s which varies $\pm 10\%$ from its nominal value. What effect will the introduction of R_s have on the ultimate state output to a constant input? Let all system resistances equal 10^5 ohms.

Solution The functional relationship in the ultimate state is

$$\phi_{us} = SE$$

where

$$S = \frac{K_1}{K}\left[\frac{1}{((R + R_s)/R_D + 1)R_C + R_I + [R_D(R + R_s)]/(R + R_s + R_D)}\right]$$

Let $R_s = R_s^\circ + \epsilon_R$; then the error in ϕ_{us} can be obtained from Equation (7.2.3)

$$\epsilon_\phi = \frac{\partial S}{\partial R_s} \epsilon_R E$$

Thus

$$\frac{\epsilon_\phi}{\phi_{us}} = \frac{1}{S}\frac{\partial S}{\partial R_S} \epsilon_R$$

$$\frac{[R_G/R_D + R_D^2/(R + R_s + R_o)^2]\epsilon_R/R_s}{[(R + R_s)/R_D + 1]R_C/R_s + R_I/R_s + [R(R + R_s)/R_s(R + R_s + R_D)]} \frac{\epsilon_s}{R_s}$$

$$= 0.1 \quad \text{and} \quad R_G = R_D = R_C = R_I = R = R_s = 10^5$$

Then $\epsilon_\phi/\phi_{us} = 1.0/42 = 0.0238$. A 10% error in R_s implies a 2.38% change in the ultimate state output. In this instance the error is reduced in propagation.

Example 7.2.2 In the galvanometer described in Section 3.5 set $\epsilon_R/R_s = 0.10$ and $R_s = 10^8$ ohms, $R_C = R_D = R_I = R = 20$ ohms and $R_G = 60$ ohms. Determine the output error.

Solution $\epsilon_\phi/R_s = 0.30$. In this instance the 10% error in the resistance is trebled by propagation.

Example 7.2.3 Let $z_o = y_1y_2y_3^n$ in which all the y carry error. Determine the error in z.

Solution Expansion of z in a series gives us

$$\begin{aligned} z &= (y_1 + \epsilon_1)(y_2 + \epsilon_2)(y_3 + \epsilon_3)^n \\ &= y_1y_2y_3^n + y_2y_3^n\epsilon_1 + y_1y_3^n\epsilon_2 + ny_1y_2y_3^{n-1}\epsilon_3 + \cdots \end{aligned}$$

and thus (if higher-order terms are neglected)

$$\frac{\epsilon_z}{z_o} = \frac{z - y_1y_2y_3^n}{y_1y_2y_3^n} = \frac{\epsilon_1}{y_1} + \frac{\epsilon_2}{y_2} + \frac{n\epsilon_3}{y_3}$$

In this instance the contribution of y_3 to the total error depends on n: the error is augmented if $n > 1$ and reduced if $n < 1$.

Example 7.2.4 Let $z_o = y_1 + y_2$, in which both values of y carry error. Find the error in z.

Solution $z = (y_1 + \epsilon_1) + (y_2 + \epsilon_2)$ and thus

$$\frac{\epsilon_z}{z_o} = \frac{z - (y_1 + y_2)}{y_1 + y_2} = \frac{\epsilon_1 + \epsilon_2}{y_1 + y_2}$$

In this case the fractional dimensionless error depends on both y_1 and y_2. If y_1 and y_2 have minimum values, then

$$\left|\frac{\epsilon_z}{z_0}\right| < \frac{|\epsilon_1| + |\epsilon_2|}{|y_1 + y_2|_{\text{MIN}}}$$

7.2.2 The literature on error propagation is ample but naive.[1] The assumption is always made that the higher-order terms are negligible and that the series expansion converges. Neither assumption can be taken for granted. Consider, for example, the error-bearing multiplication

$$z = (y_1 + \epsilon_1)(y_2 + \epsilon_2) = y_1y_2 + \epsilon_1y_2 + \epsilon_2y_1 + \epsilon_1\epsilon_2 \qquad \textbf{(7.2.4)}$$

The assumption is always made that $\epsilon_1\epsilon_2$ is negligible for all values of y. This last statement is not necessarily true. Let, for example, $y_1 = 0.1$, $y_2 = 0.10$ and $\epsilon_1 = \epsilon_2 = 0.10$. The error computed by Equation (7.2.3) is

$$\epsilon_z = y_2\epsilon_1 + y_1\epsilon_2 = 0.02$$

whereas with (7.2.4) $\epsilon_z = 0.03$. The change of 0.01 introduced by the higher-order term is as large as the error introduced by any of the low-order terms and hence cannot be neglected. If, on the other hand, $y_1 = y_2 = 10$, $\epsilon_1 = \epsilon_2 = 0.1$, then the error by Equation (7.2.2) would be equal to 2, whereas by (7.2.4) it would be 2.01. The difference between the two expressions is 0.01%, an inconsiderable amount.

The question of convergence is much more subtle than the question of higher-order terms. Consider, for example, $z = 1/(y_1 + \epsilon)$ in which ϵ is a much smaller number than y_1. Use of (7.2.2) leads to the error propagation estimate

$$\epsilon_z = \frac{1}{y_1}\left(1 + \frac{\epsilon}{y_1}\right)$$

Neither of the two error expansions discussed above is valid in the neighborhood of $\epsilon = y$. The following stratagem is used:

$$z = \frac{1}{2y + (\epsilon - y)}$$

which for ϵ near y can be expanded in the series

$$z = \frac{1}{2y} - \frac{\epsilon - y}{4y^2} + \cdots$$

and thus

$$\left|\frac{\epsilon_z}{z_0}\right| = \frac{1}{2} + \left|\frac{\epsilon - y}{4y}\right|$$

In the case of $\epsilon = y$ the error would be 50%. It is clear, therefore, that there exist three separate error expressions for a divider:

$$\begin{aligned} \left|\frac{\epsilon_z}{z_o}\right| &= 1 + \left|\frac{\epsilon}{y_1}\right| & \left|\frac{\epsilon}{y}\right| &\ll 1 \\ \left|\frac{\epsilon_z}{z_o}\right| &= \frac{1}{2} + \left|\frac{\epsilon - y}{4y}\right| & |\epsilon| &\approx |y| \\ \left|\frac{\epsilon_z}{z_o}\right| &= \left|\frac{y}{\epsilon}\right|\left(1 + \left|\frac{y}{\epsilon}\right|\right) - 1 & \left|\frac{y}{\epsilon}\right| &\ll 1 \end{aligned} \tag{7.2.5}$$

7.2.3 An interesting application of the theory discussed in the previous section is the first-order error of the "exponentiator" $z_o = ky^n$. Such a device can have an input offset error, an error in the exponent, a gain error or an output offset error. We set

$$\begin{aligned} \epsilon_y &= \text{input offset error} \\ \epsilon_o &= \text{output offset error} \\ \epsilon_k &= \text{gain error} \\ \epsilon_n &= \text{exponent error} \end{aligned}$$

and obtain the erroneous output

$$z = (k + \epsilon_k)(y + \epsilon_y)^{n+\epsilon_n} + \epsilon_o \tag{7.2.6}$$

Let us apply the theory of Section 7.2. The result for small errors is

$$\left|\frac{\epsilon_z}{z_o}\right| = \left|\frac{\epsilon_n}{k}\right| + |\epsilon_n \ln y| + \left|n\frac{\epsilon_y}{y}\right| + \left|\frac{\epsilon_o}{ky^n}\right| \tag{7.2.7}$$

The second and third terms were obtained through the expansions

$$(y + \epsilon_y)^{\epsilon_n} = y^{\epsilon_n}\left(1 + \frac{\epsilon_n\epsilon_y}{y} + \cdots\right)$$

and

$$y^{\epsilon_n} = e^{\epsilon_n \ln y} = 1 + \epsilon_n \ln y + \frac{(\epsilon_n \ln y)^2}{2!} + \cdots$$

so that it is obvious that we also need $|(\epsilon_n \ln y)| \ll 1$ to justify the use of Equation (7.2.7). Since $\epsilon_n \ll 1$, $|\ln y| < 1$ is a sufficient condition for the validity of (7.2.7), that is,

$$0.3679 = e^{-1} \leq y \leq e = 2.718$$

Outside this range, higher-order terms are needed. It is possible to rewrite Equation (7.2.7) by noting that

$$y^\epsilon = e^{\epsilon \ln \frac{y}{a} + \epsilon \ln a}$$

$$= a^\epsilon\left[1 + \epsilon \ln\left(\frac{y}{a}\right) + \frac{\epsilon^2}{2!}\ln^2\left(\frac{y}{a}\right) + \cdots\right]$$

Hence we can write

$$y^\epsilon = a^\epsilon\left[1 + \epsilon \ln\left(\frac{y}{a}\right)\right]$$
$$0.3679 < \frac{y}{a} < 2.718 \tag{7.2.8}$$

For any given values of a there is a range of y for which Equation (7.2.8) is acceptable. The values of a^ϵ for a range of a and ϵ have been computed (Table 7.1).

TABLE 7.1 Values of a^ϵ

	ϵ					
a	0	0.001	0.01	0.05	0.10	0.25
0.01	1	0.9954	0.9550	0.7943	0.6310	0.3162
0.05	1	0.9970	0.9705	0.8610	0.7411	0.4315
0.10	1	0.9977	0.9772	0.8912	0.7943	0.5623
0.50	1	0.9992	0.9931	0.9650	0.9331	0.8409
1.00	1	1.000	1.000	1.000	1.000	1.000
5.00	1	1.001	1.007	1.035	1.072	1.189
10.00	1	1.002	1.023	1.122	1.259	1.778
50.00	1	1.003	1.030	1.161	1.349	2.114
100.00	1	1.005	1.047	1.259	1.585	3.162
1000.00	1	1.007	1.072	1.413	1.995	5.623

The error can now be estimated by the equation

$$\left|\frac{\epsilon_z}{z_o}\right| = |1 - a^{\epsilon_n}| + na^{\epsilon_n}\left|\frac{\epsilon_y}{y}\right| + a^{\epsilon_n}\ln\left|\frac{y}{a}\right||\epsilon_n| + \frac{\epsilon_o}{ky_o^n} \tag{7.2.9}$$

in which a is selected so that $a = (y_{MAX} + y_{MIN})/2$ or at least so that $y_{MIN} < a < y_{MAX}$. Table 7.2 gives values of a consistent with Table 7.1.

This expression is valid only for $\epsilon/y_1 \ll 1$. If the value of ϵ were to be less than y_1 but near in value to y_1 it would be necessary to include the higher-order terms of the series. With $y_1 < \epsilon$, inclusion of the higher-order terms would not help since the infinite series for the error would diverge. For small y an alternate infinite series is used:

$$z = \frac{1}{\epsilon}\left[1 - \frac{y}{\epsilon} + \frac{1}{2!}\left(\frac{y}{\epsilon}\right)^2 + \cdots\right]$$

TABLE 7.2

a	y_{MIN}	y_{MAX}
0.01	0.0037	0.0272
0.05	0.0184	0.1359
0.10	0.0368	0.2718
0.50	0.1839	1.359
1.00	0.3679	2.718
5.00	1.839	13.59
10.00	3.679	27.18
50.00	18.39	135.9
100.00	36.79	271.8
1000.00	367.9	271.8

If $y/\epsilon < 1$ then

$$\epsilon_z = \frac{1}{\epsilon}\left(1 - \frac{y}{\epsilon}\right) - \frac{1}{y}$$

For example, if $\epsilon = 0.10$ and $y = 0.01$, then $|\epsilon_z/z| = 0.01$. [With Equation (7.2.2) the "error" would have the incorrect value of 11.00.]

Example 7.2.5 A square root extractor has an input offset error $\epsilon_y/y = 0.01$, $\epsilon_n = 0.001$, has no output offset and obeys the nominal equation $z_0 = y^{1/2}$. Compute the error for $y = 6$.

Solution Let $a = 5$ (see Table 7.2); then $a^\epsilon = 1.001$ and thus from Equation (7.2.9)

$$\left|\frac{\epsilon_z}{z_o}\right| = |1 - 1.001| + \frac{1}{2}(1.001)(0.001) + \left[1.001 \ln\left(\frac{6}{5}\right)\right]0.001 = 0.0016$$

Example 7.2.6 Determine the error at $y = 0$ for the system in Example 7.2.5

Solution It can be seen from Table 7.1 that $a^\epsilon \longrightarrow 1$ as $a \longrightarrow 0$. Let us note that $|\ln y/a| = 1$ at the lowest limit of the range. For any value of a we obtain the smallest possible value of y; hence in the limit

$$\lim_{y \longrightarrow 0}\left|\frac{\epsilon_z}{z_0}\right| = n\left|\frac{\epsilon_y}{y}\right| + |\epsilon_n| = 0.0015$$

Example 7.2.7 Obtain a complete error expression for the multiplier $z_o = y_1 y_2$.

Solution The erroneous output will have the form

$$z_o = (1 + \epsilon_o)(y_1 + \epsilon_{10})^{1+\epsilon_{11}}(y_2 + \epsilon_{20})^{1+\epsilon_{21}} + \epsilon_3$$

This can be written as

$$z_o = y_1 y_2(1 + \epsilon_o + \epsilon_{11} \ln y_1 + \epsilon_{21} \ln y_2) + \epsilon_{10} y_2 + \epsilon_{20} y_1 + \epsilon_3$$

and thus

$$\left|\frac{\epsilon_z}{z_o}\right| = |\epsilon_o| + |\epsilon_{11} \ln y_1| + \left|\frac{\epsilon_{10}}{y_1}\right| + \left|\frac{\epsilon_{20}}{y_2}\right| + \left|\frac{\epsilon_3}{y_1 y_2}\right| + |\epsilon_{21} \ln y_2|$$

7.2.4 Such theoretical results as have been discussed in the previous sections serve as guide to the form of the error to be measured experimentally. Typically an analysis such as the one in Example 7.2.7 is conducted and then calibration tests are run to obtain a range of errors for a range of inputs. For the multiplier, for example, enough data would be obtained to fit the curve

$$\epsilon_z = |y_1 y_2| A + |y_1| B + |y_2| C + D$$

with the assumption (or hope) that $\epsilon_{11} = \epsilon_{21} = 0$. Typical results for such an operation have been determined by McKenzie[2] (see Table 7.3).

TABLE 7.3 Error estimates for typical instruments

Instrument	Error estimate
Pressure Transducer	$\frac{\epsilon_z}{K\Delta p} = \frac{\epsilon_K}{K} = 0.005$
Strip Chart Recorder	$\frac{\epsilon_z}{Ky} = \frac{\epsilon_K}{K} + \frac{\epsilon_y}{y} = \frac{\epsilon_y}{y} + 0.005$
Square Root Extractor	$\frac{\epsilon_z}{Ky^{1/2}} = \frac{\epsilon_K}{K} + \frac{1}{2}\frac{\epsilon_y}{y} + \frac{1}{2}\frac{\epsilon_y}{y} + 0.005 \qquad 20 < Ky^{1/2} < 100$
Temperature Transducer	$\frac{\epsilon_z}{K\theta} = \frac{\epsilon_K}{K} = 0.005$
Multiplier	$\frac{\epsilon_z}{Ky_1 y_2} = \frac{\epsilon_{y_1}}{y_1} + \frac{\epsilon_{y_2}}{y_2} + \frac{\epsilon_K}{K} - \frac{\epsilon_{y_1}}{y_1} + \frac{\epsilon_{y_2}}{y_2} + 0.005$
Divider	$\frac{\epsilon_z}{Ky_1 y_2} = \frac{\epsilon_{y_1}}{y_1} + \frac{\epsilon_{y_2}}{y_2} + 0.005$

The results of tests carried out in the author's laboratory with more precise instruments are reasonably close to those in Table 7.3, although small differences occurred. For example, for the multipler,

$$\frac{\epsilon_z}{Ky_1 y_2} = \left|\frac{\epsilon_1}{y_1}\right| + \left|\frac{\epsilon_2}{y_2}\right| + 0.000002 \ln y_1 + 0.000002 \ln y_2 + \frac{0.004}{Ky_1 y_2} + 0.00472$$

$$100 > Ky_1 y_2 > 10 \qquad y_1 > 1 \qquad y_2 > 1$$

was obtained. Within the limits of Table 7.3, and of the equation above, the two error functions are identical.[3]

7.3 Propagation of Error in Systems with Algebraic Performance Equations and Random Errors

Let a system be represented by the algebraic performance equation

$$z = z(y_1 + \epsilon_1, y_2 + \epsilon_2, \ldots, y_m + \epsilon_m, a_1 + \eta_1, a_2 + \eta_2, \ldots, a_n + \eta_n)$$

in which $(y_1, y_2, \ldots, y_m)$ is the input vector and $(a_1, a_2, \ldots, a_n)$ is the array of system parameters. The wanted or nominal performance equation will be

$$z_o = z_o(y_1, y_2, \ldots, y_m; a_1, a_2, \ldots, a_n)$$

The error $\epsilon_z = z - z_o$ can be determined (as it was in Section 7.2) by expansion into the Taylor series

$$\begin{aligned}\epsilon_z = {} & \frac{\partial z}{\partial y_1}\epsilon_1 + \frac{\partial z}{\partial y_2}\epsilon_2 + \cdots + \frac{\partial z}{\partial y_m}\epsilon_m + \frac{\partial z}{\partial a_1}\eta_1 \\ & + \frac{\partial z}{\partial a_2}\eta_2 + \cdots + \frac{\partial z}{\partial a_n}\eta_n + \frac{1}{2!}\frac{\partial^2 z}{\partial y_1^2}\epsilon_1^2 \\ & + \frac{2}{2!}\frac{\partial^2 z}{\partial y_1 \partial y_2}\epsilon_1\epsilon_2 + \cdots \end{aligned} \qquad \textbf{(7.3.1)}$$

We assume that higher-order terms can be neglected. If ϵ_i and η_i vary randomly with each measurement, it will be necessary to replace ϵ_z by a statistic, usually the variance $\sigma^2_{\epsilon_z}$ or the standard deviation. If we square ϵ_z, sum all the terms and assume that all the errors are statistically independent, that is, the average of the products $(\epsilon_I\epsilon_J)$, $(\eta_I\eta_J)$, $(\epsilon_I\eta_J)$ $(I \neq J)$ is zero, we obtain (in the limit)

$$\begin{aligned}\sigma^2_{\epsilon_z} = {} & \left(\frac{\partial z}{\partial y_1}\right)^2 \sigma^2_{\epsilon_1} + \left(\frac{\partial z}{\partial y_2}\right)^2 \sigma^2_{\epsilon_2} + \cdots \\ & + \left(\frac{\partial z}{\partial a_1}\right)^2 \sigma^2_{\eta_1} + \left(\frac{\partial z}{\partial a_2}\right)^2 \sigma^2_{\eta_2} + \cdots \end{aligned} \qquad \textbf{(7.3.2)}$$

Equation (7.3.2) can be used to determine the propagation of error through a system whenever the input variances are known.

Example 7.3.1 A system obeys the equation

$$z = Ky_1\frac{y_2}{y_3}$$

with a constant error $\epsilon_K = 0.001$ and random errors represented by

$$\sigma^2_{y_1} = 10^{-6} \qquad \sigma^2_{y_2} = 2 \times 10^{-6} \qquad \sigma^2_{y_3} = 1.5 \times 10^{-6}$$

Determine σ_z when $k = 10.000$, $y_1 = 1.000$, $y_2 = 2.700$ and $y_3 = 3.100$ if we obtained the variances from $n = 50$ measurements.

Solution

$$\sigma_z^2 = \frac{n\epsilon_K^2}{n-1}\left|\frac{y_1y_2}{y_3}\right|^2\left|\frac{K_{y_2}}{y_3}\right|^2\sigma_y^2 + \left|\frac{K_{y_1}}{y_3}\right|^2\sigma_y^2 + \left|\frac{Ky_1y_2}{y_3^2}\right|^2\sigma_y^2$$

The first term is obtained by summing

$$\frac{1}{n-1}\sum_{i=1}^{n}(A\epsilon_K^2 + B\epsilon_K\epsilon_{y_1} + C\epsilon_K\epsilon_{y_2} + D\epsilon_K\epsilon_{y_3})$$

where A, B, C and D are nonrandom functions of y_1, y_2, y_3 and K. We assume that the mean values of the ϵ_y equal zero. Then

$$\frac{1}{n-1}A\epsilon_K^2\sum_{i=1}^{n}(1) + B\epsilon_K\sum_{i=1}^{n}\epsilon_{y_1} + C\epsilon_K\sum_{i=1}^{n}\epsilon_{y_2} + D\epsilon_K\sum_{i=1}^{n}\epsilon_{y_3} = \frac{n}{n-1}A\epsilon_K^2$$

Thus

$$\sigma_z^2 = \frac{K^2y_1^2y_2^2}{y_3^2}\left|\frac{50}{49}\frac{\epsilon_K^2}{K^2}\right| + \left|\frac{\sigma_{y_1}^2}{y_1^2}\right| + \left|\frac{\sigma_{y_2}^2}{y_2^2}\right| + \left|\frac{\sigma_{y_3}^2}{y_3^2}\right|$$

Hence $\sigma_z = 10.03 \times 10^{-3}$. The dimensionless variance (fractional error) is

$$\frac{\sigma_z}{z} = 1.187 \times 10^{-3} = 0.187\%$$

Example 7.3.2 A laboratory measurement of current is made by measuring the voltage drop across a precision resistance $R = 1 \pm 10^{-7}$ in which $10^{-7} = \sigma_R$. The following voltage measurements are obtained:

Test No.	1	2	3	4	5
Voltage	3.010	3.001	2.993	2.989	3.007

Determine the error in the current measurement.

Solution Since $E = IR$ we obtain

$$\frac{\epsilon_I}{I} = \frac{\epsilon_E}{E} + \frac{\epsilon_R}{R}$$

and thus

$$\left(\frac{\sigma_I}{I}\right)^2 = \left(\frac{\sigma_E}{E}\right)^2 + \left(\frac{\sigma_R}{R}\right)^2$$

The nominal resistance is 1 ohm and we assume that the average voltage is an estimate of the nominal voltage (or expected value). Thus $E =$ 3.000 V. The statistical computations appear in Table 7.4. Finally

$$\frac{\sigma_I}{I} = \sqrt{\frac{1.05 \times 10^{-4}}{9.00} + \frac{10^{-14}}{I}} = 0.00342$$

With such a high precision resistance, an error of 0.342% in the voltage is propagated as an error of 0.342% in the current measurement.

TABLE 7.4

E_i	$E_i - E$	$(E_i - E)^2$
3.010	0.010	1.00×10^{-4}
3.001	0.001	$.01 \times 10^{-4}$
2.993	−0.007	$.49 \times 10^{-4}$
2.989	−0.011	1.21×10^{-4}
3.007	0.007	$.49 \times 10^{-4}$
$\Sigma E_i = 15.000$	$\Sigma =$ 0.000	$\Sigma = 4.20 \times 10^{-4}$
$E = 3.000$		$\sigma_E^2 = 1.05 \times 10^{-4}$
		$\sigma_E = 1.02 \times 10^{-2}$

7.4 Propagation of Error in Differential and Integral Performance Equations

The simplicity of the theory described in Sections 7.2 and 7.3 does not extend to differential equations. The error cannot be easily estimated unless the integral-differential equation is solved first. Examine the first-order system

$$\tau\frac{dz}{dt} + z = Ky$$

in which τ is uncertain to the extent of $\tau + \epsilon_\tau$ but is not a function of time. Since it is known that

$$z = \frac{Ke^{t/\tau}}{\tau}\int_0^t e^{t'/\tau}y(t')\,dt'$$

it is possible to write

$$z = K\int_0^t \frac{e^{(t-t')/(\tau+\epsilon_\tau)}}{\tau + \epsilon_\tau}y(t')\,dt'$$

and thus

$$\epsilon_z = -\epsilon_\tau\left[\frac{t_z}{\tau^2} + \frac{z}{\tau} - \frac{K}{\tau^2}\int_0^t \frac{t'}{\tau}e^{(t-t')/\tau}y(t')\,dt'\right] \tag{7.4.1}$$

At any instant of time the error depends on time and on the input function. Consider, for example, the case of $y = 1$. We can immediately write

$$z = K(1 - e^{-(t/\tau)}) \tag{7.4.2}$$

and thus

$$\epsilon_z = Ke^{t/\tau}\frac{t}{\tau^2}\epsilon_\tau$$

This last expression can be obtained from Equation (7.4.1) but much more easily from Equation (7.4.2) through a straightforward application of the theory which appears in Section 7.2. As a general rule, the effect of a fixed change in a system parameter is best obtained by first solving the equation and the subsequent application of Section 7.2.

There are instances, however, in which the error is needed in a general form. The method used to obtain (7.4.1) can be extended to obtain the needed expression. Let the impulsive response of the system be given by $I_j(t; a_1, a_2, \ldots, a_n)$ in which $a_1, a_2, \ldots, a_n$ are system parameters and y_j $(j = 1, 2, \ldots, m)$ are the system inputs. Then

$$z = \sum_{j=1}^{m} \int_0^t I_j(t - t', a_1, a_2, \ldots, a_n) y_j \, dt'$$

If the quantities a_i and the inputs a_j all carry error, the function z can be expanded into

$$|\epsilon_z| = \sum_{j=1}^{m} \left[\sum_{i=1}^{n} |\epsilon_{a_i}| \left| \int_0^t \frac{\partial I_j}{\partial a_i} y_i(t') \, dt' \right| + |\epsilon_{y_j}| \left| \int_0^t I_j \, dt' \right| \right] \qquad \textbf{(7.4.3)}$$

The double summation represents the contribution of fixed error in the performance parameters and the single summation represents the contribution of fixed error in the inputs. The single summation also represents the sum of the products of the errors with the unit step responses of the system to the various inputs. It must be noted that (7.4.3) is completely equivalent to (7.2.3), as can be shown given that

$$\sum_{j=1}^{m} \int_0^t \frac{\partial I_j}{\partial a_i} y_i \, dt' = \frac{\partial z}{\partial a_i}$$

Example 7.4.1 Obtain a *general* error expression for the second-order system

$$\frac{d^2 z}{dt^2} + 2\zeta\omega_n^2 \frac{dz}{dt} + \omega_n^2 z = Ky$$

Solution The solution of the equation with zero initial conditions is

$$z = \int_0^t \frac{1}{\omega(1 - \zeta^2)} e^{-\zeta\omega(t-t')} \sin[\omega(t - t')] y(t) \, dt$$

and thus

$$\begin{aligned} |\epsilon_z| = {} & |\epsilon_\zeta| \left| \int_0^t \left\{ \frac{\zeta}{1 - \zeta^2} - \omega(t - t') \right. \right. \\ & \left. \left. \times [1 + \zeta(1 - \zeta^2)^{3/2} \cot \omega(t - t_1)\sqrt{1 - \zeta^2}] \right\} I(t - t') y(t') \, dt' \right| \\ & + |\epsilon_\omega| \left| \int_0^t \left\{ \frac{1}{\omega} + (t - t') \right. \right. \\ & \left. \left. \times [\zeta - \sqrt{1 - \zeta^2} \cot \omega(t - t')\sqrt{1 - \zeta^2}] \right\} I(t - t') y(t - t') \, dt' \right| \\ & + |\epsilon_y| \left| \frac{K}{\omega^2} - \frac{K}{\omega^2} e^{-\zeta\omega t} \right. \\ & \left. \times \left(\cos \omega t\sqrt{1 - \zeta^2} + \frac{\zeta\omega}{\omega\sqrt{1 - \zeta^2}} \sin \omega t\sqrt{1 - \zeta^2} \right) \right| + |\epsilon_K| |z| \end{aligned}$$

This is a fairly messy expression. For a specific function $y(t)$, ϵ_z can be computed from the above result. It would be easier, however, to compute z, then apply (7.2.2) on the result.

7.5 Propagation of Drift

Drift is a time-varying error usually introduced in the signal chain by internal changes in system properties. With algebraic performance equations drift can be handled as another error and treated by the theories of Sections 7.1 and 7.2. In differential performance equations, however, drift may impose formidable difficulties upon the designer, not the least of which is the change of ordinary differential equations with constant coefficients into equations with time-varying coefficients.

7.5.1 Drift changes the performance equation from an expression with constant coefficients to an expression with variable coefficients, or from a solvable performance equation to an equation which is either unsolvable or difficult to solve (see Sections 7.1 and 8.2).

7.5.2 When the system is represented by an algebraic performance equation, the effects of drift can be determined if the drift characteristics are known. Let a system be represented by a nominal performance equation $z_0 = z\,(a, t)$ in which a is a parameter which drifts according to the expression $a = a_0 + \epsilon t$, $\epsilon \ll 1$. Then we can write

$$z - z_o = \frac{\partial z}{\partial a}(\epsilon t) + \frac{1}{2!}\frac{\partial^2 z}{\partial a^2}(\epsilon t)^2 + \cdots \tag{7.5.1}$$

the derivatives to be computed at $a = a_0$. For "small" values of ϵt, the drift error is given by

$$|\epsilon_{zD}| = \left|\frac{\partial z}{\partial a}\epsilon\right| t$$

For larger values of time, pick a value of $t = t_1$ near the time in question so that $a = (a_o + \epsilon t_1) + \epsilon(t - t_1)$ and compute

$$|\epsilon_{zD}| = \left|\frac{\partial z}{\partial a}\epsilon\right| |t - t_1| \tag{7.5.2}$$

with the derivative evaluated at $a = (a_o + \epsilon t_1)$

7.5.3 If the coefficient of the differential equation drifts in the form $a = a_0 + \epsilon t$ in which $\epsilon \lll 1$, an estimated solution can be obtained by assuming that $z = z_o + \epsilon f(t)$ in which $f(t)$ is yet to be determined. For example, let us consider

$$\frac{d^2 z}{dt^2} + a\frac{dz}{dt} + bz = Ky(t)$$

We now substitute and obtain

$$\left[\frac{d^2 z_o}{dt^2} + a\frac{dz_o}{dt} + bz_o - Ky(t)\right] + \epsilon\left[\frac{d^2 f}{dt^2} + a\frac{df}{dt} + bf + t\frac{dz_o}{dt}\right] + \epsilon^2 f = 0$$

Thus, neglecting higher-order terms ($\epsilon^2 \approx 0$) and noting that the equation must be valid for all ϵ, we obtain

$$\begin{aligned} \frac{d^2 z_o}{dt^2} + a\frac{dz}{dt} + bz_o &= Ky(t) \\ \frac{d^2 f}{dt^2} + a\frac{df}{dt} + bf &= -t\frac{dz_o}{dt} \end{aligned} \tag{7.5.3}$$

With this operation we can solve for z_o and then use z_o to find the estimate of z. Since the solution of the second equation determines the error, we can write

$$|\epsilon_z| = |\epsilon_a| \left| \int_o^t I(t - t')t' \frac{dz_o(t')}{dt} dt' \right|$$

in which $I(t)$ is the impulsive response of the nominal system and $z_o(t)$ is the nominal response of the system to the input $y(t)$.

We can generalize the theory to higher-order systems with drifting coefficients. Set

$$\frac{d^m z}{dt^m} + [a_{m-1} + \epsilon_{m-1} g_{m-1}(t)]\frac{d^{m-1} z}{dt^{m-1}} + \cdots + [a_o + \epsilon_o g_o(t)]z = Ky(t) \tag{7.5.4}$$

Let us assume that $z = z_o + \epsilon_o f_o(t) + \epsilon_1 f_1(t) + \cdots + \epsilon_{m-1} f_{m-1}(t)$. Each of the ϵ quantities is assumed to be independent and thus, since upon substitution in (7.5.4) we obtain

$$A_o + A_1\epsilon_1 + A_2\epsilon_2 + \cdots + A_{m-1}\epsilon_{m-1} = 0$$

we must set $A_o = A_1 = A_2 = \ldots = A_{m-1} = 0$ or

$$\begin{aligned} \frac{d^m z_o}{dt_m} + a_{m-1}\frac{d^{m-1} z_o}{dt^{m-1}} + \cdots + a_o z_o &= Ky(t) \\ \frac{d^m f_o}{dt^m} + a_{m-1}\frac{d^{m-1} f_o}{dt^{m-1}} + \cdots + a_o f_o &= -g_0(t) z_o \\ \frac{d^m f_1}{dt^m} + a_{m-1}\frac{d^{m-1} f_1}{dt^{m-1}} + \cdots + a_o f_1 &= -g_1(t)\frac{dz_o}{dt} \\ \frac{d^m f_{m-1}}{dt^m} + a_{m-1}\frac{d^{m-1} f_{m-1}}{dt^{m-1}} + \cdots + a_o f_{m-1} &= -g_{m-1}\frac{d^{m-1} f_{m-1}}{dt^{m-1}} \end{aligned} \tag{7.5.5}$$

With this solvable array, the drift near $t = 0$ is determined by

$$|\epsilon_{zD}| = |\epsilon_o f_o(t) + \epsilon_1 f_1(t) + \cdots + \epsilon_{m-1} f_{m-1}(t)| \tag{7.5.6}$$

Example 7.5.1 Find the initial drift error for the system

$$(\tau + \epsilon_\tau t)\frac{dz}{dt} + z = K$$

with all initial conditions set to zero.

Solution Set $z = z_0 + \epsilon_\tau f(t)$ and obtain

$$\tau \frac{dz_0}{dt} + z_0 = K$$

$$\tau \frac{df}{dt} + f = -t\frac{dz_0}{dt}$$

We know that

$$z_0 = K(1 - e^{-t/\tau})$$

and hence

$$\frac{df}{dt} + \frac{f}{\tau} = -K\frac{t}{\tau^2}e^{-t/\tau}$$

Finally

$$f = -\frac{Kt^2}{2\tau^2}e^{-t/\tau}$$

and

$$|\epsilon_{zD}| = |\epsilon_\tau|\left|\frac{Kt^2}{2\tau^2}\right|e^{-t/\tau}$$

in which

$$|\epsilon_{zD}| \ll |z_0|$$

We note that near zero the exponential is nearly equal to unity, and thus

$$|\epsilon_{zD}|_{t=0} \approx |\epsilon_\tau|\left|\frac{K}{2}\right|\left|\left(\frac{t}{\tau}\right)^2\right|$$

Example 7.5.2 Obtain a second-order approximation for the system described in Example 7.5.2.

Solution Set $z = z_0 + \epsilon_\tau f_1 + \epsilon_\tau^2 f_2 + \ldots$; then the equation becomes

$$(\tau + \epsilon_\tau t)\frac{d}{dt}(z_0 + \epsilon_\tau f_1 + \epsilon_\tau^2 f_2 + \cdots) + z_0 + \epsilon_\tau f_1 + \epsilon_\tau^2 f_2 + \cdots$$

or

$$\left(\tau\frac{dz_0}{dt} + z_0 - K\right) + \epsilon_\tau\left(\frac{df_1}{dt} + f_1 + t\frac{dz_0}{dt}\right) + \epsilon_\tau^2\left(\tau\frac{df_2}{dt} + f_2 + t\frac{df_1}{dt}\right) + \cdots$$

Since ϵ_τ is independent of time (indeed it is at our disposal) the equation can equal zero identically only if each coefficient of ϵ^n is equal to zero. From this we obtain an array of linear ordinary differential equational equations with constant coefficients

$$\tau\frac{dz_0}{dt} + z_0 = K$$

$$\tau\frac{df_i}{dt} + f_i = -t\frac{dz_0}{dt}$$

$$\tau\frac{df_2}{dt} + f_2 = -t\frac{df_1}{dt}$$

$$\tau\frac{df_n}{dt} + f_n = -t\frac{df_{n-1}}{dt}$$

With this array it is possible to compute z to any order of ϵ_τ. In Example 7.5.1 we obtained

$$f_1 = -\frac{K}{2}\left(\frac{t}{\tau}\right)^2 e^{-t/\tau}$$

The second approximation is easily shown to be

$$f_2 = \left[\frac{K}{3}\left(\frac{t}{\tau}\right)^3 - \frac{K}{8}\left(\frac{t}{\tau}\right)^4\right]e^{-t/\tau}$$

The error is

$$|\epsilon_{zD}| = |\epsilon_\tau| K e^{-t/\tau}\left\{\frac{1}{2}\left(\frac{t}{\tau}\right)^2 + |\epsilon_\tau|\left|\left[\frac{1}{3}\left(\frac{t}{\tau}\right)^3 - \frac{1}{8}\left(\frac{t}{\tau}\right)^4\right]\right|\right\}$$

computed for the range of t

$$0 \leq \left|\frac{1}{2}\left(\frac{t}{\tau}\right)^2\right| + |\epsilon_\tau|\left|\frac{1}{2}\left(\frac{t}{\tau}\right)^3 - \frac{1}{8}\left(\frac{t}{\tau}\right)^4\right| \leq 1$$

7.6 Propagation of Noise

Noise propagates like any other error. Unfortunately, the methods discussed before cannot be used directly. Propagation of noise is understood to mean modification of system statistics as the signal is transmitted along the chain. It is necessary, therefore, to transform the performance equations so as to permit the abstraction of statistics.

7.6.1 Noise in a system can usually be treated as input noise, even when the noise is generated by a varying internal component. Consider a first-order system with a noisy time constant

$$[\tau + \epsilon_n(t)]\frac{dz}{dt} + z = Ky(t) \tag{7.6.1}$$

It is always possible to write

$$z = z_0 + z_N = z_0 + \epsilon z_{N_1} + \epsilon^2 z_{N_2} + \cdots \tag{7.6.2}$$

in which z_0 is the ideal response to the noiseless input y and z_N the noise component of the output. If we substitute (7.6.2) into (7.6.1) we obtain

$$\left(\tau\frac{dz_0}{dt} + z_0 - Ky\right) + \epsilon\left(\tau\frac{dz_1}{dt} + z_1 + n(t)\frac{dz_0}{dt}\right)$$
$$+ \epsilon^2\left(\tau\frac{dz_2}{dt} + z_2 + n(t)\frac{dz_1}{dt}\right) + \cdots$$

which implies (since ϵ is an arbitrary small parameter) that

$$\tau\frac{dz_0}{dt} + z_0 = Ky \tag{7.6.3}$$

$$\tau\frac{dz_1}{dt} + z_1 = -n(t)\frac{dz_0}{dt} \tag{7.6.4}$$

$$\tau\frac{dz_2}{dt} + z_2 = -n(t)\frac{dz_1}{dt} \tag{7.6.5}$$

$$\cdots$$

$$\tau\frac{dz_m}{dt} + z_m = -n(t)\frac{dz_{m-1}}{dt} \tag{7.6.6}$$

Equation (7.6.3) represents the nominal (ideal) response, whereas the other equations represent approximations to the noise signal output. Each equation is equivalent to the nominal system subjected to a noise input.

An interesting conclusion can be made from Equations (7.6.3) to (7.6.6): noise occurs in the system only if there exists an input signal or an initial value of the output. This is not unexpected; variations of internal components in this instance cannot generate an output—all that the change of internal parameters can do is to modulate the signal.

Example 7.6.1 A first-order system follows the equation

$$\frac{dz}{dt} + [a + f(t)]z = 0$$

in which a is a constant and $f(t)$ a randomly varying function of time. By direct integration show that output noise occurs only if $z \neq 0$ at $t = 0$.

Solution It is easy to show that if $z = z_0$ at $t = 0$

$$z = z_0 e^{-a(t)} e^{[F(0) - F(t)]}$$

in which $dF/dt = f(t)$ Obviously z exists only if $z_0 \neq 0$. It is also true, in this instance, that the noise will be suppressed in time unless

$$-at + F(0) - F(t) > 0$$

Example 7.6.2 A first-order system is governed by the equation

$$\frac{dz}{dt} + [a + f(t)]z = K$$

in which K and a are constants. Show that if $z = 0$ at $t = 0$ there exists an output noise signal only if $K \neq 0$.

Solution Let $g = at + F(t)$ (see Example 7.6.1); then the solution becomes

$$z = K\int_0^t e^{[g(\tau) - g(t)]}\, d\tau$$

Whatever the form of g, the integral will have a value as can easily be demonstrated by the mean value theorem. Hence $z \neq 0$ implies $K \neq 0$.

7.6.2 Internal noise is occasionally the result of an uncontrollable or unwanted internal energy source.[4] Such an internal source can usually be treated as an input noise source or as a noise added to the output and hence as an input to the subsequent element of the signal chain. An instance which will be discussed in the next chapter is Johnson noise in a resistor.[5] Jŏhnson noise is a randomly fluctuating voltage generated by a heated resistance. The result is not a varying resistance, but an equivalent fixed resistance with a voltage source in series. In any electrical circuit the voltage (or current source) can usually be treated as an input signal. Consider, for example, the analogue circuit in Figures 7.1a and 7.1b.

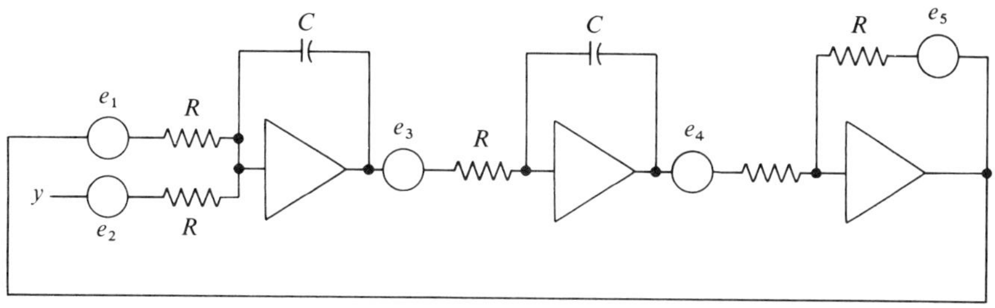

FIGURE 7.1a Second-order analogy with internally generated noise at each resistor

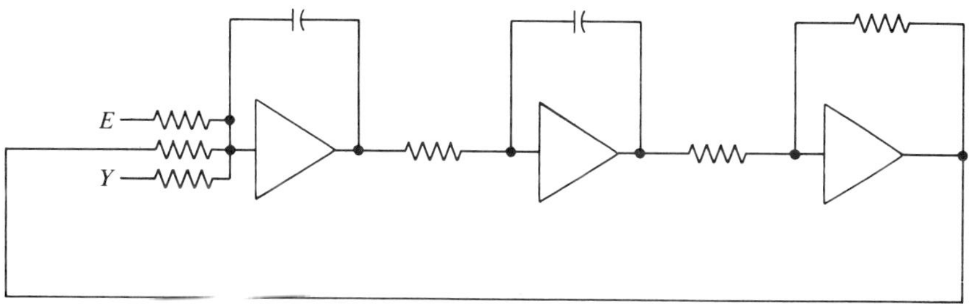

FIGURE 7.1b Equivalent circuit with the internal noise sources replaced by a single external noise source

If the resistors were to produce no noise $e_1 = e_2 = e_3 = e_4 = e_5 = 0$, the nominal performance equation would be

$$\frac{d^2z}{dz^2} + \frac{1}{(RC)^2}z = \frac{1}{(RC)^2}y(t)$$

If all the resistors were to produce noise, the equation would have the form

$$\frac{d^2z}{dt^2} + \frac{1}{(RC)^2}z = \frac{1}{(RC)^2}(y + E)$$

in which E, the equivalent input noise, is equal to

$$E = e_1 + e_2 + RC\frac{de_3}{dt} + (RC)^2\frac{d^2}{dt^2}(e_4 + e_5)$$

In this case of internal noise we can analyze the system as though there were an input signal y and an input noise signal E.

7.6.3 The description of noise as a random input signal still leaves unanswered the question of how noise may be described and manipulated. Detailed deterministic representations of noisy signals are almost out of the question; even a short period representation of a noisy signal is likely to be a rapidly varying function with nearly discontinuous derivatives[6] superposed on the useful signal. It is possible, however, to record a short segment of noisy signal and then to represent the recorded signal segment with a Fourier series.

$$\epsilon_N = y - \bar{y} = \sum_{-\infty}^{+\infty} C'_m e^{j\omega_m t} \tag{7.6.7}$$

where $\omega_m = \dfrac{\pi m}{T}$

$2T =$ time duration of signal

$\bar{y} =$ expected (or nominal) value of signal

$$C'_m = \frac{1}{2\pi}\int_{-T}^{+T} (y - \bar{y})e^{j\omega_m t}\, dt$$

and then to use Equation (7.6.7) as the input signal. If Equation (7.6.7) is introduced into the performance equation, a solution is obtained which is also a Fourier series and which can be examined for propagation, that is, for amplification and attenuation.

The difficulty with such an approach is that there is no guarantee that a short segment of noise recording is typical of the usual noise and that the coefficients C'_m and the frequencies obtained in one recording will match the coefficients and frequencies obtained in another recording. If, on the other hand, a statistically significant number of recordings (or equivalently a recording of great duration) were used, the labor of the Fourier analysis would tax the best modern computer. Luckily, however, the Fourier analysis need not be performed. The theoretical possibility of such an analysis can lead us to a principle for the estimation of the effect of noise on the system.

Suppose that Equation (7.6.7) could be applied to a linear system; then the output could be obtained directly from the theory outlined in Section 5.2. Equation (7.6.7) may be expressed in the form

$$\epsilon_N = \sum_{m=0}^{\infty} C_m \sin(\omega_m t + D_m)$$

Then the output error signal of the system will be

$$\epsilon_z = \sum_{m=0}^{\infty} C_m A(\omega_m) \sin(\omega_m t + D_m + \phi_m) \tag{7.6.8}$$

in which ϕ_m is the phase shift and $A(\omega_m)$ is the sinusoidal ultimate state sensitivity discussed in chapter 5. With Equation (7.6.8) some guidance may be had on the propagation and suppression of noise.

7.6.4 We are interested in the variance or rms error rather than the error itself. We wish to find

$$\lim_{T\to\infty}\frac{1}{2T}\int_{-T}^{+T}\epsilon_N^2\,dt=\sum_{m=0}^{\infty}\sum_{n=0}^{\infty}\lim_{t\to\infty}\frac{1}{2T}\int_{-T}^{+T}C_mC_n\sin\phi_n\sin\phi_m\,dt$$

where $\phi_n=(\omega_n t+D_n)$. If $\phi_n\neq\phi_m$ it is easy enough to show that as $T\longrightarrow\infty$ the integral vanishes. If $\phi_n=\phi_m$ the integral becomes

$$C_m^2\int_{-T}^{T}\sin^2(\omega_m t+D_m)\,dt=\frac{C_m^2}{2}$$

and hence

$$\sigma_\epsilon^2=\sum_{m=0}^{\infty}\frac{C_m^2}{2}$$

The variance of the output will be

$$\sigma_{\epsilon_z}^2=\frac{1}{2}\sum_{m=0}^{\infty}C_m^2A^2(\omega_m) \tag{7.6.9}$$

A similar operation can be performed on the signal. If the ideal, noise-free input signal has a mean value $\bar{y}$, then it will be possible to write

$$\lim_{T\to\infty}\frac{1}{2T}\int_{-T}^{+T}(y-\bar{y})^2\,dt=\frac{1}{2}\sum_{n=0}^{\infty}G_n^2(\omega_n)$$

where it has been assumed that in the interval $(-T, T)$

$$y-\bar{y}=\sum_{n=0}^{\infty}G(\omega_n)\sin(\omega_n t+H_n)$$

The ideal output signal will have a variance of

$$\sigma_{z_0}^2=\frac{1}{2}\sum_{m=0}^{\infty}G_n^2(\omega_n)A_n^2(\omega_n^2)$$

whereupon a signal to noise ratio (N_s) can now be defined as the ratio of the rms input to the rms noise or

$$N_s=\frac{\sigma_{z0}}{\sigma_{\epsilon z}}=\left[\frac{\sum_{n=0}^{\infty}A_n^2(\omega_n)G_n^2(\omega_n)}{\sum_{m=0}^{\infty}A_m^2(\omega_m)C_m^2(\omega_m)}\right]^{1/2} \tag{7.6.10}$$

Example 7.6.3 A first-order system with $\tau=0.01$ and $K=1.0$ is subjccted to a noise input for which

$$C_n=0.03e^{-\omega_n^2}\qquad \omega_n=\frac{n\pi}{T}$$

Determine the variance of the output noise.

Solution With Equation (7.6.9) we obtain

$$\sigma_{\epsilon_z}^2=0.00045\sum_{n=0}^{\infty}\frac{e^{-2\omega_n^2}}{1+(0.01\omega_n)^2}$$

7.6.5 The Fourier series representation permits us to compute derivatives of the noise signal and hence the variance of the derivatives to obtain estimates of equivalent input noise (chap. 8).

The pth derivative of the input noise will be

$$\sum_{n=0}^{\infty} (-1)^{n/2}\omega_n^p C_n \sin(\omega_n t + D_n) \qquad p \text{ even}$$

$$\sum_{n=0}^{\infty} (-1)^{(n-1)/2}\omega_n^p C_n \cos(\omega_n t + D_n) \qquad p \text{ odd}$$

from which it can be shown that

$$\sigma_p^2 = \sum_{n=0}^{\infty} \omega_n^{2p} C_n^2 \tag{7.6.11}$$

Example 7.6.4 The system discussed in Section 7.6.2 is subjected to an input noise for which $C_n^2 = 0.02^2$ and $\omega_n = 0.1, 0.3, 0.3, 0.5, 0.7, 0.9$ radian/second for $n = 0$ to 4 and $C_n = 0$ for all other values of n. Determine the variance of the equivalent noise input if $RC = 1$.

Solution The equivalent input was shown to be

$$E = e_1 + e_2 + RC\frac{de_3}{dt} + (RC)^2\frac{d^2}{dt^2}(e_4 + e_5)$$

It is easy to show that

$$\sigma_{e_1}^2 = \sigma_{e_2}^2 = \frac{1}{2}(0.02)^2$$

$$\sigma_{\dot{e}_3}^2 = \frac{1}{2}(0.02)^2[(0.1)^2 + (0.3)^2 + (0.5)^2 + (0.7)^2 + (0.9)^2]$$

$$\sigma_{\ddot{e}_4} = \sigma_{\ddot{e}_5} = \frac{1}{2}(0.02)^2[(0.1)^4 + (0.3)^4 + (0.5)^4 + (0.7)^4 + (0.9)^4]$$

The square of E has the form

$$E^2 = e_1^2 + e_2^2 + (RC)^2\left(\frac{de_3}{dt}\right)^2 + (RC)^4\left(\frac{d^2e_4}{dt^2}\right)^2$$
$$+ (RC)^4\left(\frac{d^2e_5}{dt^2}\right)^2 + \text{Cross Products}$$

Since the noise from each resistance is produced independently of the other resistances, we assume that the noise signals are uncorrelated and that the means of the cross products are zero. (See Section 6.6.) Thus $\sigma_E^2 = 7.23 \times 10^{-4}$ volts.

7.7 Relationship to Spectral Density

The approximate theory described in Section 7.6 bears a direct relationship to the theory of spectral density discussed in Section 6.7. If

$$y(t) = \sum_{n=0}^{\infty} C_n \sin(\omega_n t + D_n) \tag{7.7.1}$$

we can obtain the correlation function

$$R(\tau) = \sum_{n=0}^{\infty} \frac{C_n^2}{2} \tag{7.7.2}$$

Hence $R(0) = \sigma_y^2$. The spectral density is given by

$$S(\omega) = \frac{\pi}{2} \sum_{n=0}^{\infty} C_n^2 \delta(\omega - |\omega_n|) \tag{7.7.3}$$

Thus

$$\frac{1}{2\pi} \int_{-\infty}^{+\infty} S(\omega)\, d\omega = \sum_{n=0}^{\infty} \frac{C_n^2}{2} = \sigma_y^2 \tag{7.7.4}$$

σ_y^2 represents the integral of the spectral density. We have obtained a practical theorem which allows us to compute the variance if the spectral density is known. This last result will be of great use whenever the spectral density is a continuous function.

PROBLEMS

7.1 Derive Equation (7.1.1) and show that τ becomes $R_0 C$ in the limit as A approaches zero.

7.2 (This is a difficult problem but not impossible.) A galvanometer obeys the usual second-order equation

$$\ddot{z} + 2b\dot{z} + \omega_n^2 z = Kf(t)$$

As a result of external vibration the spring constant ω_n^2 is replaced by

$$\omega_n^2(1 + \epsilon \cos \alpha t)$$

in which ϵ is a small number much less than one. Show that if $f(t) = \cos \alpha t$, the output will not oscillate about zero. (HINT: Expand z in a cosine Fourier series and compare terms.)

7.3 A system obeys the performance equation

$$z = a_1 a_2^2 y_1 + a_3 y_2$$

Set $y_1 = 10$ and $y_2 = 7$. Compute the relative error in z if a_1, a_2 and a_3 each can vary $\pm 6\%$ and $2a_1 = 4a_2 = a_3 = 8$.

7.4 A strain gage with a nominal gage factor of 2.00 is replaced with a similar strain gage in which the gage factor may be in error $\pm 1\%$. What is the estimated output error if the input voltage is also in error 1%?

7.5 A bourdon tube pressure gage can be shown to obey the empirical equation

$$\phi = \frac{0.05}{E}\left(\frac{R}{d}\right)^{0.20}\left(\frac{A}{B}\right)^{0.33}\left(\frac{A}{D}\right)^{3}\sqrt{\Delta_p}$$

in which

ϕ = angular deflection of element tip (degrees)
E = modulus of elasticity of tube material (psi)
R = radius of curved tubing (inches)
d = thickness of tube material (inches)
A = major axis length of tube cross section (inches)
B = minor axis length of tube cross section (inches)

In manufacturing, E can vary $\pm 0.1\%$. R varies $\pm 3\%$, d may change $\pm 0.1\%$, whereas B and A can vary 1%. Determine the error in ϕ which can occur as a result of manufacturing imperfections.

7.6 In an amplified pH measuring system, the output voltage E_0 is related to the input voltage E_i through the equation

$$\frac{E_0}{E_i} = \frac{1}{1 + (1/G)}$$

in which G is the gain of the amplifier. Ideally $E_0 = E_i$. Determine the value of G for which E_0 differs from E_i no more than 0.01%.

7.7 The Alnor Velometer® reads air velocity accurately at the calibration point of 20°C, 50% relative humidity and pressure of 1,000 millibars. For other conditions the following equation is used

$$V = V_R\left(\frac{\rho_0}{\rho}\right)^{1/2}$$

in which

V = actual velocity (m/s)
V_R = velocity read on device (m/s)
ρ = density of air under measurement conditions (K/m^3)
ρ_0 = density of air at calibration point (K/m^3)

Determine the actual error in the reading V_R in m/s if $V = 2$ m/s and ρ differs 1% from ρ_0.

7.8 The Ostvald Viscosimeter is a capillary viscosimeter in which the viscosity of a liquid is measured by comparing the rate of flow of a fluid in a tube to the rate of flow of a known fluid. The viscosity is computed by

$$\nu_1 = \nu_0 \frac{t_1}{t_0}$$

in which ν_1 is the wanted kinematic viscosity, ν_0 the kinematic viscosity of the reference liquid, and t_1 and t_0 are the times for standard volumes of the two fluids to pass through the device. It is known that the absolute viscosity of water is given by

$$\frac{1}{\mu} = 2.148[(\theta - 8.43) - \sqrt{8078.4 + (\theta - 8.435)^2}] - 120$$

in which μ is in poises and θ is the temperature in degrees Celsius. Determine the relative error in the measurement t_1 if a test for the viscosity of water were conducted at an erroneous temperature of 20.1°C instead of 20°C.

7.9 In problem 7.6 the quantity $1/G$ can be considered an error ϵ. Under what circumstances is the expression

$$\frac{E_0}{E_i} = a_0 + a_1\epsilon + a_2\epsilon^2 + \cdots \qquad \text{(a)}$$

valid? Generate an expansion for the case in which the expansion (a) fails to converge.

7.10 A device carries out the operation

$$z = Ky_1^{n_1}y_2^{n_2}$$

Both n_1 and n_2 can be in error. Obtain an expression for the relative error propagated to the output.

7.11 The output of a pressure transducer is connected to a multiplier so as to be multiplied by a function which can be in error 0.35%. The product is introduced into a square root extractor and then recorded with a strip chart recorder. Determine the error propagated to the output.

7.12 A system obeys the performance equation

$$z = f(a_1, a_2, a_3, a_4; y)$$

The following information is known

$$a_1 = 1.03 \qquad a_2 = 10.1 \qquad a_3 = 107 \qquad a_4 = 14.2$$

$$\frac{\partial f}{\partial a_1} = 3.71 \qquad \frac{\partial f}{\partial a_2} = 3.79 \qquad \frac{\partial f}{\partial a_3} = 11.1 \qquad \frac{\partial f}{\partial a_4} = 7.91$$

and the following measurements are made of the parameters a_i:

Test No.	a_1	a_2	a_3	a_4
1	1.03	10.3	107	14.0
2	0.98	10.7	106	14.2
3	1.10	11.1	105	13.7
4	1.01	11.0	107	14.2
5	0.92	9.72	109	13.1
6	0.97	10.3	108	13.1
7	0.88	9.61	108	13.9
8	1.02	9.64	104	14.5
9	0.97	10.1	106	13.7
10	1.04	9.91	109	14.5

Determine the average values of the a_i and compare with the nominal values. Is there a systemic error? Using the nominal values of a_i, compute the variance of the output error. Show that this differs from the variance computed from the average values.

7.13 Let a system parameter be subjected to a systemic error and to a random error. Show that if the errors are small, the output variance is given by

$$\sigma_{\epsilon_z}^2 = A\alpha^2 + B\sigma_{\epsilon_a}^2$$

in which α is the systemic error and $\sigma_{\epsilon_a}^2$ is the variance of the difference of the parameter a from its expected value.

7.14 A flow meter obeys the equation

$$Q = K\sqrt{\Delta p}\,(\text{m}^3/\text{s})$$

It is known that $\sigma_{\epsilon_k}^2 = 4 \times 10^{-6}$ and that $K = 0.027$. The following set of measurements has been recorded for a fixed Q:

Test No.	1	2	3	4	5	6	7
Δp	30,000	24,700	29,900	31,500	29,900	30,300	30,700

Estimate Q and the variance of the errors in Q.

7.15 By direct solution obtain an expression for the error of a second-order system in which ω_n^2 can be in error but in which all other parameters and the input are perfect. If the error in ω^2 is $\pm 3\%$ and $\zeta = 0.3$, $K = 1$ and $y = 1$, compute the output error.

7.16 A third-order system is represented by the transform

$$\frac{\mathcal{L}(z)}{\mathcal{L}(y)} = \frac{K}{(s + a_1)(s + a_2)(s + a_3)}$$

Let $a_1 = 2.14$, $a_2 = 0.351$, $a_3 = 0.972$ and let each parameter be subjected to an error of 0.8%. Use Equation (7.4.3) to determine the output error if the input is $y = 1$.

7.17 Carry out problem 7.16 by direct integration and the methods of Section 7.2.

7.18 In precision current measurement, a voltage drop is measured across a precision resistor and the current computed by Ohm's law. If the resistance of 1 ohm drifts 0.0023 ohm per hour, determine the drift in the computed current.

7.19 In a second-order system $\omega = 81$ and $\zeta = 0.707$. If the quantity ω drifts according to the equation

$$\omega = 81 + 0.00006t$$

compute the output drift error of the lowest order (small t).

7.20 Let the third-order system in problem 7.16 be subjected to drift according to the rules

$$a_1 = 2.14 + 0.0005t$$
$$a_2 = 0.351 + 0.0005t$$
$$a_3 = 0.972 + 0.0005t$$

Let the input be $y = 0$ and let the following initial conditions exist: $z(\text{o}) = 1.21$, $\dot{z}(\text{o}) = 0$. Express the system in the form of Equation (7.5.4) and evaluate the drift error of the output in the neighborhood of $t = 0$.

7.21 Obtain a second-order estimate to the system described in problem 7.19.

7.22 A second-order system is subjected to internal noise

$$\frac{d^2z}{dt^2} + [a_1 + \epsilon_1 n_1(t)]\frac{dz}{dt} + [a_2 + \epsilon_2 n_2(t)]z = Ky(t)$$

Obtain the family of equivalent equations needed for the estimation of the effects of noise.

7.23 In Equation (7.6.1) let $\epsilon = 0.01$ and

$$n(t) = u(t) - u(t - 0.4) + 0.7u(t - 0.4)$$
$$- 0.9u(t - 0.5) + 0.2u(t - 0.6)$$

Let $y = 0$ and let $z(\text{o}) = 1$. If $\tau = 0.01$ determine the noise output signal from time $t = 0$ to $t = 1$ second to the order of ϵ^2.

7.24 Show that the noisy circuit in Figure 7.1a is equivalent to the noiseless circuit with an equivalent noise input. HINT: Note that the performance equation of the first integrator is

$$\text{OUTPUT} = -\frac{1}{RC}\int_{0+}^{t} (y + e_2 + e_1 + z)\, dt$$

and that the inverter follows the performance equation

$$-(\text{INPUT} + e_4) = (\text{OUTPUT} + e_5)$$

7.25 A second-order system, $\omega = 20$, $\zeta = 0.5$, $K = 3$, is subjected to a signal

$$y = 3 + 0.5 \sin 20t$$

and to a noise input signal

$$n(t) = 0.21 \sin t + 0.17 \sin 3t + 0.05 \sin 5t + 0.01 \sin 7t$$

Compute the signal-to-noise ratio.

7.26 A first-order system, $\tau = 0.01$, $K = 2.5$, is subjected to the input signal

$$y = 4 \cos 2t + 2 \cos 4t$$

and a noise signal for which $\omega_m = 2.4(2m + 1)$ and

$$C_m = \frac{m}{(9 - m)!} \qquad m = 1, 3, 7, 5, 9$$

Compute the signal-to-noise ratio.

7.27 Compute the variance of the derivative of the noise signal given in problem 7.25.

7.28 Compute the variance of the second derivative of the noise signal given in problem 7.26.

NOTES

[1]The principles of professional self-defense and the laws of libel being what they are, it is best to give no references here. The reader can find examples of naiveté without the author's help.

[2]Ref. 94. Table 7.4 includes some modifications since some of McKenzie's expressions do not seem to be complete.

[3]It is best to calibrate each design.

[4]See chapter 8.

[5]Ref. 107, 108, 109.

[6]The phrase "nearly discontinuous" is rather imprecise. If n is noisy, then dn/dt consists of a set of continuous functions pieced together with randomly distributed very large slopes, i.e., $|d^2n/dt^2| > M > 0$ where M is a very large number. In the limit, as M approaches infinity, dn/dt becomes piecewise continuous with discontinuities on a set of measure zero. If the quantities $|d^2n/dt^2|$ are too great for the instrument to follow, the input signal can be replaced by a discontinuous input without loss of generality. Such signals with continuous derivatives which appear as discontinuous to the user system are said to have nearly discontinuous derivatives.

8

ERROR IN PRACTICE

8.1 Sources of Error

The designer of instruments finds himself in the situation of optimizing an instrument with respect to an unknown form of error in an unknown location. Ideally, one should design the instrument to avoid all sources of error known to man: practically, such a super-safe approach is restrictive on measurement function and wasteful of time and materials. The designer is forced by circumstances to design systems for the avoidance of the known error inherent in measurement and then to redesign the instrument after additional error is discovered in the prototype.

The experience of other engineers and scientists is a valuable, though confused, source of information on error elimination. The instrumentation literature is about equally divided into articles which describe new instruments and articles which discuss design practice or minimization of error. Some journals, for example, *The Review of Scientific Instruments*, specialize in the description of the theory and function of individual instruments with moderate discussion of error and the sources thereof.[1] On the other hand, engineering journals, such as

Instrumentation Technology, are inclined to publish articles on general instrument classes and the design problems involved therein. The distinction between "scientific" instrumentation and "engineering" instrumentation is elusive. The scientist will agonize over his instrument, adjusting, changing, modifying, praying and cursing until the one perfect measurement can be made. Experimental physicists are often exquisite instrument-makers with infinite patience and seemingly infinite time. Robert Millikan had his first inkling of a method for the determination of the charge of the electron in 1906. He spent three years refining his procedure and was able to give some crude answers in 1909. The idea for the "perfect" experiment came to him late in 1909—by 1912 he had data good enough to convince the world that he had obtained experimental proof of the atomic theory of matter.[2]

Engineers do not have the time granted to physicists, even though academic research engineers have been known to undertake measurement problems which may require a decade to solve.[3] Usually the engineering measurement "lead-time" can be no more than a few months to a year—a customer can rarely wait ten years for a solution to an individual problem. The answer must be obtained with known components and the error must be estimated in advance.

The point to be noted is that scientific experimentation and engineering design are not the result of a single intellectual exercise, but rather the result of many careful, small essays with novelty and many false steps. Design is an accumulative learning experience; each design error restricts the options and widens the knowledge available to the designers. When all options but one are removed, a final design occurs even though the specification may not have been met.

In the elimination of error from an instrument the steps to be followed are reasonably clear:

1.0 The available signals should be examined for meaning and for noise or unwanted signals.
   - 1.1 The character of the signal and of the noise should be studied to determine the basic difference between them. If there is no discernible difference between the signal and the noise there can be no way to separate them.
   - 1.2 The signal-to-noise ratio should be found.

2.0 A number of possible solutions to the measurement problem should be studied for capacity to discriminate the signal from the noise. Each solution should be optimized to give the best signal-to-noise ratio at the output.

3.0 Each of the possible solutions should be examined for possible instrument error.
   - 3.1 The elements of the instruments should be isolated to determine the possible causes of systemic errors.
   - 3.2 The elements of the instruments should be studied to determine the possible causes of internally generated noise.

3.3. The instruments should be studied to determine the possibility of drift and performance changes.

4.0 A judgment is now made, largely personal and intuitive, as to which choice is best to meet system performance specifications. In general, the designer picks the most promising proposal and studies it for possible improvement or combination with other systems.

5.0 A model is built and tested. The results of the tests are used as an additional information input for a repetition of the design procedure. The repetition is usually complete. Tests often show that the characteristics of the signal out of the input noise were imperfectly understood.

At the end of the first "go-around" a curious psychological hangup occurs. Instead of repeating the design procedure, the engineer becomes enamored with his original work and spends his time trying to make the child of his first love function properly. Such intellectual trauma is rather ordinary and is a compound of arrogance, laziness and the inability to admit mistakes. Engineers and scientists should protect themselves from the "final design syndrome" and should always try to answer the question, "Where did we go wrong?" rather than try to answer the related question, "Can we fix our gadget?" The answer to the latter question is a result of the answer to the former, and design improvement is always achieved by someone who considers the basic mistakes made in the initial work.

8.2 Drift and Its Compensation

The author does not wish to imply that elimination of measurement error becomes simple whenever the proper state of grace is achieved by the experimenter. Error, performance, specifications and system selection are so involved with each other that no person can understand the effects of even a single change on the totality. Consider, for example, the simple *RC* circuit (Fig. 8.1) with a thermally drifting resistance. The

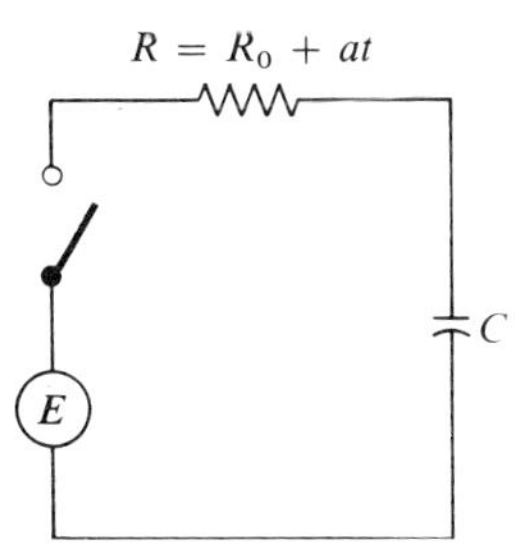

FIGURE 8.1 Simple RC series circuit with drifting resistance

equation for the circuit charge (q) is

$$RC\frac{dq}{dt} + q = CE \tag{8.2.1}$$

If R drifts linearly with time in the form

$$R = R_0 + At$$

it is easy to see that the solution of Equation (8.2.1), with $q(0^+) = 0$ is

$$\frac{q}{EC} = \left[1 - \left(\frac{R_0}{R_0 + At}\right)\right]^{1/AC} \tag{8.2.2}$$

If A approaches zero, that is, if the system approaches the ideal or nominal performance, Equation (8.2.2) becomes

$$\frac{q}{CE} = 1 - e^{-t/R_0C} \tag{8.2.3}$$

as can be established with the aid of the well-known theorem

$$\lim_{A\to 0}\left(\frac{1}{1 + (At/R_0)}\right)^{1/A} = e^{-t/R_0}$$

The basic difference between the stable system and the drifting system can be demonstrated by a typical numerical example. Let us choose $R_0 = 100$, $A = 100$, $C = 10^{-2}$ and plot the result in Figure 8.2.

Obviously the net effect is to change the time constant τ. Since τ is the time at which 63.21% of the transient has vanished, it can be seen that in the drifting system

$$\tau = \frac{R_0}{A}(0.3679^{-AC} - 1)$$

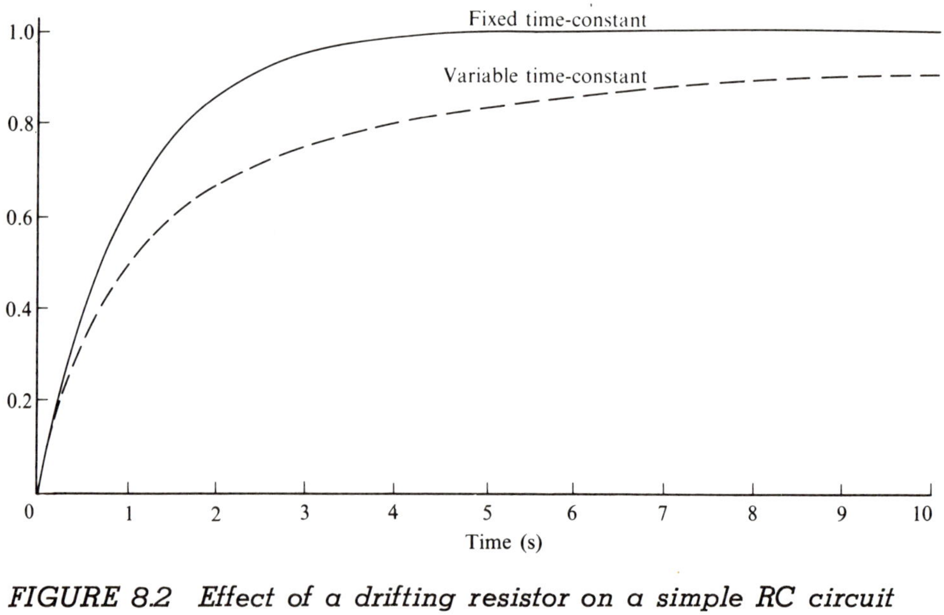

FIGURE 8.2 Effect of a drifting resistor on a simple RC circuit

which in the limit as $A \longrightarrow 0$ becomes the nominal $\tau = R_0 C$. In Figure 8.2 the nominal time lag is 1 second, whereas the actual time lag is 1.72 seconds.

The effects of the change of resistance have been determined. How shall this error be corrected? The temptation here is to buy a better resistor or to supply some method for the maintenance of the resistance near R_0. Both solutions will clear the problem after a fashion but may be rather expensive. It is best, as mentioned before, to determine the reason for the thermal drift and then to take appropriate measures.

The resistance of a material changes with temperature according to the basic equation[4]

$$R = R_0(1 + \alpha\theta + \alpha_1\theta^2 + \cdots) \tag{8.2.4}$$

where R is the resistance at any temperature θ measured above a reference temperature θ_0 for which the resistance is R_0. Drift can occur if the temperature changes. The question therefore is to determine how the temperature changes:

(1) The resistance is immersed in an environment in which the temperature changes.
(2) The resistance cannot dissipate the energy lost throughout its length.

In the first instance, correction of the drift can be accomplished by compensation or by making the heat capacity of the resistor so large that the temperature varies little and varies slowly. In the second instance, improvement of the heat transfer characteristics of the resistor is one solution; reduction of the current in the circuit, and hence of the power lost, $P_W = i^2R$, is another solution.

8.2.1 The concept of performance equation modification for the compensation of error can be formalized.

Definition 8.2.1 IF AN ERROR-PRODUCING ELEMENT IN AN INSTRUMENT IS CONNECTED TO A DEVICE (A) WHICH PRODUCES A SIGNAL EQUAL TO THE ORIGINAL ERROR IN SUCH A FASHION THAT THE ERRORS CANCEL, THEN THE ERROR IS SAID TO BE COMPENSATED AND THE DEVICE A IS CALLED A COMPENSATOR.

A simple illustration is afforded by the compensation of thermal drift in the resistor discussed in Section 8.1.1. The quantity α in Equation (8.2.4) is of the order of 4×10^{-3}, whereas α_1 is of the order of 10^{-8} and may be neglected in the first design.

Compensation is achieved by means of a device called a *thermistor*, a ceramic resistor with the property of decreasing resistance with increasing temperature.[5]

$$R_T = R_{T0}(1 - \beta\theta + \cdots) \tag{8.2.5}$$

For small variations in temperature the thermistor can be placed in series with the resistance to give an equivalent resistance.

$$R_e = R_o + R_{T0} + (\alpha R_o - \beta R_{T0})$$

Let $R = R_e$ be the resistance required in the circuit, and let

$$R_{To} = \frac{\alpha}{\beta} R_o$$

Thus

$$R_o + R_{T0} = R_o\left(1 + \frac{\alpha}{\beta}\right) = R \tag{8.2.6}$$

If we now select components with

$$R_o = \frac{R}{1 + \alpha/\beta}$$

$$R_{T0} = \frac{R}{1 + \beta/\alpha}$$

we shall have the resistance of constant value R in the circuit as long as the temperature remains in the range of validity of Equations (8.2.4) and (8.2.5).

Another simple example is the procedure for the temperature compensation of a resistance strain gage bridge (see Section 3.5.2). If the input impedance of the strain gage amplifier is extremely high, the output voltage e_o of the bridge (Fig. 8.3) is given by

$$e_o = \frac{(R_1 R_4 - R_2 R_3)e_1}{(R_1 + R_2)(R_3 + R_4)} \tag{8.2.7}$$

in which e_1 is the constant excitation voltage. Let R_1 be the active gage $R_1 = R + \delta + \epsilon_T$. Let all other resistances be identical inactive strain gages with $R_2 = R_3 = R_4 = R + \epsilon_T$. The quantity δ is the small change in resistance induced by change of length in the active gage, and ϵ_T is the change in resistance induced in all the gages by changes of temperature in the environment. In most strain gages, ϵ_T is of the order of δ and is considered small. Expansion of Equation (8.2.7) by the binomial theorem gives

$$\frac{e_o}{e_1} = \frac{\delta}{4R} - \left(\frac{\delta\epsilon + \delta^2}{R^2}\right) + \cdots \tag{8.2.8}$$

Obviously the thermal error $\delta\epsilon/R^2$ is of the same order as the nonlinearity error δ^2/R^2. The error is of order 10^{-8} as compared to $\delta/4R$ which is of order 10^{-4}. With $e_1 = 10$ volts and the amplifier gain $K = 100$, the useful signal from the amplifier will be of the order 0.1 volt whereas the error will be of the order 0.00001 volt. For many measurement purposes the thermal drift is "fully compensated." Comparison with the uncompensated system should convince the reader that immersing the inactive gages in the same thermal environment as the active gages is a method of compensation. It is common to place a single gage at the active measurement point and to place the three bridge resistances, not usually gages but some less expensive components, in the neighborhood of the output instrumentation. Then the active gage will change with temperature an amount ϵ_{T1} and the other gages an amount ϵ_{T2}.

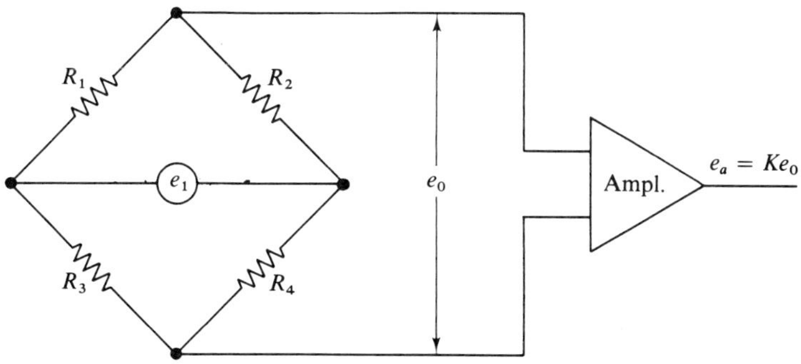

FIGURE 8.3 Full resistance bridge connected to a high impedance amplifier

Instead of Equation (8.2.8) we obtain

$$\frac{e_o}{e_1} = \frac{\delta}{4R} + \frac{\epsilon_{T1} - \epsilon_{T2}}{4R} + \text{higher-order terms} \qquad \textbf{(8.2.9)}$$

If the active gage temperature were to increase and the bridge temperature to decrease or remain unchanged, the error would be of the same order as the useful signal.

8.2.2 Thermal drift is common in mechanical and hydraulic instruments as well as in electrical instruments. It is a reasonable design rule to assume that all properties of information chain elements and components are affected by temperature change and that all instruments should be checked for thermal drift. A well-known example is the bourdon tube, the most common fluid pressure-measuring device.[6] A flat elastic tube is formed in a semicircle with one end sealed and one end open to the working fluid, the pressure of which is to be measured (Fig. 8.4). Increase of pressure in the working fluid tends to circularize the tube cross section which in turn tends to straighten out the semicircle. The

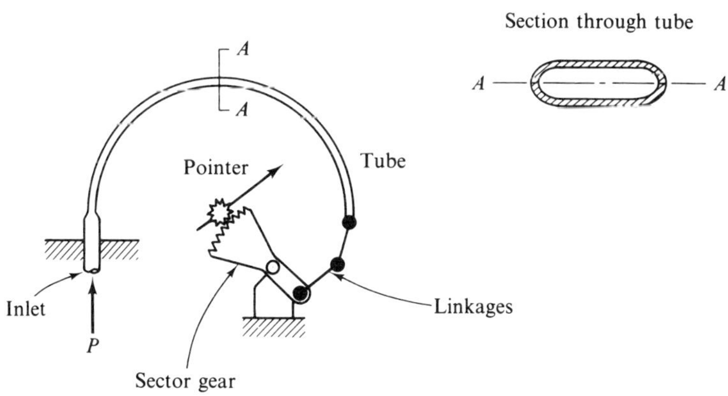

FIGURE 8.4 Bourdon tube pressure gage

motion of the tube tip is transmitted to a pointer through linkages and a sector-gear-pinion combination. The performance equation for tip motion is a function of the shape and material of the tube as well as of the difference in pressure between the interior of the tube and the exterior.[7]

Two types of drift can be discovered:

(1) Thermal Drift: the physical dimensions of the tube change with temperature.

(2) Barometric Drift: for a given measurement (p) the bourdon indication is proportional to $p - p_A(t)$ in which $p_A(t)$ is the time variable ambient pressure (occasionally called *case* pressure).

Both forms of drift can be compensated by introducing another bourdon tube in all ways identical to the active element but not connected to the working fluid (Fig. 8.5). The compensator is attached to the working bourdon so as to rotate the indicator needle in the opposite direction to the active element. Since the compensator responds only to thermal drift, the effect is to compensate the error.

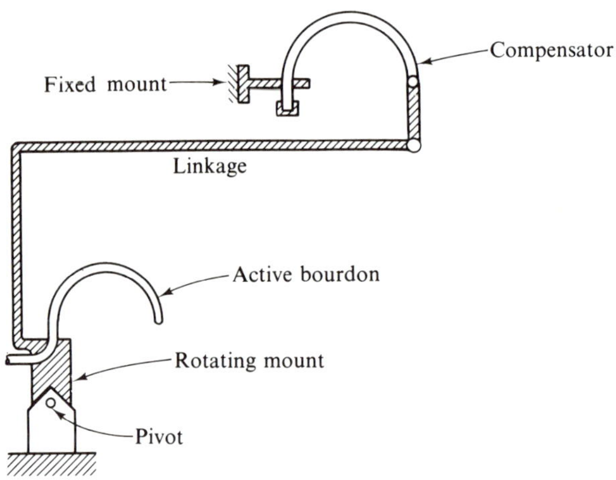

FIGURE 8.5 Bourdon tube compensated by an inactive gage

8.2.3 In some instances thermistors exhibit an irreversible change of response as a function of time. The thermistor is known to obey the law[8]

$$R = R_0 e^{B(1/\theta - 1/\theta_0)} \tag{8.2.10}$$

in which R_0 is the resistance at temperature θ_0 and B is a parameter related to the particular type of thermistor (θ is in absolute degrees).*

---

*Equations (8.2.10) and (8.2.5) correspond to each other for a small variation in temperature. Let $\theta = \theta_0 + \theta_1$ in which $\theta_1 < \theta_0$. Then Equation (8.2.10) can be expanded into the infinite series

$$R = R_0 - \left(\frac{B}{\theta_0^2}\right)\theta + \cdots$$

This last equation is equivalent to Equation (8.2.5).

If the thermistor is installed in a bridge with inactive legs of resistance R_0, the voltage presented to the amplifier will be (Fig. 8.3)

$$e_0 = \left\{\frac{1 - e^{B(1/\theta - 1/\theta_0)}}{2[1 + e^{B(1/\theta - 1/\theta_0)}]}\right\} e_1 \tag{8.2.11}$$

The response is nonlinear for wide variations in θ. For small changes of temperature above the reference temperature, the response is close enough to a straight line to prevent scale crowding (Fig. 8.6). The quantity B is itself a function of temperature.

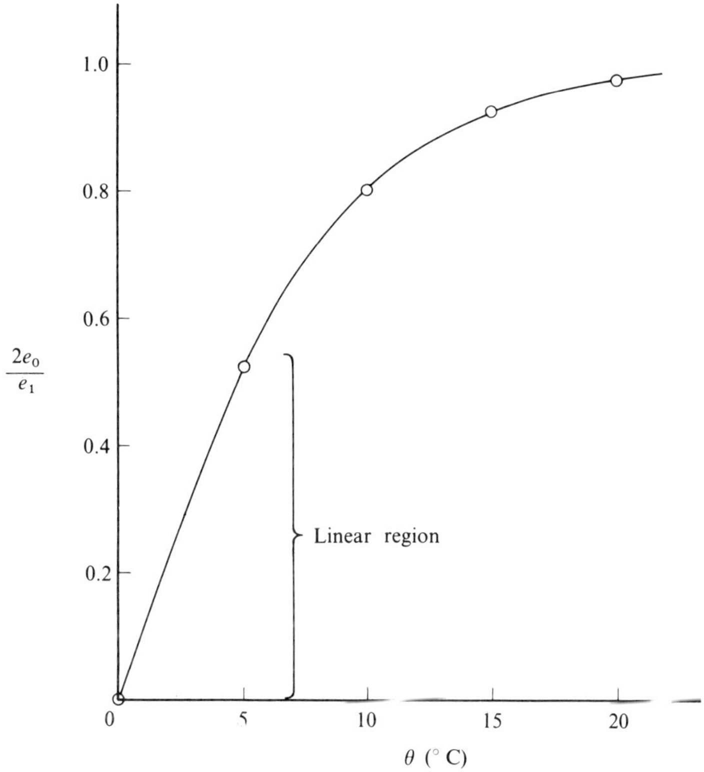

FIGURE 8.6 Behavior of a thermistor with B = 16,000 in the neighborhood of 0°C

When a thermistor is new, the quanity B changes slowly with time, approaching a steady ultimate state after many months of use. The long-term drift propagates through the instrument. Let $B = B_0 + \epsilon_B$ where B_0 is the nominal value and ϵ_B the drift; then

$$|\epsilon| = \left|\frac{\partial e_0}{\partial B}\right|\left|\frac{dB}{dt}\Delta t\right| = \left|\frac{\partial e_0}{\partial B}\right||\epsilon_B|$$

$$\left|\frac{\epsilon}{e_0}\right| = \left|\frac{1/\theta_0 - 1/\theta}{\sinh(1/\theta_0 - 1/\theta_0)B}\right|\left|\frac{\epsilon_B}{B}\right| \tag{8.2.12}$$

The quantity ϵ_B/B is 0.2% per year and thus $|\epsilon/e_0| < 0.2\%$ year (Fig. 8.7). Unless we were interested in high-precision thermometry, the drift could be neglected.

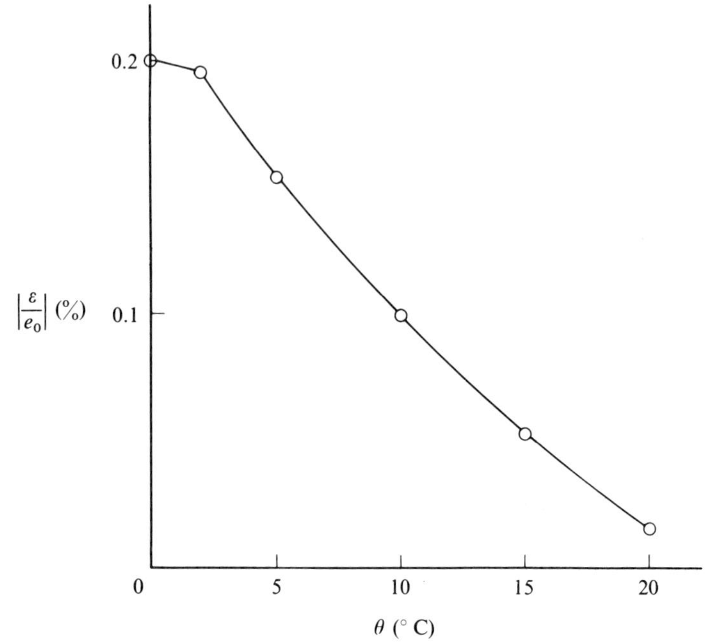

FIGURE 8.7 Output drift per year in a thermistor circuit

8.2.4 The direct current electronic amplifier drifts continuously. (Some typical numbers are given in Table 8.1.) The drift is not cumulative, however. The input voltage error wanders slowly within its range between positive and negative values. The long-time average of the drift is near zero. An unstabilized dc operational amplifier may drift 20 μV in one hour but, at worst, drift 25 μV from the nominal input value in 24 hours. The causes of amplifier drift are various and ill defined. Thermal drift, aging, changes in bias and supply voltage all contribute to drift. Proper design keeps drift within the limits of Table 8.1. Stabilization, a form of compensation,[9] can reduce drift to less than 2 μV per hour with a long-term drift at the input of less than 1 μV per day. Chopper stabilized instrument and computing amplifier with less than 1 μV/day drift can be purchased "off-the-shelf."

TABLE 8.1 Typical dc amplifier drift (input voltage)

Vacumn tube dc amplifier		60 μV/h
Stabilized VT dc amplifier		10 μV/h
Solid state dc amplifier	Maximum	50 μV/h
	Minimum	2 μV/h

The use of an amplifier can sometimes create drift, even though the basic amplifier is nearly driftless. Consider, for example, the use of high-gain amplifier as an integrator (Fig. 8.8). The performance equation for the circuit is

$$(1 - A)RC\frac{de_0}{dt} + e_0 = Ae_i \tag{8.2.13}$$

If $A \longrightarrow \infty$, Equation (8.2.13) reduces to

$$RC\frac{de_0}{dt} = -e_i$$

and thus

$$e_0(t) - e_0(0) = -\frac{1}{RC}\int_0^t e_i\, dt \tag{8.2.14}$$

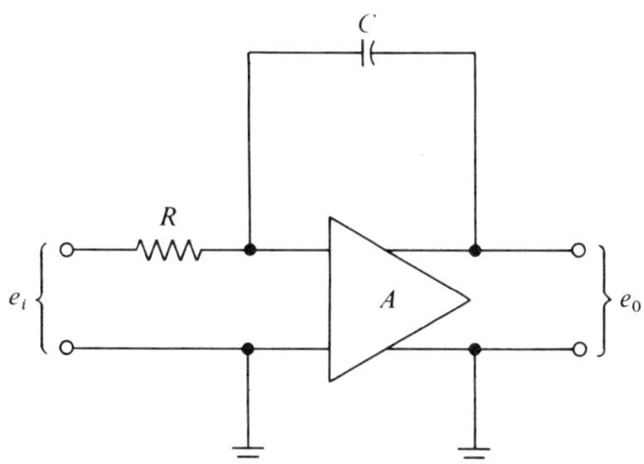

FIGURE 8.8 The elementary integrator

There is no such device as an amplifier with infinite open-loop gain, and, in effect, Equation (8.2.14) should be

$$e_0(t) - e_0(0) = -\frac{1}{[1 - (1/A)]RC}\int_0^t e_i\, dt' - \frac{1}{[1 - (1/A)]RC}$$

$$\int_0^t e_i \sum_{m=1}^{\infty}\left[\frac{t' - t}{(1 - A)RC/m!}\right]^m dt' \tag{8.2.15}$$

The second term is a drift error. Consider the simple case of integration of the constant $e_i = 1$ with the initial condition $e_0(0) = 0$. Then

$$e_0(t) = \frac{-t}{[1 - (1/A)]RC} - A\sum_{m=1}^{\infty}\left[\frac{t}{(1 - A)RC}\right]^m$$

The term multiplied by A represents drift.

Example 8.2.1 Let $RC = 1$ and $A = 10^4$. At what time will the error be 1.0% if $e_i = 1$?

Solution

$$\begin{aligned} e_0 &= -A(1 - e^{t/(A-1)RC}) \\ &= [-10^4(1 - e^{t/10^4}) + t] - t \end{aligned}$$

The time is obtained by setting

$$\left|\frac{t_1 - 10^4(1 - e^{t/10^4})}{t_1}\right| = 0.01$$

The answer is $t_1 = 198.7$ seconds. In less than 3.50 minutes the integrator shows significant error. In ten minutes the error will be 3.33%. In 2.8 hours the error will be 100%.

8.3 Systemic Error

Most of the errors discussed up to now are the result of inherent qualities of nature and tend to exist in all instruments. Systemic errors, on the other hand, are always the result of human failure and must be corrected by redesign.

8.3.1 Systemic error can occur as a result of one or more improprieties:

(1) improper design of instrument,
(2) improper manufacture of instrument and
(3) improper use of instrument.

Improper design arises from

(1) gross error in computation,
(2) neglect of nonlinearities,
(3) neglect of shop practice and
(4) neglect or ignorance of a relevant input.

Improper manufacturing changes the performance equations and may be a true manufacturing failure or may be the design of an engineer who is unfamiliar with the limits of manufacturing accuracy.

The improper use of an instrument cannot be summarized in a few words. There is no end to the dumb-headed ingenuity of the laboratory worker who fails to read or follow the instructions. The principal causes of systemic error through improper use are listed below:

(1) failure to level or balance a mechanical system
(2) failure to ventilate an electrical system
(3) use of instrument in an extreme environment
(4) use of instrument on measurements outside the design range
(5) failure to maintain and clean the instrument
(6) failure to calibrate at proper intervals
(7) failure to null or balance between measurements
(8) failure to apply the proper correction to the measurements
(9) failure to reproduce the specified conditions for a specific measurement

8.3.2 A classical simple example of systemic error which can be of manufacturing origin is that of the error-laden screw micrometer.[10] This device consists of a fine screw connected to a tapped frame (Fig. 8.9). Each turn of the screw corresponds to linear movement of the screw

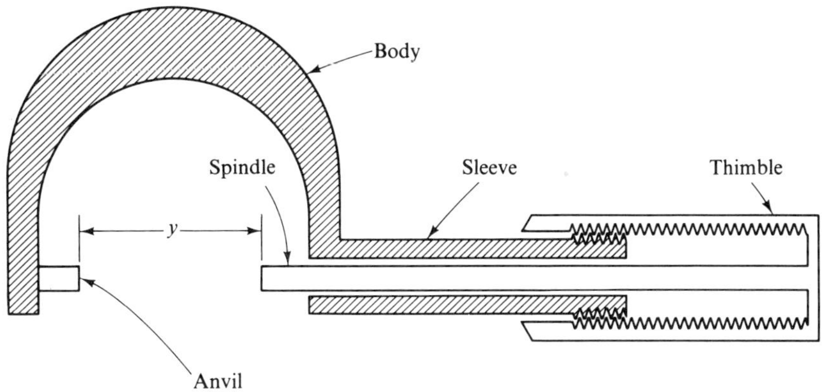

FIGURE 8.9 Simplified section through a screw micrometer

end. A vernier scale, mounted circumferentially on the screw and on a sleeve fixed to the frame, is calibrated in units related to the distance between the end of the screw and the limit stop anvil. If the screw is perfect, a change of position z of the vernier should correspond exactly to a change of position y along the axis of the screw. The performance equation will then be

$$z = ky \tag{8.3.1}$$

in which k is a constant related to the lead or to the pitch of the screw.[11]

Systemic error can be introduced in the following ways:

(1) The anvil and the end of the screw do not come together when the vernier reads zero. This can be the result of poor initial adjustment or of wear or damage.

(2) The screw can have a uniformly wrong pitch. Although such an error could conceivably be the result of the use of improperly set screw-cutting equipment, it is most likely to be the result of manufacture and calibration of the instrument at a temperature much different from the temperature for which it is designed.

(3) The screw and the thread in the body can be loose with respect to each other and the screw is at an angle to the line of advance. Such an error can occur through wear and can be considered a form of drift. If the screw becomes very loose, the alignment will be different with every measurement and will generate a random error.

Let us consider the case of a micrometer with a zero-offset error and a uniformly wrong-pitch screw. Let the gain equal k_1 mm/mm and let the zero error be z_0. Then the performance equation will be

$$z_E = k_1 y + z_0 \tag{8.3.2}$$

The output error will be

$$|\epsilon| = |z_E - z| = |(k_1 - k) + z_0|$$

Occasionally a micrometer is found in practice which appears to be untrue. On-the-spot calibration can be accomplished by measurement of two precisely manufactured length standards such as Johansen blocks. To the two measurements, y_1 and y_2 will correspond two measurements z_1 and z_2. The performance parameters in Equation (8.3.2) are then found by the obvious equations

$$k_1 = \frac{z_1 - z_2}{y_1 - y_2}$$

$$z_0 = \frac{z_2 y_1 - z_1 y_2}{y_1 - y_2}$$

The resulting equation can be used to compute y from our measurement z.[12]

Modern micrometers purchased from reliable manufacturers are not likely to be burdened with design error. Equation (8.3.2) can nevertheless be used to "calibrate out" thermal drift, whenever the micrometer is used in an environment which differs from the specified environment.

8.3.3 Systemic errors which are the result of the neglect of nonlinearities and of the neglect of important inputs are commonly *accepted* in the measurement of quantities in fluid flow. Systemic errors of 10% in mass flow measurement are considered normal in commerical practice and systemic errors of 1% are assumed to be inevitable. Such fatalism is bred of long experience; outside the scientific laboratory, mass flow is accepted as an approximate measurement.

Precise methods of mass transport rate measurements do exist; precise methods, unfortunately, are expensive in money, time and labor and demand the use of high-level personnel. Engineers and their clients, therefore, are resigned to cheap, sloppy measurement or expensive, accurate measurements.

In a pipe, a common technique for the measurement of fluid volume rate or fluid mass flow rate is to introduce an orifice in the pipe and to determine the flow rate indirectly through the pressure change from upstream to downstream of the orifice (Fig. 8.10). Commonly the

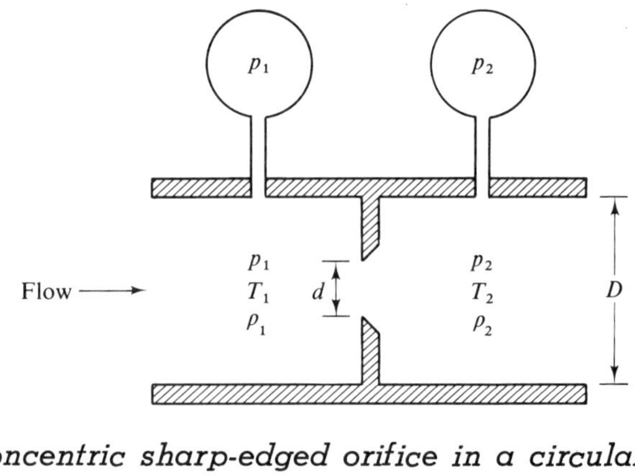

FIGURE 8.10 Concentric sharp-edged orifice in a circular pipe

calibration is accomplished through use of the equation

$$Q_M = CA\sqrt{\Delta p} \tag{8.3.3}$$

in which

Q_M = mass flow rate
A = area of the orifice opening
Δp = pressure drop
C = an experimentally derived calibration constant, the discharge coefficient

The theory is well known and the practice voluminously documented.[13] Error may be introduced in four ways:

(1) incorrect coefficient,
(2) pressure transduction error,
(3) electronic processing error and
(4) recording error.

Error in the coefficient is the primary cause of system error. The coefficient is usually obtained for a specific fluid, a specific temperature, a given density and viscosity.[14] Deviation from any of the design quantities (or neglect of any of the relevant inputs) can cause a systemic error. The coefficient has never been estimated theoretically to acceptable accuracy.[15] In practice, engineers combine theoretical analysis with experimental corrections in what is known as a "rational" procedure.

The American Society of Mechanical Engineers proposes, for example, a rational equation for the volume rate of flow of gases moving through orifices at low subsonic speeds:[16]

$$Q = K_1 Y y d^2 \sqrt{\left(\frac{S}{G}\right)\left(\frac{p_1}{T_1}\right)}\sqrt{\Delta p} \tag{8.3.4}$$

$$Y = \left[R^{2/\Gamma}\left(\frac{\Gamma}{\Gamma - 1}\right)\left(\frac{1 - R^{(\Gamma-1)/R}}{1 - R}\right)\left(\frac{1 - \beta^4}{1 - \beta^4 R^{2/\Gamma}}\right)\right]^{1/2}$$

where

Q = volumetric flow rate
K_1 = quantity dependent on units of measurement
K = quantity dependent on the Reynolds number and the geometry of the system
S = ratio of the specific weight of the gas to the specific weight computed by Boyle's law
G = specific gravity of gas referred to air
d = diameter of orifice
Y = compressibility factor
T_1 = upstream absolute temperature
Γ = ratio of specific heats
β = d/D diameter ratio
R = p_1/p_2
D = diameter of pipe

p_1 = upstream pressure
p_2 = downstream pressure
R_e = Reynolds number of pipe flow

The coefficient C in Equation (8.3.4) is a function of many parameters:

$$C = C(R_e, S, G, T_1, \Gamma, \beta, R)$$

and the error contributed by C is

$$\epsilon_C = \frac{\partial C}{\partial R_e}\epsilon_{R_e} + \frac{\partial C}{\partial S}\epsilon_S + \frac{\partial C}{\partial G}\epsilon_G + \frac{\partial C}{\partial T_1}\epsilon T_1 + \frac{\partial C}{\partial \Gamma}\epsilon_\Gamma + \frac{\partial C}{\partial R}\epsilon_R$$

Consider the case in which only Y changes. Then it can be shown that

$$\frac{\epsilon_C}{C} = \frac{\epsilon_Y}{Y}$$

Numerical data for the customary changes in Y are given in Table 8.2.[17] With this table it is possible to estimate the maximum possible error in the system. Let the nominal values of the quantities be $\Gamma = 1.20$, $R = 0.90$, $\beta = 0.65$. The largest changes occur by changing the quantities to $\Gamma = 1.40$, $R = 0.98$, $\beta = 0.25$. Then

$$\frac{\epsilon_Y}{Y} = \frac{0.9893 - 0.9201}{0.9201} = 0.0751$$

7.50% error is possible in this system by neglect of the changes in the compressibility ratio.

TABLE 8.2 Values of compressibility factor Y

Γ	R	$\beta = 0.25$	$\beta = 0.5$	$\beta = 0.65$
1.40	0.98	0.9893	0.9884	0.9863
	0.96	0.9782	0.9765	0.9724
	0.94	0.9672	0.9646	0.9585
	0.92	0.9560	0.9526	0.9447
	0.90	0.9447	0.9405	0.9309
1.30	0.98	0.9883	0.9874	0.9851
	0.96	0.9767	0.9747	0.9703
	0.94	0.9647	0.9620	0.9555
	0.92	0.9526	0.9491	0.9407
	0.90	0.9405	0.9361	0.9259
1.20	0.98	0.9873	0.9863	0.9839
	0.96	0.9747	0.9727	0.9679
	0.94	0.9618	0.9589	0.9512
	0.92	0.9489	0.9450	0.9360
	0.90	0.9358	0.9311	0.9201

A properly calibrated orifice flow meter can be error laden as a result of the external instrumentation. Equation (8.3.4) can be written as

$$Q = C_1\sqrt{\frac{p_1\,\Delta p}{T_1}} \qquad \textbf{(8.3.5)}$$

in which C_1 is a system "constant." Let us put together an instrument to read Q directly. The electronic signal chain shown in Figure 8.11 is capable of introducing systemic error in each operation so that even if p_1 and T_1 were measured perfectly, the ratio p_1/T_1 would be wrong. With the theory developed in the previous chapter the dimensionless error can be represented by

$$\left|\frac{\epsilon_Q}{Q}\right| = \left|\frac{\epsilon_{C_1}}{C_1}\right| + \frac{1}{2}\left|\frac{\epsilon_{p_1}}{p_1} + \frac{\epsilon_\Delta}{\Delta p} + \frac{\epsilon_{T_1}}{T_1}\right| + \begin{Bmatrix}\text{information}\\ \text{chain errors}\end{Bmatrix}$$

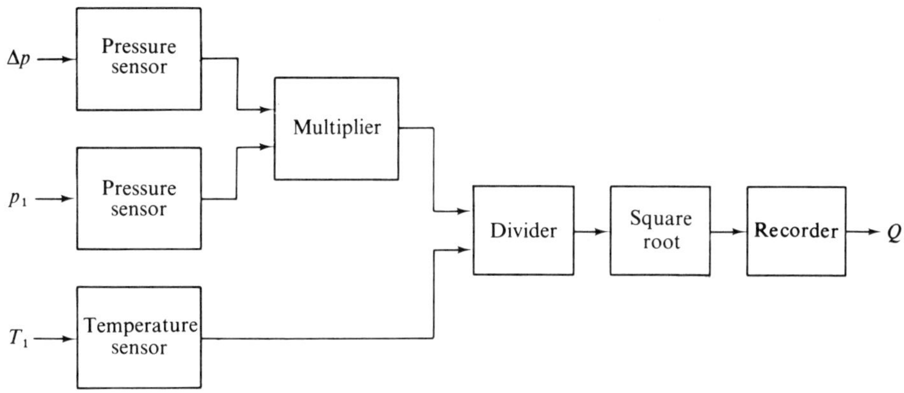

FIGURE 8.11 Signal chain for an electronic flow rate computer

The first term represents the systemic orifice calibration error; the three terms multiplied by 1/2 represent transduction errors. The determination of component and element error is a function of cost and state of the art. At any time in history the error should be determined with respect to the current equipment. Equipment commercially available in 1972 was shown to develop the following dimensionless error at the output:[18]

Device	Dimensionless Error
Recorders	0.0050
Pressure transducers	0.0050
Differential pressure transducers	0.0050
Temperature transducers	0.0070
Multipliers and dividers	0.0075
Square root extractor	0.0075

If the quantity C_1 remains constant and is properly chosen, the estimated maximum possible error can be computed to be equal to 0.0360. An error of 3.6% can be generated by the signal chain alone.

8.3.4 Withdrawal of energy by a transducer occasionally disturbs the measurand beyond an allowable minimum (see Section 1.4). Such a withdrawal is the cause of one of the most insidious systemic errors: the instrument measures correctly the properties of the disturbed measurand; no discernible error exists unless the result wanted is the value of the properties of the undisturbed system.

The measurement of heat flux at the boundary of a fluid presents a particularly extreme instance of measurand disturbance: in the measurement of heating rate, errors of 25% are possible.[19] The heat flux sensor is usually placed flush with, but insulated from, the boundary of a system and hence may achieve a temperature different from the temperature of the boundary. The boundary temperature discontinuity affects the local fluid density, viscosity and velocity in the neighborhood of the boundary and hence changes the local heat transfer characteristics.

The most common technique in the measurement of heat flux is to introduce a thermal capacity in the system, that is, a slug of highly thermally conductive material in an insulated housing (Fig. 8.12). A thermocouple at the base of the capacity element is used to record the temperature history. The heat flux $\dot{Q}$ is determined by the rate of change of the temperature through the equation

$$\dot{Q} = c_1 \frac{d\theta}{dt} + q_L \tag{8.3.6}$$

in which c_1 is a calibration constant, θ is the temperature at the thermocouple and q_L is the rate of heat loss from the calorimeter. The quantity c_1 can be estimated from the well-known equation[20]

$$c_1 = c_S \mathcal{V} \rho$$

where

ρ = density of material in slug
c_S = specific heat of receiver material
$\mathcal{V}$ = volume of receiver material.

It is assumed that c_S is a constant and that the temperature of the slug is uniform throughout. If the losses from the calorimeter are negligible,

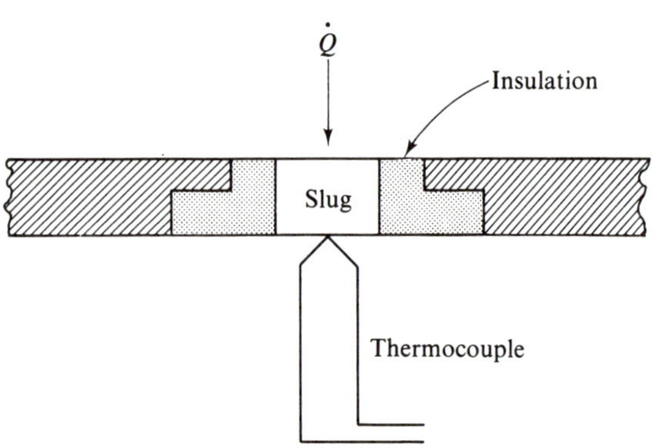

FIGURE 8.12 Slug (or slope) calorimeter

the heat energy will be stored in the capacity according to the first integral $Q = c_1\theta$, and heat flow will occur according to the law of Fourier, proportionally to the difference in temperature between the slug and the fluid, and dependent solely on the heat transfer characteristics of the fluid. In practice, however, c_S varies with temperature and the slug temperature has a measurable temperature gradient so that both c_1 and q_L must be determined experimentally.[21] In the laboratory, θ is recorded as a function of time and the slope is determined graphically. c_1 and q_L are obtained from calibration charts as a function of θ, and $\dot{Q}$ is computed. An alternate procedure is to use an on-line digital computer or a fairly complicated analogue computing circuit similar to the one shown in Figure 8.13. The digital computer is slightly more accurate and considerably faster than manual computation, whereas the analogue processor introduces systemic error. That the electrical differentiator is likely to introduce noise in the system is hardly a problem; the heat transfer fluctuations are so slow compared to the noise that filtering is fairly simple. Low-noise long-term differentiators for slow inputs are built easily from stock components.[22]

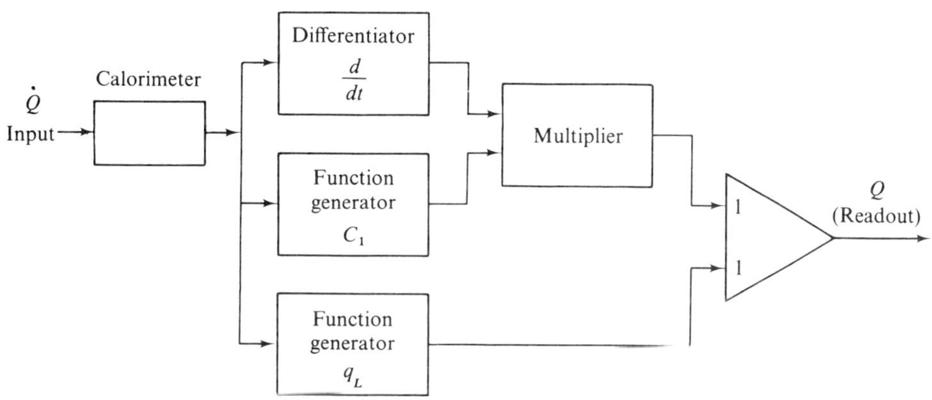

FIGURE 8.13 Electrical processing of the output of a slug calorimeter

We have already discussed the two common sources of systemic error in the calorimeter: neglect of the variation of c_1 with temperature, and neglect of q_L. The latter is often the largest contributor to error. A 1/8-inch thick 1/4-inch diameter copper slug insulated with MinK 2000 insulation[23] has been shown to suffer from a heat loss given approximately by[24]

$$q_L = 5.22 \times 10^{-5}\theta^{1.7}$$

At a temperature of 700°F the system heat transfer rate was measured to be 32 Btu/ft²/s, whereas $q_L = 4$ Btu/ft²/s. In a well-insulated calorimeter, neglect of q_L would generate an error of 12.5%.

Yet to be discussed is the disturbance of the measurand by the calorimeter. Let the calorimeter be embedded in a plate over which a heated gas is passing at a steady rate. It is assumed that the temperature of the plate is θ_p, that a steady state gas temperature distribution and hydrodynamic velocity distribution have been established and that a steady heat transfer rate occurs into the plate. The heat flux transducer is not likely to reproduce the conditions which occur on the plate, and thus is likely to have a different temperature from the plate. The gas which passes over the boundary temperature discontinuity will be affected. A first-order analysis (somewhat too lengthy to reproduce) has been made with results which can be used to estimate the error.[24]

Let gas flow over a flat plate of length b (Fig. 8.14) and then pass over the transducer. The insulation between the transducer and the plate is of negligible thickness and is "perfect." Let x be the distance from the leading edge of the plate to some point on the calorimeter slug. It can be shown[24] that

$$\frac{\dot{Q}_p - \dot{Q}_T}{\dot{Q}_p} = \frac{\theta_T - \theta_p}{\theta_g - \theta_p}\left[1 - \left(\frac{b}{x}\right)^{9/10}\right]^{-1/9} \tag{8.3.7}$$

where

$\dot{Q}_p$ = heat flux through the plate
$\dot{Q}_T$ = heat flux through the transducer
θ_p = temperature of the plate
θ_T = temperature of the transducer
θ_g = temperature of the gas

x varies from b to $b + d$ where b is a large number and d, the diameter of the transducer, is very much smaller than b. We replace x by $b + \eta$, $0 < \eta < d$, and noting that $b \gg \eta$, we write

$$\left|\frac{\epsilon_\theta}{\dot{Q}}\right| = \left|\frac{\theta_T - \theta_p}{\theta_g - \theta_p}\right|\left|\frac{10b}{9\eta}\right|^{1/9}$$

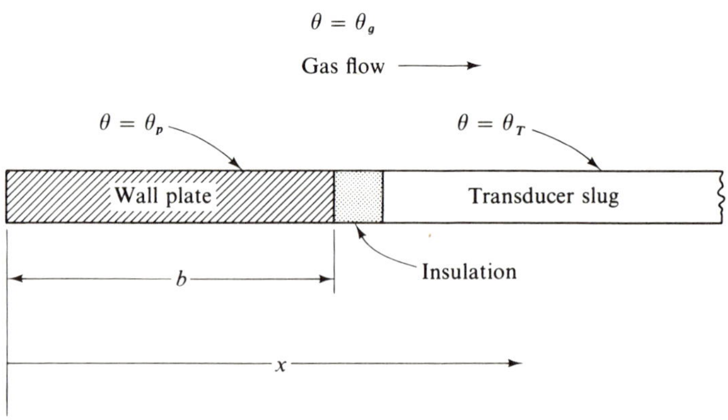

FIGURE 8.14 Simplified physical model for the determination of transducer interference in the vicinity of a slug calorimeter

At the center of the transducer ($\eta = d/2$), the most likely location of the thermocouple, the error will have the form

$$\left|\frac{\epsilon_{\dot{Q}}}{\dot{Q}}\right| = \left|\frac{\theta_T - \theta_p}{\theta_g - \theta_p}\right|\left|\frac{20b}{9d}\right|^{1/9} \tag{8.3.8}$$

In a fairly common case[24] in which $\theta_T - \theta_p = 100°\text{F}$ and $\theta_g - \theta_p = 1{,}500°\text{F}$, $d = 0.5$ inch and $b = 12$ inches, we obtain $\epsilon_{\dot{Q}}/\dot{\theta} = 0.104$, or 10.4% error.

8.3.5 Precise measurement of mass is achieved by comparison on the analytical balance (Fig. 8.15). A beam, or truss, presumably rigid, is mounted on a rigid frame which supports the beam at a height above a fixed, horizontal base. Two pans, each pivoted on a knife edge, are suspended at equal distances from the fulcrum. A known mass is placed in one pan and an unknown mass in the other pan. If the two masses are nearly equal, the indicator needle will oscillate about an ultimate state deflection ϕ_0. The difference between the known mass and the unknown mass will be equal to the angle ϕ_0 multiplied by the ultimate state sensitivity S. A good balance can be used to measure a mass change of one millionth of the maximum allowable load. A 1.5-kilogram balance can be designed inexpensively to measure a difference of 0.0015 gram. Commonly available commerical balances can be read to 0.0005 gram and high-precision balances with a maximum allowable load of 10 grams can be read to 10^{-12} grams, a 10^{-11}th part of the maximum load[25].

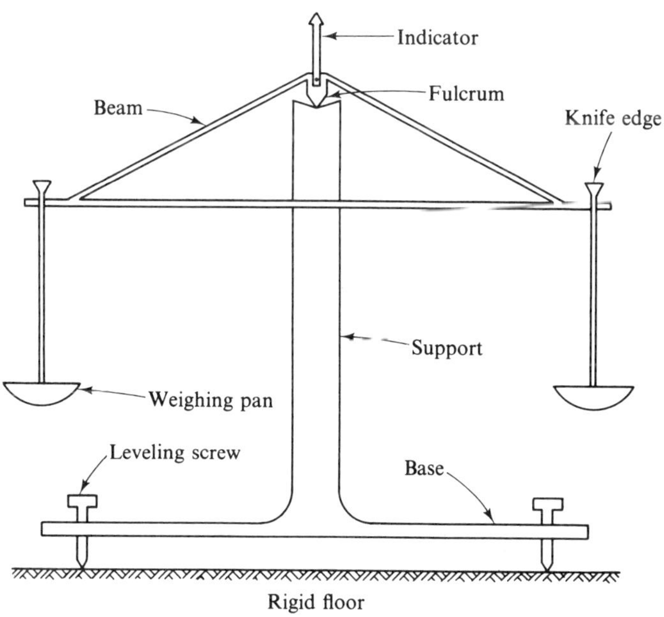

FIGURE 8.15 The basic components of the analytical balance

The limiting factor in balance accuracy and threshold is always of a systemic nature. If the balance is properly leveled, mounted on a massive, isolated, vibration-free mounting, and if the whole mechanism is encased in a vacuum chamber, then the lowest limit of mass discrimination will depend on the rigidity of the beam and on the ability of an operator or of an electrical transducer to detect a change of fulcrum center-line angular position with respect to a zero reference line. Improvement of system design always hinges on improvements of manufacturing precision and position detection.

Extremely minute changes become important near the threshold of measurement. A good example of small effects which can be of great significance if neglected is the buoyancy of bodies in air. If two apparently balanced masses M_1 and M_2 have different volumes V_1 and V_2, the actual balance would be

$$M_1 - \rho V_1 = M_2 - V \rho_2$$

in which $\rho = 0.001240$ g/cc at 20°C and sea level and ρ is the density of air. The difference between the masses, or measurement error, would be

$$|\epsilon_M| = |M_2 - M_1| = \rho |V_2 - V_1|$$

If M_1 were 1 g and the difference in volume 0.001 cc, M_2 would be 1.00000124 g. With a scale designed to detect a difference of 1 microgram, the buoyancy error would be of the order of the threshold, or 1.24 micrograms. To obtain that last significant figure the balance should be mounted inside a vacuum chamber[26] or else the masses should be enclosed in containers of nearly identical volumes such that $|\rho(V_2 - V_1)|$ would be less than the acceptable error.

The systemic error introduced by manufacturing inprecision or structural defect can be described in terms of the perfectly balanced beam with unequal arms. When suspended on the fulcrum, the unloaded beam and pans can be balanced by moving a small counterbalance weight to the point that the indicator needle is exactly on zero. If the distances between the central fulcrum and the pan fulcri contact points differ slightly, an error is introduced in the measurements even though the beam was balanced without load.

In Figure 8.16 the two beam components have lengths l_1 and l_2 whereas $l_1 \neq l_2 \neq l$, and l is the nominal design length. The two angles α_1 and α_2 are unequal and not equal to their nominal value α. Let the pans be held in place so that $\phi = 0$ at time $t = 0$. Then one pan will be loaded with the mass M_1 which is to be compared to the mass M_2 on the other pan. If the pans are released and allowed to move, the balance output will follow the performance equation

$$J\frac{d^2\phi}{dt^2} + f\frac{d\phi}{dt} + M_1 l_1 \cos(\alpha_1 + \phi) - M_2 l_2 \cos(\alpha_2 - \phi) - mh \sin\phi = 0 \qquad \textbf{(8.3.9)}$$

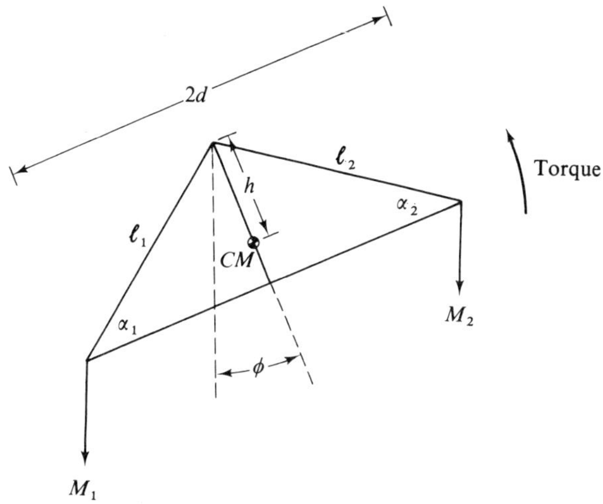

FIGURE 8.16 Displaced balance beam with unequal arms

in which

J = moment of inertia of beam assembly
f = damping (extremely small)
m = mass of beam
h = position of center of mass of the beam below the fulcrum contact point.

The beam will now oscillate about its ultimate state position:

$$\tan \phi_u = \frac{M_1 l_1 \cos \alpha_1 - M_2 l_2 \cos \alpha_2}{M_1 l_1 \sin \alpha_1 + M_2 l_2 \sin \alpha_2 - mh} \qquad \textbf{(8.3.10)}$$

In measurement we change the reference mass M_2 to set $\phi_u = 0$, so that at the point of apparent balance we would have unequal masses

$$\frac{M_1}{M_2} = \frac{l_2 \cos \alpha_2}{l_1 \cos \alpha_1}$$

The mass error can be found by setting $M_2 = M_1 + \epsilon_M$:

$$\left| \frac{\epsilon_M}{M_1} \right| = \left| \frac{l_1 \cos \alpha_1}{l_2 \cos \alpha_2} - 1 \right| \qquad \textbf{(8.3.11)}$$

From Figure 7.15 it can be seen that if d is the nominal half-length of the beam base then

$$l_1 \cos \alpha_1 = d + \epsilon_1$$
$$l_2 \cos \alpha_2 = d + \epsilon_2$$

in which ϵ_1 and ϵ_2 are the changes in length of the beam base from the nominal. Expansion of (8.3.11) with the assumption that ϵ_1 and ϵ_2 are small gives the error equation

$$\frac{\epsilon_M}{M_1} = \frac{2(\epsilon_1 + \epsilon_2)}{d}$$

If a scale with a 20-cm half-beam were to be designed with a maximum error of 10^{-6} gram per grams, the allowable error in the beam half-length would be 10^{-5} centimeters. Such manufacturing precision can be achieved with some tedious hand-finishing.

8.3.6 Systemic errors are occasionally the result of the misuse of a component outside its design range (see Section 1.5). A common mechanical example of a limited performance device is the dashpot viscous damper.[27]

Mechanical devices designed to operate at low frequencies are occasionally subjected to high-frequency inputs or noise. Such noise is reduced by means of mechanical filters which are combinations of springs, masses and viscous dampers. Dampers are devices which absorb energy and dissipate it in the form of heat. Such devices develop forces or torques proportional to some power of the driving velocity and hence modify the performance equation by changing the damping factor ζ, thereby reducing the cutoff frequency. (See chapter 5.) The design method for the damper is to select the needed value of ζ and then to generate a damping coefficient $f = 2\zeta\omega$ in which ω is the natural undamped frequency of the system. The damper is then designed to maintain f within the operating range of the device.

For the damping of linear motion it is customary to introduce a piston into a cylinder filled with a viscous fluid (Fig. 8.17). In the design of such dampers it is assumed that

(1) the fluid is maintained at a reasonable constant viscosity and is nearly incompressible;

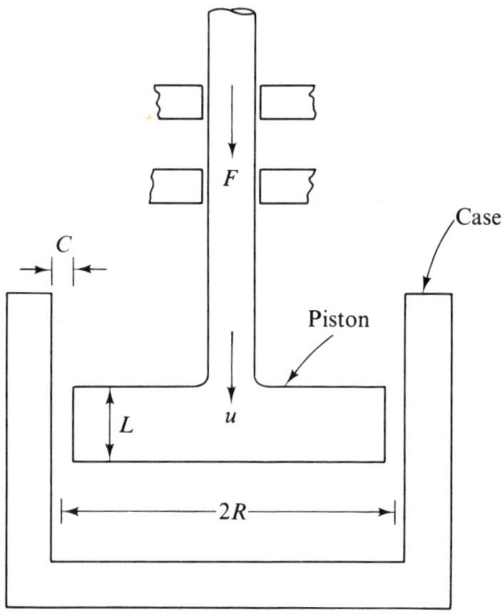

FIGURE 8.17 A simple linear viscous damper or dashpot

(2) the clearance (C) between the piston and the cylinder is small relative to the piston diameter;
(3) the piston length (L) is great enough for the fluid velocity pattern to remain constant over the major portion of the piston length;
(4) the piston is maintained concentric with the cylinder at all times.

If all the assumptions are satisfied, it can be shown that for a range of velocities the flow of the fluid between piston and wall is in the laminar mode and that the velocity distribution is given by a simple modification of the Hagen-Poiseuille law:[28]

$$\left(\frac{dp}{dl}\right)\frac{r^2}{4} = \mu u + A \ln r + B \tag{8.3.12}$$

in which

p = pressure
l = distance coordinate along piston measured from bottom edge of piston
r = distance coordinate from center of dashpot
μ = dynamic viscosity coefficient
u = fluid velocity at point (r, l)
A, B = constants to be determined

The boundary conditions are

$$u = U \quad \text{at } r = R$$

$$u = 0 \quad \text{at } r = R + C$$

$$dp/dl = 0 \quad \text{for } U = 0$$

where U is the velocity of the piston. Application of the first two conditions yields

$$\left(\frac{dp}{dl}\right)\frac{R^2}{4} = \mu U + A \ln R + B$$

$$\left(\frac{dp}{dl}\right)\frac{(R + C)^2}{4} = A \ln (R + C) + B$$

and thus

$$\frac{dp}{dl}\left[\frac{R^2}{4} - \frac{(R + C)^2}{4}\right] = \mu U + A \ln \frac{R}{R + C}$$

Application of the third condition then tells us that $A = 0$. B is found to be

$$B = \frac{dp}{dl}\frac{(R + C)^2}{4}$$

The pressure drop per unit length is

$$\frac{dp}{dl} = -\frac{4\mu U}{C(2R + C)}$$

Over the length L there will be a pressure drop Δp, whence

$$\Delta p = \frac{-4\mu UL}{C(2R + C)}$$

The force on the driving piston is

$$F = \pi R^2 \, \Delta p = fU$$

whence the damping factor becomes

$$f = \frac{4\pi\mu L}{C/R(2 + C/R)} \tag{8.3.13}$$

Equation (8.3.13) is surprisingly accurate as long as the flow is laminar, that is, as long as

$$\frac{\rho UC}{\mu} < 1{,}000 \tag{8.3.14}$$

where ρ is the density of the fluid.

Example 8.3.1 A dashpot is designed to operate with SAE 10 lubricating oil at 25°C. The clearance between cylinder and wall is 0.10 cm. Determine the maximum operating velocity.

Solution $\mu/\rho = 1 \text{ cm}^2/\text{s}$ and thus from inequality (8.3.14) we find a maximum velocity of

$$U_M = 10{,}000 \text{ cm/s}$$

A velocity of 100 m/s is not too great a limitation. Typical vibrations have velocities less than 10 m/s.

Example 8.3.2 A second-order undamped system has an unwanted linear oscillation of 10.0 Hz superposed on a natural frequency of 100 Hz. The oscillations have an amplitude of 0.25 cm. Design a dashpot with SAE 10 lubrication oil at 25°C to achieve $\zeta = 0.5$.

Solution The unwanted oscillations exhibit a displacement $y_N = 0.25 \sin \omega_0 t$ and a velocity $0.25\, \omega_0 \cos \omega_0 t$, The maximum velocity will occur at $\omega_0 t = 0$ and then $U_{\text{MAX}} = 15.7$ cm/s. Let us introduce a safety factor by the use of 20 as an upper limit, so that $C \leq 50$ cm.

We now use Equation (8.3.14) and note that

$$f = 2\zeta\omega = 628.0$$

whereupon it follows that

$$628 = \frac{4\pi\mu L}{C/R(2 + C/R)}$$

or

$$L = 84.4\frac{C}{R}\left(2 + \frac{C}{R}\right)$$

Any combination of L, C and R which satisfies this equation is an acceptable design. Set $C/R = 0.01$; then $L = 1.69$ cm. Let $R = 5$ cm; then $C = 0.05$ cm.

The sources of systemic error in such a system are various. The most common source, however, is the result of the application of the dashpot to a system in which the input velocity exceeds U_M. For values of Reynolds number $\rho UC/\mu$ between 1,000 and 5,000 the damper becomes erratic and noisy. Above 5,000 the resultant force becomes roughly proportional to U^2 and the element becomes false.

8.3.7 Design by linearized performance equations of highly nonlinear systems is a common source of distortion (a systemic error) in vacuum tube and transistor amplifiers. Consider, for example, the simple vacuum tube triode (Fig. 8.18). e_c is the input voltage and e_b the output voltage with i_b the plate output current. (Note that the current is assumed to flow from plate to cathode, a quaint conceit of the electrical engineer.[29]) If a specific plate voltage and a specific grid voltage are applied to the vacuum tube, a specific plate current is obtained. The variation of the current over the permissible range of plate and grid voltage can be plotted in a family of curves known as the characteristic curves of the triode (Fig. 8.18). The curves are nonlinear, but for a sufficiently large plate voltage the output current varies linearly with the grid voltage and with increments of plate voltage above the minimum. The characteristic curves are replaced as a first approximation by a set of parallel, equally spaced straight lines, with some warning about a minimum plate voltage or dc level, and minimum grid voltage or bias (Fig. 8.19). Then we can write an idealized equation

$$i_b = \frac{1}{r_p} e_b - \mu e_g$$

or

$$r_p i_b = e_b - \mu r_p e_g = e_b - g_m e_g \tag{8.3.15}$$

Equation (8.3.15) could also be obtained by a series circuit with a

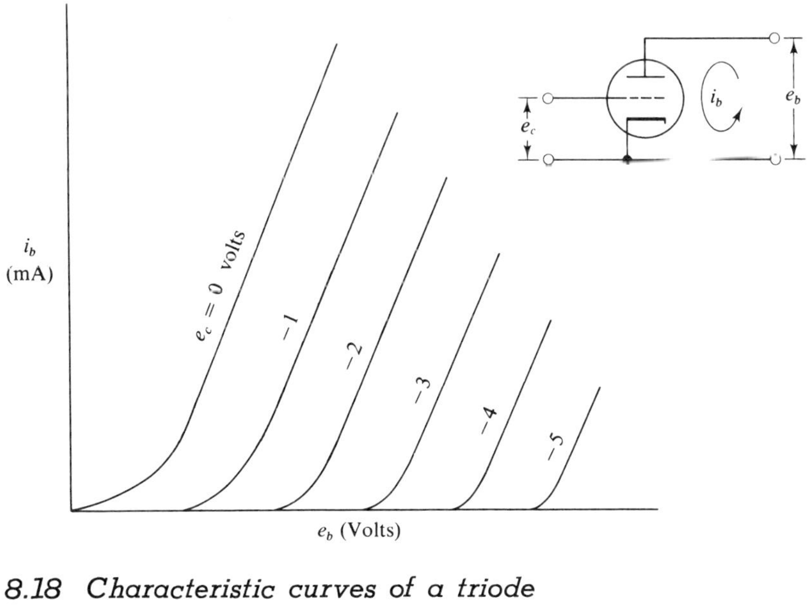

FIGURE 8.18 Characteristic curves of a triode

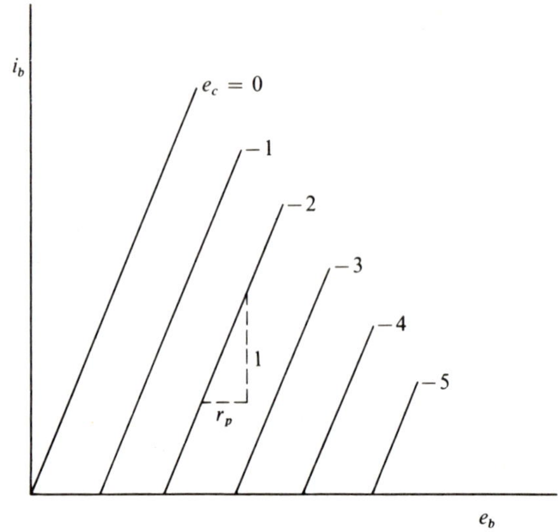

FIGURE 8.19 Linearized vacuum tube characteristic curves

voltage source $-g_m e_g$ and a series resistance r_p. For this reason the quantity r_p is called the *plate resistance*. g_m is called the *gain of the vacuum tube*. Triodes, therefore, can be replaced in design by the equivalent circuit shown in Figure 8.20a. If the vacuum tube output were to be connected to a load resistance R_L, the equivalent circuit would be as in Figure 8.20b. Application of the circuit laws gives us the performance equation for the plate-loaded amplifier (Fig. 8.20b):

$$e_o = -\left(\frac{g_m r_p R_L}{r_p + R_L}\right)e_g = -Se_g \qquad \textbf{(8.3.16)}$$

The quantity S is the ultimate state sensistivity or gain of the amplifier.

The simple performance Equation (8.3.16) is based on assumptions

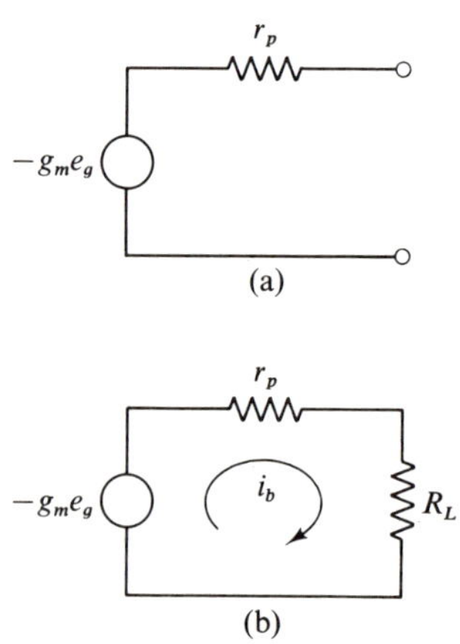

FIGURE 8.20 Elementary equivalent vacuum tube triode circuit

which are all violated in practice:

(1) The characteristics are linear in all variables.

(2) There exists no capacitance or inductance between the elements of the triode

(3) The grid circuit is a pure voltage source

In practice, however, the characteristics are nonlinear; the plate circuit has both inductance and capacitance so that all sorts of errors occur. The capacitance and inductance actually change the performance equation to

$$M_1 \frac{d^2 e_o}{dt^2} + M_2 \frac{de_o}{dt} + M_3 e_o = -Se_g$$

whereas the neglect of nonlinearities changes the sensitivity of the system. The first statement can be established by computing an equivalent circuit with r_p replaced by a resistance, capacitance and inductance in parallel, whereas the second statement requires examination of the function i_p.

Let i_p be expanded in a Taylor series about the grid bias e_g^o and the applied plate voltage e_b^o. Then if e_g' and e_b' are the deviations from e_g^o and $e_{b'}^o$

$$i_p(e_g^o + e_g' e_b^o + e_b') = i_p(e_g^o e_b^o) + \left.\frac{\partial i_p}{\partial e_g}\right|_{o,o} e_g'$$
$$+ \left.\frac{\partial i_p}{\partial e_b}\right|_{o,o} e_b' + \frac{1}{2!}\left(e_g' \frac{\partial}{\partial e_g} + e_b' \frac{\partial}{\partial e_b}\right)^2 i_b + \cdots$$

The current above the base current is i_p'

$$i_p' = i_p - i_p(e_g^o e_b^o) = -\mu e_g' + \frac{1}{r_p} e_b' + \frac{1}{2!} \left.\frac{\partial^2 i_b}{\partial e_g^2}\right|_{o,o} (e_g')^2$$
$$+ \frac{1}{2!} \left.\frac{\partial^2 i_b}{\partial e_b^2}\right|_{o,o} (e_b')^2 + \left.\frac{\partial^2 i_p}{\partial e_g\, \partial e_b}\right|_{o,o} e_g' e_b' + \cdots$$

so that, in effect, the performance equation becomes

$$R_p i_p' = -G_m e_g' + e_b'$$

where

$$R_p = \frac{r_p}{1 + r_p(\partial^2 i_p/\partial e_b^2)e_b + r_p(\partial^2 i_p/\partial e_g\, \partial e_p)e_g' + \cdots} \tag{8.3.17}$$

and

$$G_m = \left(\mu - \frac{\partial^2 i_p}{\partial e_g^2} e_g' - \frac{\partial^2 i_p}{\partial e_g\, \partial e_p} e_p' - \cdots\right) R_p$$

The performance equation of the amplifier now shows a new form

$$e_o = -\left(\frac{G_m R_p R_L}{R_p + R_L}\right) e_g \tag{8.3.18}$$

which can be shown to be expressible in the form

$$e_o = -Se_g + \epsilon(e_{g'} e_o)$$

in which ϵ represents a systemic error which depends on the extent of misuse of the amplifier outside its linear characteristics.

One must be careful, however, to not overemphasize the problem. Whereas in sound amplification and in signal recording the distortion cannot be tolerated, in instruments with output indication dials moderate distortion either crowds the markings together, or spreads them farther apart. Such scale distortion, if not excessive, and if the scale spacings do not fall below the design threshold, can be tolerated.

8.4 Noise

Both the discovery and the elimination of noise are difficult tasks. The noisiness of a signal is not discernible without an instrument and the noisiness of an instrument is not easily distinguishable from the noise of the signal. If the output signal is noisy, the designer will attempt to discover the source of the noise by diagnostic filtering. This process consists of introducing filters at the output and changing the filter characteristics until an acceptable signal is received. The filter is next placed at the input and then, sequentially, after each element of the information chain. As soon as the output with an intermediate filter location yields the same signal as the output-filtered instrument, the noise source has been discovered. Improvement of the noise characteristics of the instrument can now proceed.

Experienced designers avoid some trouble by avoiding inherently noisy elements and components. Mechanical vibration, slip ring and commutator noise, resistance thermal noise, ground-loop interference and "crosstalk" are all in the experience memory bank and are to be avoided or at least filtered.

8.4.1 Rotating elements which need electrical connection with fixed elements can achieve such a contact through flexible wires, through slip rings or through radio signals. Flexible wires limit the rotation to less than one full turn;[30] slip rings are noisy and radio cannot carry large amounts of power and is open to atmospheric noise and man-made interference. In practice, flexible wires are used in galvanometric type equipment and slip rings in electric motor type equipment with radio restricted to special applications. Gyroscopes, servomechanisms, most inertial guidance devices, radar antennas and radio telescopes all have at least one set of slip rings and brushes and one or more flexible connection. The flexible pigtails will eventually fail in fatigue (a point to be considered in design), whereas the slip rings are sources of noise.

Ideally the slip ring connection should act as a constant low resistance; unfortunately in most slip ring assemblies the resistance varies randomly.[31] Three factors influence the resistance of the connection between the brush and the ring:

(1) type of material in ring and brush,
(2) shape of the contact surface and
(3) pressure between the ring and the brush.

Most metals become covered with a surface film which has high resistance compared to the resistance of the basic metal. The film is punctured by a sufficiently high voltage and the resistance of the contact drops drastically. The film will rebuild and the resistance will rise only to be suddenly decreased again. Various combinations of materials seem to minimize the film effect; noble metals, for example, appear to be the best materials for low-noise slip rings. Silver slip rings or gold slip rings with silver-graphite brushes are common in high-precision inertial navigation devices; and platinum (though less successful in noise abatement) is favored by some designers.

The contact surface shape between slip ring and brush is highly variable: only the high points of the brush and the slip ring are in contact. During high-speed motion, heat is generated, the high points melt and the contact surface changes shape. Other high points appear as a result of heating, vibration and arching so that the actual contact area is a random quantity and consequently so is the resistance. For unknown reasons noble metals again seem to be best for constancy of surface shape.

The contact pressure between brush and ring cannot be too high, or else the friction will be too high for a lightly damped instrument. In addition, the heat generated by the friction will enhance film buildings. The pressure cannot be too low or else the brush will not remain in steady contact with the ring. The brush pressure spring is to be adjusted with care and much experimentation; manufacturers' instructions about brush pressure should be followed religiously.

8.4.2 The correction and the elimination of many types of electrical noise are fairly simple and the methodology well known.[32] The common sources of electrical noise are

(1) power equipment and lines,
(2) earth ground currents,
(3) common mode voltages,
(4) acoustic and other mechanical interference,
(5) thermoelectric noise,
(6) electrochemical noise,
(7) Johnson noise (see Section 7.4.3).

The noise enters the system in two different fashions. For items 1, 2 and 3 the means of transmission are

(a) common impedance,
(b) magnetic coupling,
(c) capacitative or electrostatic coupling,
(d) electromagnetic radiation.

For item 4 the noise is transmitted mechanically to some element or component which reacts to mechanical deformation with a change of the electrical performance equation. Items 5, 6 and 7 are inherent to the components and are classified as internal noise. Thermoelectric noise can be removed by having no dissimilar metals in contact throughout the measuring circuit. Electrochemical noise requires that there be no dissimilar metals and no electrolytes of any sort, including acid atmospheres in the neighborhood of the circuit. Johnson noise will be discussed in Section 8.4.3.

Coupling (magnetic, electrostatic and electromagnetic) can be reduced by shielding and removing the coupled circuits from the proximity of each other. For complete information the reader is referred to the literature.

Common impedance noise is interesting by itself since it is a common error of the inexperienced designer. This type of noise is the result of two independent devices connected to the same power sources or impedances. If the line is sufficiently long, a voltage drop is created which can disturb signals. Consider, for example, the thermocouple circuit shown in Figure 8.21. A thermocouple is connected to a cascade of two amplifiers so that the output voltage e_o is obtained between the output lead of the second amplifier and the "negative" return wire of the power lines. An entirely different circuit, a switch used to energize a relay coil which draws 0.5 ampere, draws its power from the same source. The lead wire 1-2 consists of 20 meters of AWG #12 wire which has a resistance of 5.25 ohms per thousand meters. Whenever the relay is energized, a voltage drop of $iR = 0.5 \times 20/1{,}000 \times 5.25 = 0.0525$ volt occurs in the line. Let the voltage of the thermocouple referred to point 1 be e_1. If the first amplifier has a gain of 100, the voltage presented to the second amplifier will be (referred to point 2)

$$e_2 = 100e_1 + 0.0525 = 100(e_1 + 0.000525)$$

In other words, the relay coil current would create a spurious increment of 0.525 mV at the input. This noise could be equivalent to an error of 50°C with some thermocouples. It is best, therefore, to energize the relay from a separate power source.

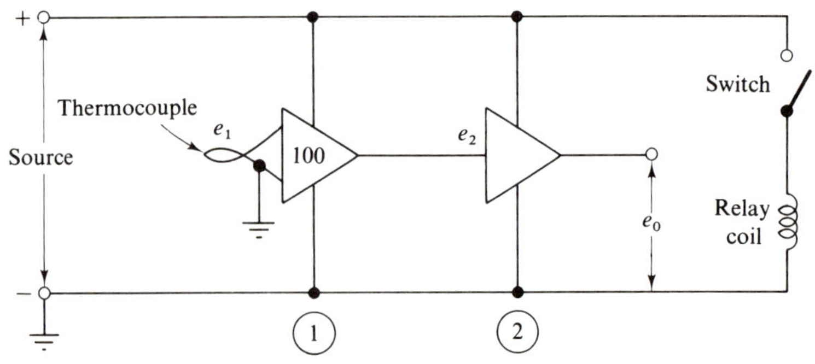

FIGURE 8.21 Thermocouple amplifier system in common impedance connection with a relay

8.4.3 An excellent example of the application of the theory outlined in chapter 7 and the difficulty inherent in even the simplest noise computation is the estimation of the signal-noise ratio in a hot-wire anemometer.[33] As a preliminary to the discussion let us examine the performance equation of a resistance bridge with noisy resistors (Fig. 8.22a). This equation is

$$\left(\frac{R_4}{R_2 + R_4} - \frac{R_3}{R_1 + R_3}\right)E = U \qquad \textbf{(8.4.1)}$$

If the nominal resistances of R_1, R_2 and R_3 are R and $R_4 = R + \delta$ and if each resistance changes by a small "noisy" amount $r_i(t)$, the performance equation can be written as

$$\left[\frac{R + r_4}{(2R + \delta) + (r_2 + r_4)} - \frac{R + r_3}{2R + (r_1 + r_3)}\right]E = U$$

This expression can be simplified if it is assumed that $r \ll R$, as is normally the case, and if δ is considered a small number, which is not necessarily true but which is true in most applications. Then it can be assumed that all products of r and δ are negligible so that by the binomial theorem we can obtain

$$\left(\frac{R}{2R + \delta} - \frac{1}{2}\right)E + \frac{E}{4R}(r_1 - r_2 - r_3 + r_4) = U$$

Consider now a bridge with noiseless resistances but with small noise voltage sources in series with the resistance (Fig. 8.22b). The bridge performance equation becomes

$$\left(\frac{R}{2R + \delta} - \frac{1}{2}\right)E + \frac{1}{2}(-e_1 - e_2 + e_3 + e_4) = U \qquad \textbf{(8.4.2)}$$

Thus it is possible to set

$$e_1 = \frac{Er_1}{2R}, \qquad e_2 = \frac{Er_2}{2R}, \quad \text{etc.}$$

and to state that a noisy resistance in a bridge can be replaced by a series voltage source as long as the resistance changes and the voltage

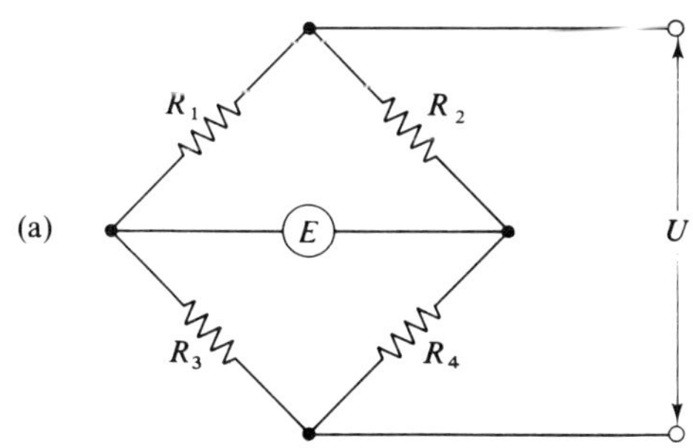

FIGURE 8.22a Elementary resistance bridge

(b)

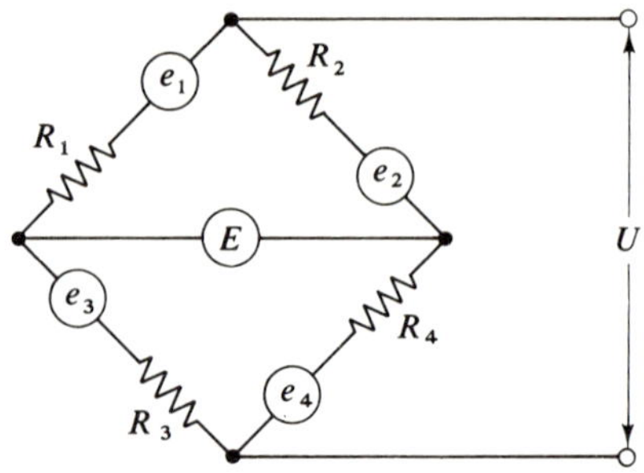

FIGURE 8.22b Equal arm bridge with series voltage noise sources

source are assumed to be small. The signs of the voltages do not matter since we will use the variance of the output signal. If δ and E are constant, it immediately follows that

$$\sigma_U^2 = \frac{1}{4}(\sigma_{e_1}^2 + \sigma_{e_2}^2 + \sigma_{e_3}^2 + \sigma_{e_4}^2) \qquad \textbf{(8.4.3)}$$

It is known[33] that if the operating temperature of the resistance is T, and ω_{MAX} is the maximum frequency of the noise, then

$$\sigma_e^2 = \int_0^{\omega_{MAX}} 4k_B TR \, d\omega$$

where k_B is Boltzman's constant.[34] The resistance R operates at T and since R_4 is not equal to R, R_4 is likely to operate at temperature $T + \Delta T$. Thus

$$\sigma_U^2 = k_B TR\omega_{MAX}\left(1 + \frac{\Delta T}{4T}\right) \qquad \textbf{(8.4.4)}$$

Example 8.4.1 A resistance bridge with $R = 10\ \Omega$, $T = 300°\text{K}$, $\Delta T = 20°\text{K}$ is known to generate 100 μV of rms noise. If Boltzman's constant is $k_B = 1.38 \times 10^{-23}$ W/°Ks, determine the maximum frequency of the system noise.

Solution

$$\sigma_V^2 = (100 \times 10^{-6})^2 = 10^{-8}$$

and thus

$$4.14 \times 10^{-20}\omega_{MAX} = 10^{-8}$$

Hence $\omega_{MAX} = 2.42 \times 10^{11}$ rad/s $= 3.86 \times 10^7$ kHz.

8.4.4 The resistance bridge is the fundamental element of the hot-wire anemometer. One of the resistance arms is replaced by a thin wire of some uniform material. The wire is immersed in a moving gas and is heated to a temperature above the ambient temperature by passage of a current. Heat energy will be carried away from the wire by convection,[35] thereby changing the resistance of the wire. A feedback mechanism is used to return the wire to a constant temperature (or sometimes to a constant current). The voltage change needed to compensate the system can be shown to be a function of the heat energy convected from the hot wire and thus a function of the gas velocity, since it is well known that the heat transfer Q from the wire is given by[36]

$$Q = f(v)(T_W - T_A) \tag{8.4.5}$$

in which

$$\begin{aligned} Q &= \text{heat transferred (watts)} \\ v &= \text{gas velocity (m/s)} \\ T_W &= \text{temperature of wire (°K)} \\ T_A &= \text{temperature of gas (°K)} \end{aligned}$$

The output from the bridge is connected to a high input-impedance amplifier and the amplifier output is connected to the input diagonal of the bridge (Fig. 8.23). The amplifier input is usually connected to an active filter and to an rms amplifier and hence to some form of display transducer.

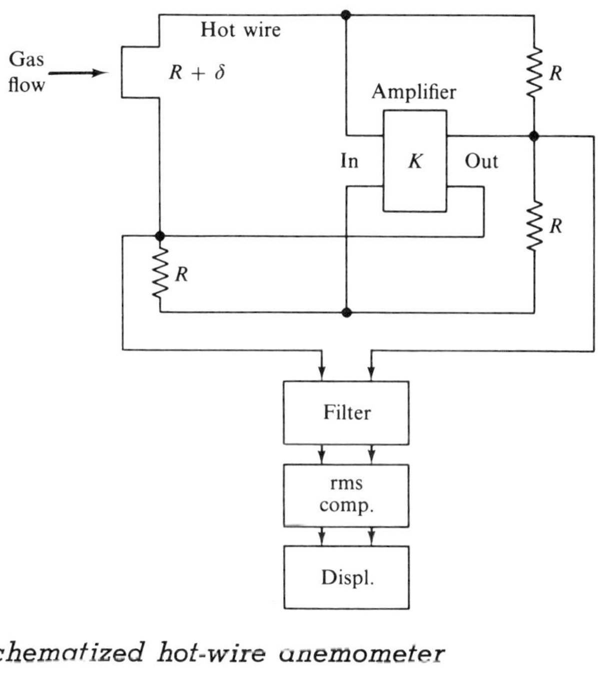

FIGURE 8.23 Schematized hot-wire anemometer

The bridge consists of the active hot wire, usually a very small piece of platinum (or some alloy like wollanston), and three balance resistors. The hot wire has very small heat capacity and hence changes temperature rapidly with changes of heat according to the well-known equation

$$c\frac{dT_W}{dt} = \text{net heat transfer}$$

and hence will also change resistance quickly according to

$$(R_W - R_0) = \alpha R_0(T_W - T_0) \tag{8.4.6}$$

where

R_W = resistance of wire at temperature T_W
R_0 = resistance of wire at some reference temperature T_0
α = temperature coefficient of resistance

The balancing resistances are selected to have a very large heat capacity so that their resistance changes little if at all. In high-quality systems the balance resistors are embedded in large heat sinks or are kept in a temperature controlled area. The active wire is initially balanced against the remainder of the bridge and then is "overheated" sufficiently to insure proper heat transfer. In a well-designed system it is possible to read velocity fluctuations in millimeters per second superposed on velocities of meters per second. Obviously the instrument is ideal for the measurement of turbulence in moving gases.[37]

The elements of the basic system are

(a) the hot wire transducer,
(b) the bridge and
(c) the feedback amplifier.

Noise is generated in each resistance of the bridge and also between the input and the output of the amplifier. If we replace the amplifier by a voltage source U and a series resistance R_A (see Section 8.3.7) and then place noise sources in series with the amplifier and the resistances [see Figure 8.24 and Equation (8.4.2)], we can immediately write the bridge equation

$$\frac{1}{2}\left[e_1 + e_2 + (e_3 + e_4)\left(1 + \frac{\delta}{2R}\cdot\frac{2R + R_A}{R + R_A}\right)\right] + \frac{(U + e_2 + e_4)\delta}{4(R + R_A)} = u \tag{8.4.7}$$

The voltage u is introduced into an amplifier of gain K and time constant τ. This amplifier has an extremely high input impedance so that negligible current is drawn from the bridge. The amplifier output is

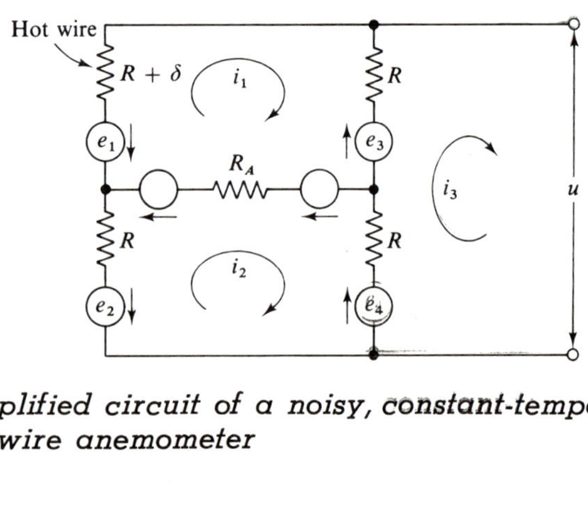

FIGURE 8.24 Simplified circuit of a noisy, constant-temperature, hot-wire anemometer

the voltage source U. Thus the amplifier equation is

$$\tau \frac{dU}{dt} + U = ku$$

The performance equation for the wire is written simply as

$$C \frac{dT_W}{dt} = \text{power added by circuit} - \text{heat dissipated per unit time}$$

or

$$C \frac{dT_W}{dt} = i_1^2(R + \delta) - f(v)(T_W - T_A)$$

The quantity i_1 is easily found by Kirchhoff's law so that the wire equation becomes

$$C \frac{dT_W}{dt} = \frac{(U + e_0)^2(R + \delta)}{2(R + R_A) + (1 + R_A/R)\delta} - f(v)(T_W - T_A) \qquad \textbf{(8.4.8)}$$

in which

C = thermal capacity = $C_S \rho \pi (D^2/4) l$(watt-s/deg)
C_S = specific heat (W-s/°g)
ρ = density of wire (g/m³)
D = diameter of wire (m)
l = length of wire (m)
$e_0 = e_A + (e_1 + e_3)(1 + R_A/2R) + (e_2 + e_4) R_A/2_R$ (volts)

The fourth system equation is the previously discussed relationship

$$R_W - R_0 = \alpha R_0(T_W - T_0)$$

Usually T_0 is taken to be the temperature of melting ice so that $T_W - T_0$ is simply T_W in degrees Celsius.

The speed of response of the anemometer depends on the speed of response of the hot wire and the speed of response of the amplifier; all other system components react at the speed of electricity, nearly instantaneously with respect to most measurements.[38] If the quantities C and τ are made small enough, the transients will vanish rapidly and the instrument will essentially operate in the transient-free, ultimate-state mode. In an ideal system the time derivative is nearly equal to zero and the noise sources are negligible, whereupon the ideal system equations are

$$\frac{U_0^2(R + \delta)}{4[(R + R_A) + 1 + (R_A/2_R)\delta]^2} = f(v)[(T_W - T_0) - (T_A - T_0)]$$

$$R_W - R_0 = \alpha R_0(T_W - T_0) \quad \text{and} \quad ku = U$$

$$\frac{U\delta}{4(R + R_A)} = u\left\{1 + \frac{\delta}{R}\left[\frac{2R + R_A}{2(R + R_A)}\right]\right\}$$

It follows that

$$\delta = \frac{4R(R + R_A)}{(k - 2)R - R_A}$$

The system tends to keep δ constant and thus, since $R_W = R + \delta$, the wire temperature is also maintained at the constant value

$$T_W = \frac{1}{\alpha R_0}\left[R + \frac{4R(R + R_A)}{(k-2)R - R_A} - R_0 + \alpha R_0 T_0\right]$$

We can see, therefore, that since δ and T_W tend to be invariant with time then

$$U_0^2 = (\text{constant}) f(v)$$

The constant is easily determined.

It is not necessary to solve the equations to obtain an estimate of the relationship between signal and noise. If it were possible to eliminate the excess variables and write the array in a single equation

$$F\left(\frac{d^2U}{dt^2}, \frac{dU}{dt}\right) + U = Av + B_1 e_1 + B_2 e_2 + B_3 e_3 + B_4 e_4 + B_A e_A$$

it could be argued that if the noise and the signal had similar spectra then

$$\frac{|A|\sigma_v}{(B_1^2\sigma_{e_1}^2 + B_2^2\sigma_{e_2}^2 + B_3^2\sigma_{e_3}^2 + B_4^2\sigma_{e_4}^2 + B_A^2\sigma_{eA}^2)^{1/2}}$$

would represent an estimate of the signal-noise ratio. It is assumed, somewhat intuitively, that the effects of the instrument depend only on the input frequency and that with similar frequency range the dimensionless sinusoidal sensitivity (see chapter 5) is the same for signal and for noise. This last assumption is true with linear systems but fails in nonlinear systems.

We do not have a chance to carry out the work, however. If after monumental labor we were to write a single differential equation in U, v and e, we would obtain a nonlinear expression of the form

$$F\left(\frac{d^2U}{dt^2}, \frac{dU}{dt}, U, U^2, f(v), e_0, e_e\right) = 0$$

At this point we would be unable to proceed: no practical method for the general solution of nonlinear differential equations exists.

A partial rescue is effected by use of the so-called variational procedure for the linearization of a set of equations.[39] The practical impetus for the procedure arises from the method of operation of the instrument. The hot-wire anemometer is usually used to measure turbulence, small fluctuations of velocity over the mainstream velocity. Thus it is possible to think of the system variables as composed of two parts: (1) the response to the mainstream velocity in the absence of noise and (2) a small variation to the mainstream velocity response as a result of turbulence input and system noise.

$$V = v_0 + V_1$$

$$\delta = \delta_0 + \delta_1$$

$$U = U_0 + U_1$$

$$u = u_0 + u_1$$

The variations (subscript 1) are assumed to be quite small and the ratio of the variation to the mainstream velocity response is assumed to be much less than 1. The performance equations can be symbolized by

$$f_1(e_0, e_a, \delta_0 + \delta_1, U_0 + U_1, u_0 + u_1) = 0$$

$$\frac{dU}{dt} + f_2(e_0, e_a, \delta_0 + \delta_1, U_0 + U_1, v_0 + v_1, u_0 + u_1) = 0$$

$$\frac{d\delta}{dt} + f_3(e_0, e_a, \delta_0 + \delta_1, U_0 + U_1, v_0 + v_1, u_0 + u_1) = 0$$

Each quantity f is now expanded into a Taylor series

$$\frac{\partial f_1}{\partial e_0} e_0 + \frac{\partial f_1}{\partial e_a} e_a + \frac{\partial f_1}{\partial \delta} \delta_1 + \frac{\partial f_1}{\partial U} U_1 + \frac{\partial f_1}{\partial U_0} u_1 + \cdots = 0$$

$$\frac{\partial U_1}{\partial t} + \frac{\partial f_2}{\partial e_0} e_0 + \frac{\partial f_2}{\partial e_a} e_a + \frac{\partial f_2}{\partial \delta} \delta_1 + \cdots = 0$$

$$\frac{\partial \delta_1}{\partial t} + \frac{\partial f_3}{\partial e_0} e_0 + \cdots = 0$$

The derivatives are computed with the quantities $e_0 = e_a = \delta_1 = U_1 = u_1 = v_1 = 0$ and are only functions of the reference state U_0, u_0, δ_0, v_0. The higher-order terms in the series contain products of the variational quantities, δ_1^2, $\delta_1 e_a$, δu_1, δ^3, u_1^3, etc., and are small enough to be neglected. The resulting expressions then take the form

$$\begin{aligned} A_{11}u_1 + A_{12}U_1 + A_{13}\delta_1 &= B_{11}e_1 + B_{12}e_2 + B_{13}e_3 \\ &\quad + B_{14}e_4 + B_{15}e_a = e_e \end{aligned}$$

$$\frac{dU_1}{dt} - A_{21}u_1 - A_{22}U_1 = 0$$

$$\frac{d\delta_1}{dt} - A_{32}U_1 - A_{33}\delta_1 = B_{31}v_1 + B_{32}e_0$$

As a further simplification it is noted that δ_0 itself is a small quantity. In a typical system $R \approx 6\,\Omega$, $R_A \approx 40$, $K \approx 1{,}000$, from which it follows that $\delta_0 \approx 0.178$. δ_0 can be treated as a small quantity, and whenever it appears in the denominator in the form $(1 + \delta/R)$ its effects are neglected.[40] The arbitrary function $f(v)$ is also expanded in a Taylor series:

$$f(v) = f(v_0 + v_1) = f(v_0) + \frac{df(v_0)}{dv} v_1 + \cdots$$

and the higher-order terms neglected. $f(v)$ has been given[41] as

$$f(v) = lk\left[0.43 + E\left(\frac{vD}{\nu}\right)^m\right] \tag{8.4.9}$$

in which

l = length of wire (m)
k = thermal conductivity of wire (W/m deg)
D = diameter of wire (m)

ν = kinematic viscosity of gas (m^2/s) and E and m are found from the table below.

vD/ν	E	m
1 – 4,000	0.48	0.5
4,000 – 40,000	0.174	0.618
40,000 – 400,000	0.0239	0.805

The linearized system equations become

$$0 = a_{11}u_1 + 0U_1 + a_{13}\,\delta_1 + e$$

$$\frac{dU_1}{dt} = a_{21}u_1 + a_{22}U_1 + 0\,\delta_1$$

$$\frac{d\delta_1}{dt} = 0u_1 + a_{32}U_1 + a_{33}\,\delta_1 + b_{31}e_e + b_{32}v_1$$

where

$$a_{11} = 1 \qquad a_{13} = \frac{-U_0}{4(R + R_A)}$$

$$a_{21} = \frac{k}{\tau} \qquad a_{22} = -\frac{1}{\tau}$$

$$a_{33} = 0f(v_0)(R - R_0) \qquad a_{32} = \frac{\alpha U_0}{2C(R + R_A)^2}$$

$$b_{31} = a_{32} \qquad b_{32} = \frac{-f'(v_0)(R - R_0)}{C}$$

$$e = \frac{1}{2}(e_1 + e_2 - e_3 - e_4)$$

$$e_e = \left(1 + \frac{R_A}{2R}\right)(e_1 + e_3) + \frac{R_A}{2R}(e_2 + e_4) + e_A$$

It is quite simple to eliminate all variables except U_1:

$$\frac{d^2U_1}{dt^2} - (a_{22} + a_{33})\frac{dU_1}{dt} + (a_{22}a_{33} + a_{21}a_{13}a_{32})U_1$$

$$= -a_{21}\frac{de}{dt} + a_{21}a_{33}e - a_{21}a_{13}b_{31}e_e + a_{21}a_{13}b_{32}v_1 \qquad \textbf{(8.4.10)}$$

As was discussed in chapter 7, the noise can be described by a Fourier series

$$e = \sum_{-\infty}^{+\infty} C_m e^{j\omega_m t} \qquad \sigma_e^2 = \sum_{0}^{\infty} |C_m|^2$$

According to the theory of the sinusoidal ultimate state (chap. 5)

$$U_1 = \sum_{-\infty}^{+\infty} A(j\omega_m)C_m e^{j\omega_m t}$$

and

$$\sigma_{U_1}^2 = \sum_{0}^{\infty} |A(j\omega_m)|^2 |C_m|^2$$

It can be shown from Equation (8.4.10) that

$$|A(j\omega_m)|^2 = [-(\omega_m^2) + (a_{22}a_{33} + a_{21}a_{33} + a_{21}a_{13}a_{32})^2 + (\omega_m a_{22} + \omega_m a_{33})^2]^{-1}$$

If $\omega_m \ll |a_{22}a_{33} + a_{21}a_{13}a_{32}|$, that is, if the driving frequency is much smaller than the natural undamped frequency of the system, as would be expected to occur in a well-designed instrument, we would obtain as a first estimate

$$\sigma_{U_1}^2 = \frac{\sigma_e^2}{(a_{22}a_{33} + a_{21}a_{13}a_{32})^2}$$

The response derivative of a noise component would appear as

$$U_1' = \sum_{-\infty}^{+\infty} jA(j\omega_m)C_m\omega_m e^{j\omega_m t}$$

and thus

$$\sigma_U^2 = \sum_0^{\infty} \omega_m^2 |A(j\omega_m)|^2 |C_m|^2$$

This expression is written in integral form

$$\sigma_U^2 = \int_0^{\omega_M} \frac{\omega^2 C^2(\omega)}{\omega_N^4} d\omega$$

in which $C^2(\omega) = C_m^2(\omega)\delta(\omega - \omega_m)/2\pi$ is the spectral density, $\omega_N^2 = a_{22}a_{33} + a_{11}a_{13}a_{32}$ and ω_{M} is the maximum expected frequency in the system. But the noise from a resistance is usually "white," that is, exhibits a spectral density C^2 independent of frequency, whence

$$\sigma_{U'}^2 = \frac{\omega_{\text{M}}^3}{3}\frac{C^2}{\omega_N^4} = \frac{\omega_{\text{M}}^3}{3}\frac{\sigma_e^2}{\omega_N^4}$$

The response to the noise itself can be shown to be $\omega_{\text{M}}(C^2/\omega_N^4)$, and thus

$$\sigma_{U''}^2 = \frac{\omega_{\text{M}}^2}{2}\frac{\sigma_e^2}{\omega_N^4}$$

Computation of the variances is straightforward if it is assumed that the noise components are uncorrelated (chap. 6) and thus

$$\sigma_e^2 = \frac{1}{4}(\sigma_{e_1}^2 + \sigma_{e_2}^2 + \sigma_{e_3}^2 + \sigma_{e_4}^2)$$

$$\sigma_{e_e}^2 = \left(1 + \frac{R_A}{2R}\right)^2(\sigma_{e_1}^2 + \sigma_{e_e}^2) + \left(\frac{R_A}{2R}\right)^2(\sigma_{e_2}^2 + \sigma_{e_4}^2) + \sigma_{e_A}^1$$

The noise generated by a resistance R has been given as

$$\sigma_R^2 = 4k_B T_R R\omega_M$$

If the temperature of the bridge resistance is T_A and that of the wire is $T + \Delta T$, then

$$\sigma_e^2 = 4k_B TR\omega_m\left(1 + \frac{\Delta T}{4T}\right)$$

$$\sigma_{e_0}^2 = 4k_B TR\omega_m\left[\left(1 + \frac{R_A}{2R}\right)\left(2 + \frac{\Delta T}{T}\right) + 2\left(\frac{R_A}{2R}\right)^2 + \frac{\sigma_{e_A}^2}{\sigma_R^2}\right]$$

A final combination of all the material in the previous paragraphs yields the variance σ_N^2 of the noise output:

$$\sigma_N^2 = \left(\frac{a_{21}\sigma_R^2}{\omega_N^2}\right)^2 \left\{\frac{\omega_M^3}{3}\left(1 + \frac{\Delta T}{4T}\right) + (a_{13}b_{31})^2 \right.$$
$$\left. \times \left[\left(1 + \frac{R_A}{2R}\right)^2\left(2 + \frac{\Delta T}{T}\right) + 2\left(\frac{R_A}{2R}\right)^2 + \frac{\sigma_{eA}^2}{\sigma_R^2}\right]\right\}$$

The reader can now show easily that the variance of the signal output is

$$\sigma_S^2 = \left(\frac{a_{21}a_{31}b_{32}}{\omega_N^2}\right)^2 \sigma_V^2$$

and that the signal-noise ratio is

$$n = \frac{\sigma_S}{\sigma_N}$$

Example 8.4.2 A constant-temperature hot-wire anemometer is designed to function with a platinum wire 5 mm long and 0.006 mm in diameter. The air temperature is 20°C, the output impedance of the amplifier is 40 Ω and the resistance of the platinum at 0°C is 17.7 Ω. Determine the signal-noise ratio of the instrument if the stream velocity is 5 m/s and the turbulence is on the order of 0.01 m/s rms. The following handbook information is needed:

$$\begin{aligned}
\alpha &= 3 \times 10^{-3} \\
k_B &= 1.38 \times 10^{-22}(\text{W/deg s}) \\
v &= 10^{-6}(\text{m}^2/\text{s}) \\
k &= 79.8\ (\text{W/m deg}) \\
\rho &= 2.14 \times 10^4\ (\text{kg/m}^3) \\
C_s &= 157\ (\text{W/deg}) \\
K &= 1{,}000 \\
\tau &= 10^{-3}\ (\text{s}^{-1}) \\
\omega_M &= 2{,}000\ (\text{Hz})
\end{aligned}$$

Solution The following quantities are computed:

$$\begin{aligned}
C &= 4.74 \times 10^{-7} \\
R &= 18.7\ \Omega \\
Dv/v &= 30, \text{ from which } m = 1/2E = 0.48 \\
f(v_0) &= lk[0.43 + 0.48(5D/v)^{1/2}] = 1.22(\text{W/deg}) \\
f(v_0) &= lkE(D/v)^m mV_0^{m-1} = 0.104\ (\text{W/deg/m/s}) \\
U_0 &= 3.1 \times 10^{-2} \\
\delta_0 &= 0.236 \\
\Delta T &= 4.45°\text{K} \\
T &= 293°\text{K}
\end{aligned}$$

From the above data we obtain

$$a_{11} = 1 \qquad a_{13} = -1.32$$
$$a_{21} = 10^6 \qquad a_{22} = -10^3 \qquad a_{33} = -2.68 \times 10^6$$
$$a_{32} = b_{31} = 1.14 \times 10^3 \qquad b_{32} = -2.19 \times 10^5$$

Also

$$\sigma_{e_A}^2 = 10^{-4} \qquad \sigma_{v_1}^2 = 10^{-4}$$

and finally

$$\sigma_N^2 = 2.82 \times 10^{-4}$$

$$\sigma_v^2 = 2.48 \times 10^{-2}$$

$$\sqrt{\frac{\sigma_v^2}{\sigma_N^2}} = 9.34$$

This is not a particularly advantageous ratio. A real improvement would be obtained if the amplifier were designed to generate only 10^{-8} volts rms of noise. In this case the ratio would be near 530. Such quiet amplifiers are very expensive. It can be shown that increasing the amplifier output impedance is a cheapter method of reducing noise.

8.5 Noise as a Measurement Principle

Noise can be useful. Since resistance noise depends on the temperature, it follows that the noise output of a resistor could be used to measure temperature.[42] Although no commercially available noise thermometer is known to the author, the method has some attractiveness for high-temperature measurements. A natural extension of the noise formula presented in Section 7.9 is the Nyquist formula

$$d(\sigma_N^2) = 4k_B[R_e(Z)]T\, d\omega$$

where $R_e(Z)$ is the real part of the complex impedance Z of an electrical network. If the noise is processed through an amplifying system, the result is

$$\sigma_N^2 = \int_0^\infty 4k_B A(\omega) R_e(Z) T\, d\omega$$

A reference circuit is used to measure the variance of the noise at an accurate, known, fixed reference temperature T_1:

$$\sigma_{N_1}^2 = 4k_B T_1 \int_0^\infty A(\omega) R_e(Z_1)\, d\omega$$

A similar circuit is introduced into another medium at temperature T_2 and processed through the same amplifier

$$\sigma_{N_2}^2 = 4k_B T_2 \int_0^\infty A(\omega) R(Z_2)\, d\omega$$

The impedance Z_2 is varied so that $\sigma_{N_1}^2 = \sigma_{N_1}^2$ and thus

$$\frac{T_2}{T_1} = \frac{\int_0^\infty A(\omega) R_e[Z_1(\omega)]\, d\omega}{\int_0^\infty A(\omega) R_e[Z_2(\omega)]\, d\omega}$$

If the impedances Z_1 and Z_2 can be selected to be independent of frequency and of the amplifier characteristics then

$$\frac{T_2}{T_1} = \frac{R_e(Z_1)}{R_e(Z_2)}$$

and T_2 can be computed in terms of T_1.

A possible, though very crude, setup for measurement is shown in Figure 8.25. In this instrument the rms noise of circuit R_1, T_1, is measured first. Then the switch is shifted to R_2 and the resistance varied until the rms output from R_2 is the same as the rms output of R_1. The temperature is obtained directly by computation or by a vernier dial on R_2.[43]

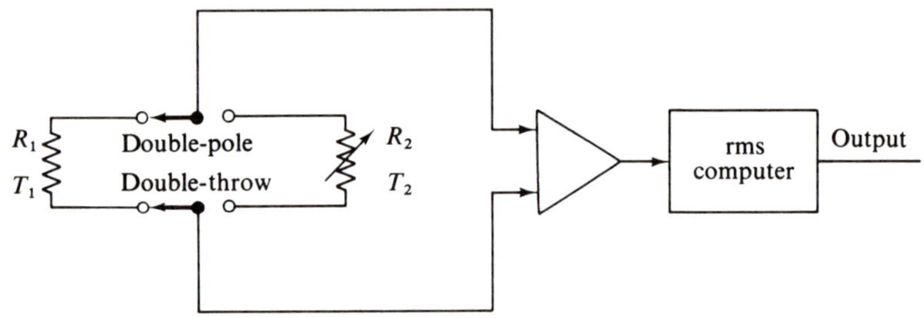

FIGURE 8.25 Crude noise thermometer

PROBLEMS

8.1 An amplifier subjected to sinusoidal vibrations picks up oscillations mechanically through a plate capacitor and follows the performance equation

$$\tau\frac{dz}{dt} + k_0(1 + \mu \sin \omega_0 t)z = Sy \quad (\mu \text{ is small})$$

The input to the amplifier is from a strain gage bridge which measures the vibrations of the structure on which the amplifier is mounted; hence

$$y = K \cos \omega_0 t$$

Show that the output of the amplifier is not proportional to y and compute the systemic error under the assumption that the ideal measurement is obtained when $\mu = 0$.

8.2 In the circuit shown in Figure 8.3, R_1 is subjected to a large temperature variation, and thus

$$R_1 = R_0(1 + 4 \times 10^{-3}\theta + 7 \times 10^{-6}\theta^2)$$

Select a thermistor $R_T = R_{T0}(1 - \beta\theta - \beta_1\theta^2)$ so that $R_e = R_0$ with neglected errors of order θ^3 or smaller.

8.3 Obtain a performance equation for the system in Figure 8.5 with the following assumptions.

(a) The bourdon tubes have performance equations in the form

$$f_1(\ddot{\delta}_1, \dot{\delta}_1, \delta_1) = A_1(p - p_a) + A_2(\theta - \theta_0)$$
$$f_2(\ddot{\delta}_2, \dot{\delta}_2, \delta_2) = B_1(p - p_a) + B_2(\theta - \theta_0)$$

δ_1 = displacement of endpoint of active bourdon
δ_2 = displacement of endpoint of compensating bourdon

(b) Adopt the following notation:

p = working pressure (input)
p_A = atmospheric pressure
θ = working (case) temperature
θ_0 = reference temperature
J_1 = moment of inertia of bourdon tube when the semicircle changes through an angle ϕ_1
R_1 = radius of bourdon element
$\delta_1 = R_1\phi_1$ = displacement of active bourdon tip
J_0 = rigid moment of inertia of tube rotating about its pivot
δ_0 = motion of active tip for a rotation of bourdon element about the pivot $\delta_0 = R_0\phi_0$
δ_2 = motion of compensator tip
ϕ_2 = angular displacement of compensator element
R_2 = radius of compensator element
$\delta_2 = \phi_2 R_2$
J_2 = moment of inertia of compensator for angular acceleration in ϕ_2
n = linkage ratio $n\delta_2 = \delta_0$
f_1 = angular damping in active element
f_2 = angular damping in compensator
f_0 = angular damping in pivot

(c) Assume that all phenomena are linear and that the active and compensating tubes show an elastic (or spring) moment $k_1\phi_1$ and $k_2\phi_2$, respectively. See Figure 8.26 for nomenclature.

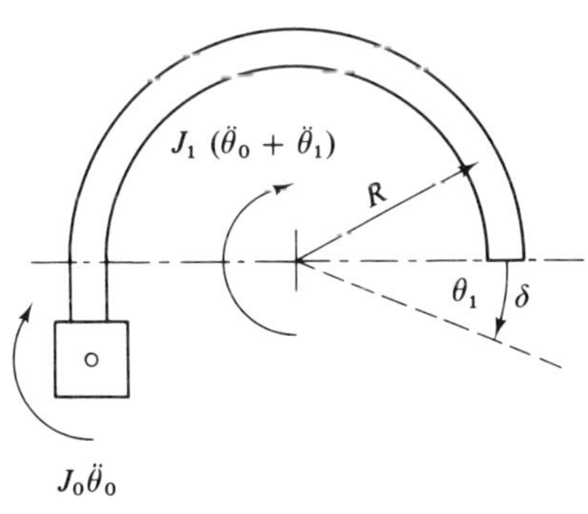

FIGURE 8.26 Diagram for problem 8.3

8.4 Evaluate B_1 and B_2 in terms of A_1 and A_2 and the other parameters in problem 8.3 to fully compensate for a constant change in temperature and a constant change in pressure.

8.5 What should B_2 be to compensate the bourdon tube in problem 8.3 for a sinusoidal temperature variation $\theta - \theta_0 = a \sin \omega_0 t$?

8.6 Obtain an estimate of the error for an amplifier with a steady input $e_1 = 1$ [(Equation 8.2.13)] in which the gain drifts. $A = A_0(1 - \alpha t)$.

8.7 Compute the drift at the end of five minutes for an integrator in which $RC = 10$, $A = 200$ and the input voltage $e_i = 10$.

8.8 In an integrator $RC = 0.01$. Find the value of A for which e_0/e_i is 0.99 at the end of ten minutes, i.e., the gain for which the drift is 0.1% per minute.

8.9 A micrometer has a pitch of 0.1 mm per turn. If the barrel is 2 cm in diameter, determine the gain k. The metal of the screw expands (in this configuration) with temperature at a rate of 0.002 mm/°C. What measurement error is introduced per degree C?

8.10 An orifice meter wears with use so that the area (A) becomes larger. It follows that A and β change with time. Let the diameter $d = d_0(1 + bt)$ mm in which d_0 is the nominal initial diameter and $b = 10^{-2}$ mm/year. Compute an estimate of the flow error after 20 years for the case in which $R = 0.94$, β at $t = 0$ is 0.5, $\Gamma = 1.30$, $p_1 = 25{,}000$ dynes-cm^2, $T_1 =$ 300°K and $d_0 = 1$ cm.

8.11 A heat flux transducer is 0.28 cm in diameter and is embedded in a conducting wall which is maintained at a steady 50°C whenever the gas passing over the wall is held at a steady 180°C temperature. The slug is made of platinum, $C_s = 157$ W/°C and $\rho = 2.14 \times 10^4$ kg/m^3 and is 3 mm thick. Determine C_1. If $q_L = 7.31 \times 10^{-5}\,\theta^{1.9}$ watts and $\dot{Q} =$ 1,000 watts steadily at the transducer, determine $\dot{Q}$ on the plate if $b = 1$ meter. [Note: use Equation (8.3.6) to find θ_T. Assume the initial temperature of the transducer to be 50°C.]

8.12 Two samples weigh 0.0112 gram and 0.0117 gram, respectively. The samples are placed in opposite pans of an analytical balance open to the air at sea level. The first sample has a density of 10^{-1} grams/cc and the second 10^{-2} grams/cc. The difference in masses in air will be given by the scale as $0.0117 - 0.0112 + \epsilon_M$. Find ϵ_M.

8.13 Two masses have identical densities but the volume of one mass is different from the volume of the other. The balance will settle at some angle ϕ not equal to zero. Assume a perfect balance and show how ϕ can be used to determine the density of the air.

8.14 As a first approximation it can be shown that in an analytical balance the quantity h (Fig. 8.16) changes with the applied load according to the law

$$\frac{\epsilon_h}{h_0} = \frac{h - h_0}{h_0} = \frac{(M_1 + M_2)g \cos \phi}{AE}$$

in which $h_0 = h$ for $M_1 + M_2 = 0$, $\phi = 0$ and

g = gravitational acceleration constant
A = cross-sectional area of bar
E = modulus of elasticity of material in bar

Determine the ultimate state error induced in the measurement of an angle ϕ given an otherwise "perfect" balance.

8.15 A second-order system has the performance equation

$$\ddot{z} + 10\dot{z} + 2500z = Ky$$

Design a linear damper using SAE 10 lubrication oil at 25°C ($\mu/\rho =$ 1 cm²/s) to make $\zeta = 1$. Let the maximum value of $\dot{z}$ be 20 m/s.

8.16 A rotary damper is composed of a fixed cylinder with another cylinder inside it. An oil is placed between the two cylinders. Obtain a performance equation for the system. (Note: the basic applicable fluid motion equation is the equation of viscous resistance in laminar flow between two parallel plates. Look it up in the literature.)

8.17 Obtain an equation similar to (8.3.18) for the transistor amplifier. (Refer to the literature for the characteristics of transistors.)

8.18 From Equation (8.3.18) compute an approximation of the error

$$e_o + Se_c = \epsilon(e_c, e_o)$$

in the form

$$\epsilon = a_1 e_c + a_2 e_o + a_3 e_c e_o + a_4 e_c^2 + a_5 e_o^2$$

all higher-order terms to be neglected.

8.19 In Figure 8.21 the gain of the second amplifier is assumed to be 1. Let the transducer be an iron-constantan ® thermocouple and let the wire between (1) and (2) be 100 feet of AWG 10 copper wire. Find the characteristics of the system from the literature and determine the noise introduced into the system in terms of an equivalent temperature error at the input.

8.20 Obtain a second-order noise equation for a resistance bridge containing e^2 and δ^2 and $e\delta$ and compare to Equation (8.4.2). If $\delta = 0.200\ \Omega$ and $R = 20\ \Omega$ and $e_1 = e_2 = e_3 = e_4 = 2 \times 10^{-2}$ volts, determine the value of the second-order error. Is it negligible?

8.21 The maximum white-noise frequency in a system is 20,000 Hz. If in a bridge the resistance $R = 20$ ohms, temperature $T = 320°$K and the temperature change in the active resistance is zero, determine the rms Johnson noise in the circuit.

8.22 Repeat Example 8.4.2 with the stream average velocity equal to 20 m/s.

8.23 In the noise thermometer (Fig. 8.25) let $R_1 = 10.002$ and $R_2 = 9.725$. $T_1 = 204°$K. Find T_2.

NOTES

[1]Do not, however, dismiss RSI as a source of design information. Most new, clever ideas in measurement usually appear first in RSI. If error sources are discussed in RSI, the treatment is likely to be more thorough than in engineering journals. See, for example, Ref. 105.

[2]Ref. 85, pp. 334–35. R. Millikan subsequently won the Nobel Prize for his determination of Plank's constant within 0.5%. This measurement took him 3 years to perform.

[3]The author has used as many as three years to design one instrument and spent a fruitless decade in search of a nonsonic method for the measurement of the adiabatic compressibility of hydraulic oils at high temperature and high pressure. Ref. 86, 87, 88.
[4]Ref. 6, pp. 2–48.
[5]Ref. 6, pp. 2–62.
[6]Ref. 6, pp. 2–70 and pp. 3–16.
[7]The design details are too numerous to be discussed here. See Ref. 6, 3 and 66, as well as the promotional literature of bourdon manufacturers.
[8]Ref. 6, pp. 2–60. For additional drift and aging phenomena see Ref. 96 and 97.
[9]Ref. 103.
[10]Ref. 15, pp. 11–14.
[11]Pitch in a single thread screw is equal to the lead or distance traveled by the nut in a 2π turn. In micrometer design the lead is the relevant parameter.
[12]The discussion in Ref. 15 for the calibration of a micrometer leads to a crude approximation of the error. If the micrometer were to be used to measure a large number of identical parts against a standard, the approximate procedure prescribed in p. 13 of Ref. 15 would be quite adequate.
[13]Ref. 89, pp. 33–34; also Ref. 3, 4, 6, 8, 90, 91, 92, 93, 94 and 95.
[14]Ref. 89 and 96.
[15]Ref. 93. The derivation requires little more than a first course in thermodynamics. See, for example, Ref. 99, pp. 648–78.
[16]Ref. 92 and Ref. 93. The irrationality of "rational" equations gives us pause. The possibility of writing $K_1 K_1 Y \sqrt{S/G}$ as a single constant to be obtained from a single table for a single fluid has been ignored.
[17]Ref. 93, p. 63.
[18]Ref. 94, p. 25.
[19]Ref. 98.
[20]Ref. 99, Chapter 1 and passim.
[21]Ref. 100.
[22]Ref. 103, p. 48 and p. 90.
[23] ® Johns-Manville Co.
[24]Ref. 100.
[25]Beam balances with manual readout can be read to 10^{-5} grams. Vacuum-mounted beam balances with feedback nulling and electronic readout can measure 10^{-12} grams.
[26]Analytical balances designed to be used in vacuum are available commercially.
[27]Ref. 101.
[28]See any elementary textbook in fluid mechanics, for example, Ref. 102, pp. 184–86.
[29]The reason for current moving opposite to electron drift is buried in antiquity. Upon this arcane and asinine concept several lawsuits hinged: two claimants for a patent on the triode vacuum tube based the distinction between two identical devices on the distinction between the real electron drift and the imaginary plate to cathode "current."
[30]By means of a rather complicated gimbal, within gimbal, within gimbal combination and a flexible wire, it is possible to connect a single rotating element to ground without slip rings. The servomechanism needed to move

all the gimbals as functions of a single output makes the cost of the device quite high and replaces one source of noise with several others.

[31]Ref. 82 and Ref. 83. Reference 82 is rather old (1940) but is at this moment (1972) the most current single reference on the design of electrical contacts.

[32]Ref. 109.

[33]Ref. 104, 105. The results presented in Section 8.4.3 differ somewhat from those given in References 104 and 105. In the present work it is assumed that there exists an amplifier output noise (rather than the customary assumption that internal amplifier noise is equivalent to an input noise) and that the amplifier output impedance is not zero. The author of References 104 and 105 makes slightly different assumptions. In addition, a theoretical estimate of the wire response is made in this text.

[34]Ref. 105, Ref. 108 (Johnson noise).

[35]Ref. 104.

[36]Ref. 106.

[37]Hot-wire anemometer probes are too fragile to use in liquids. For water, the hot-film anemometer is commonly used. The theory is identical to the theory of the hot-wire system. In practice, however, hot-film anemometers are touchy and difficult to use.

[38]Most resistors also behave as small capacitors and small self-inductors. Thus the actual impedance of the resistance is not R but

$$\frac{RLs}{R + Ls + RCLs^2}$$

Care must be taken in the design of bridges to extremize stray transient in resistive components.

[39]Ref. 14, 23.

[40]This simple fiction is not needed for the theory. The coefficients change a little if δ_0 is retained. In some commercially available anemometers δ_0 is quite large since K is small. The author uses a device built in his own laboratory with $K = 10{,}000$ and $R_A = 500$ and hence $\delta \approx 0.22$. This keeps the wire at 101°C, an overheat ratio of 4.

[41]Ref. 106, p. 242.

[42]Ref. 107.

[43]Ref. 107, 108 and 109.

9

FILTERS

9.1 Introduction

Real systems tend to amplify, suppress or otherwise distort sinusoidal signals (chap. 5). Advantage is taken of predictable distortion to suppress unwanted signals (noise) or enhance desired inputs.

Example 9.1.1 Show that a system described by

$$\tau_L \frac{dz_1}{dt} + z_1 = Ky_1 \tag{9.1.1}$$

tends to suppress high-frequency inputs, whereas

$$z_2 = k\left(\tau_H \frac{dy_2}{dt} + y_2\right) \tag{9.1.2}$$

tends to emphasize or amplify high-frequency signals.

Solution Let the input be $y = C \sin \omega t$. Then in the ultimate state

$$z_1 = A_1 KC \sin (\omega t + \phi_1)$$

$$z_2 = A_2 kC \sin (\omega t + \phi_2)$$

in which (see Section 5.4)

$$A_1 = 1/\sqrt{1 + (\omega\tau_L)^2}$$

$$A_2 = \sqrt{1 + (\omega\tau_H)^2}$$

It is clear that A_1 approaches zero and A_2 approaches infinity as ω becomes large.

Example 9.1.2 The outputs of the two devices described in Example 9.1.1 are connected to elements of the chain which have an input threshold amplitude of 0.01. Select the parameters so that the first filter will suppress all input signals with frequencies greater than 10 rad/s, whereas the second filter suppresses input signals with frequencies lower than 10 rad/s.

Solution Let $A_1K = A_2k = 0.01$ and $\omega = 10$ to obtain

$$(1 + 100T_L^2)^{1/2} = 100K$$

$$100k(1 + 100T_H)^{1/2} = 1$$

Any combination of K with T_L and k with T_H which satisfies the equations is a solution.

The "low-pass" first-order filter described by Equation (9.2.1) in Example 9.1.1 tends to pass the low-frequency inputs relatively undisturbed, that is, $A_1 \approx 1$ for $\omega\tau_L \approx 0$. The system described by Equation (9.1.2), on the other hand, has no region of flat response (chap. 5) and passes distorted high-frequency signals. Hence it cannot be used as a true high-pass filter. It is necessary to "flatten out" the high-frequency Bode plot. A simple scheme is to introduce a cascaded first-order low-pass filter in such a fashion that the increase in amplitude of A_2 is counteracted. The simplest technique is to use the system

$$\tau_L\frac{dz}{dt} + Z = K\left(\tau_H\frac{dy}{dt} + y\right) \qquad \textbf{(9.1.3)}$$

and to pick τ_L, τ_H and K so as to meet the filtering criteria. The output ratio is

$$A_3 = \sqrt{[1 + (\tau_H\omega)^2]/[1 + (\tau_L\omega)^2]}$$

For $\omega \approx 0$, $A_3 \approx 1$ and for large values of ω

$$A_3 \approx \frac{\tau_H}{\tau_L}$$

The output signal will have the form

$$z = KA_3C \sin(\omega t + \phi_3)$$

so that the amplitude of z can be shown in a Bode plot as (Fig. 9.1)

$$20 \log |z| = 20 \log (KC) + 20 \log A_3$$

$$|z|_{\text{dB}} = (KC)_{\text{dB}} + (A_3)_{\text{dB}}$$

The quantities K, τ_L and τ_H can be selected so that the low-frequency inputs are suppressed. It is required that $\tau_H > \tau_L$.

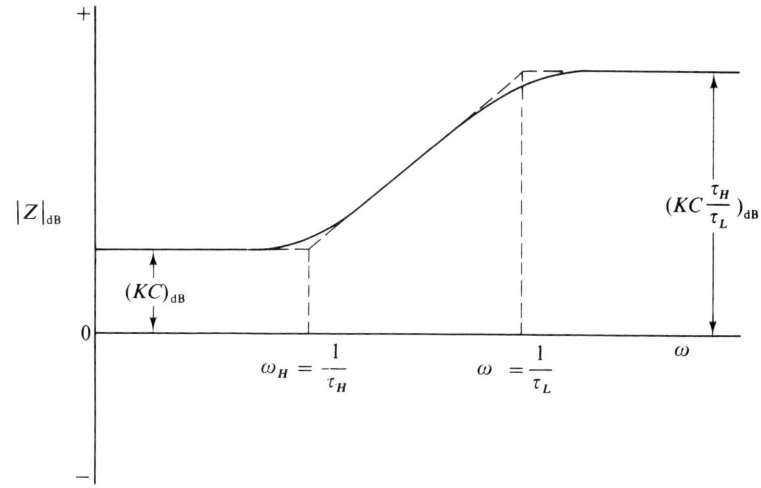

FIGURE 9.1 Gain characteristics of high-pass filter

Example 9.1.3 A high-pass filter is connected to a system which suppresses input signals of amplitude 10^{-3}. Design a filter which eliminates all signals with frequency 60 rad/s and which passes high-frequency signals with an amplitude of 1, given that $C = 1$.

Solution If no regard is given to signal amplitude distortion and phase shift, the specifications can be translated into

$$K\frac{\tau_H}{\tau_L} = 1$$

and

$$\sqrt{[\tau_L^2 + 3{,}600(\tau_H\tau_L)^2]/[\tau_H^2 + 3{,}600(\tau_H\tau_L)^2]} = 10^{-3}$$

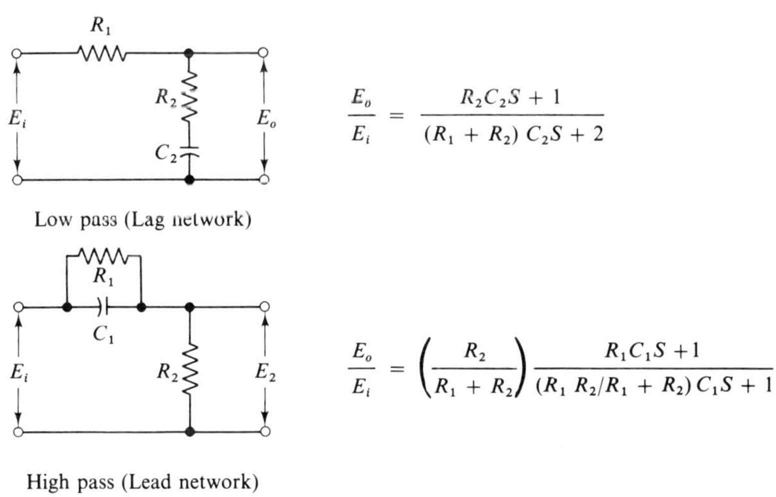

FIGURE 9.2 Realization of two simple filter-networks

Any combination of τ_H, τ_L and K which satisfies the equation is a solution to the problem.

The system described by Equation (9.1.3) is physically realizable and acts as a high-pass filter if $\tau_H < \tau_L$ and as a low-pass filter if $\tau_L < \tau_H$. Electrical networks with the required performance equation are shown in Figure 9.2. Mechanical or pneumatic equivalent filters can be found in the literature.[1]

9.2 Approximate Filter Equations

Filter designers decide in advance what amplitude characteristics are needed for a system and then construct the filter by the superposition of first-order and second-order elements. A Bode diagram is drawn to match the filter amplitude ratio specifications and then the roll-off slopes and the flat regions are approximated by a simple extension of the previous methods.

9.2.1 Let it be required to obtain the approximate performance equation of a system with a prescribed Bode diagram. Such an approximation may be built up by means of the graphical summation of a first-order system. Let us note from chapter 5 that

$$A^2 = |G(j\omega)|^2 = \frac{[(\tau_{H1}\omega)^2 + 1][(\tau_{H2}\omega)^2 + 1]\cdots[(\tau_{Hm}\omega)^2 + 1]}{[(\tau_{L1}\omega)^2 + 1][(\tau_{L2}\omega)^2 + 1]\cdots[(\tau_{Ln}\omega)^2 + 1]}$$

is a representation of the amplitude of the output of a system with the transformed performance equation

$$z = KG(s)\mathcal{L}(y)$$

in which

$$G(s) = \frac{(\tau_{H1}s + 1)(\tau_{H2}s + 1)\cdots(\tau_{Hm}s + 1)}{(\tau_{L1}s + 1)(\tau_{L2}s + 1)\cdots(\tau_{Ln}s + 1)}$$

If we generate the nomenclatures

$$A_{Hi} = [1 + (\tau_{Hi}\omega)^2]^{1/2}$$

$$A_{Li} = [1 + (\tau_{Li}\omega)^2]^{-1/2}$$

We see immediately that

$$A_{\mathrm{dB}} = 20\log\left(\frac{\prod_i A_{Hi}}{\prod_j A_{Lj}}\right) = \sum_{i=1}^{m}(A_{Hi})_{\mathrm{dB}} - \sum_{j=1}^{n}(A_{Lj})_{\mathrm{dB}} \tag{9.2.1}$$

Equation (9.2.1) is the key to a simple graphical procedure for the construction of the approximate arbitrary Bode diagram of a cascade of first-order systems. As a first step let us replace the Bode plot of the first-order systems

$$A_H = \sqrt{1 + (\tau_H \omega)^2}$$
$$A_L = 1/\sqrt{1 + (\tau_L \omega)^2}$$

by approximate lines, the tangent near $\omega = 0$ and the tangent to the curve for large ω (Fig. 9.3). The approximate curve is now built up of a line parallel to $A = 0$ up to the "breakpoint" from which the approximation becomes a line with a positive slope of 20 dB/dec for (A_H) dB and a negative slope of 20 dB/dec for (A_L) dB. The equations of the sloping lines are

$$(A_H^\infty)_{\text{dB}} = 20 \log (\tau_H \omega) \qquad (A_L^\infty)_{\text{dB}} = 20 \log (\tau_L \omega) \qquad \textbf{(9.2.2)}$$

The breakpoint occurs at the intersection of the lines (9.2.2) with the $A = 0$ axis so that the break frequencies ω_H, ω_L are given by

$$\omega_H = \tau_H^{-1} \quad \text{and} \quad \omega_L = \tau_L^{-1}$$

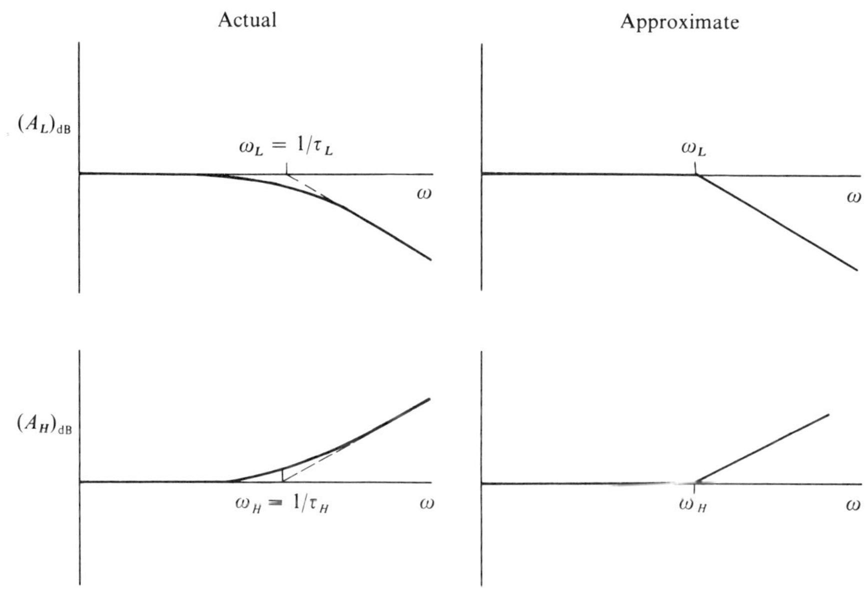

FIGURE 9.3 Replacement of first-order Bode plots by straight-line approximations

9.2.2 The construction of the approximate Bode plot is quite simple. The graph is approximated by straight line segments and then replaced by a set of first-order curves. It was demonstrated in chapter 5 (Section 5.8) that the high-frequency slope of the Bode diagram for linear systems is always an integral multiple of (n20) dB/dec, and thus the best fit linear system to the given gain characteristics is given by the closest (n20) dB/dec slope at any point.

Example 9.2.1 A system measures signals of all frequencies. A set of unwanted signals lies between 20 rad/s and 30 rad/s. Find the transfer function and performance equation of a filter which tends to attenuate the signals in the given band. The signals in the range between 24 and 26 rad/s should be down -3 dB.

Solution A straight-line approximation to the sinusoidal gain characteristics is shown in Figure 9.4a. The initial graph is built up by the system shown in 9.4b (breakpoint $\tau_1 = 0.05$ s) and the system shown in 9.4c (breakpoint $\tau_2 = 0.0417$ s). The sum of (a) and (c) is given by (d). Here the two slopes are selected to cancel and generate a straight-line response. To find (n) it is necessary to compute the slope. From (a) and (d) we note that the first negative slope is equal to $-3/(\log 24 - \log 20)$ so that

$$\frac{-3}{\log 24/20} = -20n = -38$$

and thus $n \approx 2$. We use a slope of 40 dB/dec. Now the curve can be completed by adding a positive 40 dB/dec slope at the breakpoint 26 rad/s, and a negative slope of 40 dB/dec at the breakpoint 30 rad/s.

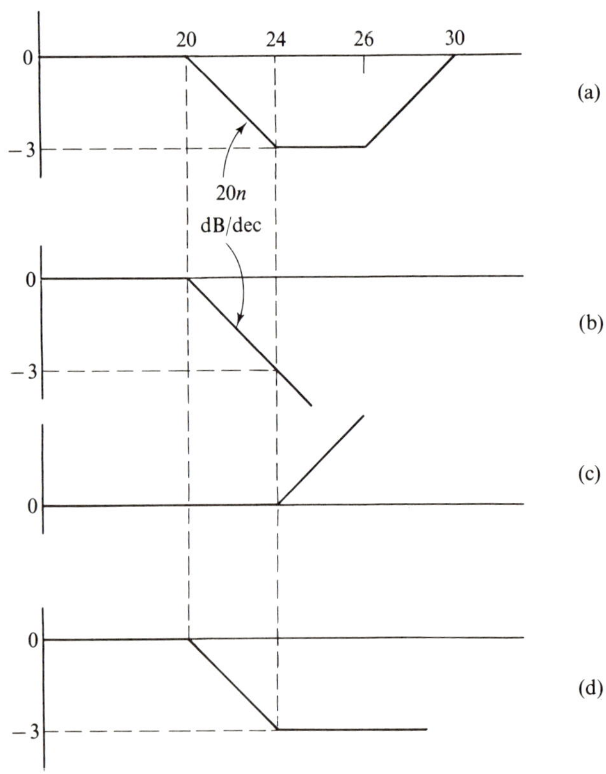

FIGURE 9.4 First step in the synthesis of a band-pass filter

40 dB/dec slopes can be obtained by superposing two first-order systems so that the transfer function is now given by

$$G(s) = \frac{(0.0417s + 1)^2(0.0387s + 1)^2}{(0.050s + 1)^2(0.0333s + 1)^2}$$

The performance equation is found by inverting

$$(0.0500s + 1)^2(0.0333s + 1)^2 Z = K(0.0417s + 1)^2(0.0387s + 1)^2 Y$$

K is determined from other conditions such as the threshold of the instrument.

9.2.3 The typical second-order Bode plot "rolls off" with a slope of -40 dB/dec. It is possible, therefore, to use a single second-order system in place of a cascade of two first-order systems. The breakpoint is found by noting that

$$\lim_{\omega \to \infty} A(\omega) = K\omega^{-2}$$

and that therefore the break occurs at

$$K = \omega_B^2$$

if it is noted that the performance equation must be set in the form

$$\frac{d^2z}{dt^2} + 2\zeta K^{1/2}\frac{dz}{dt} + Kz = Ky$$

so that $A \longrightarrow 1$ as $\omega \longrightarrow 0$. Examination of Figure 5.7a shows that the second-order system which approaches the two tangents $A_{\text{dB}}(0) = 0$ and $A_{\text{dB}}(\infty) = K_{\text{dB}} - 40 \log \omega$ is a system in which $\zeta = 0.707$. The break is much sharper than in the first-order system or in any combination of two first-order systems. The last statement is obvious once it is realized that a cascade of two first-order systems is equivalent to a single second-order filter with real characteristic roots, that is, with $\zeta \geq 1$ where the equality applies to the case of equal real roots. It is possible, therefore, to design second-order filter systems with $\zeta = 0.707$, which can be used for filter design in place of a pair of first-order elements. From the equations developed in chapter 5 (Example 5.4.2) we obtain for $\zeta = 0.707$

$$A_L = \left[1 + \left(\frac{\omega^2}{K_L}\right)^2\right]^{-1/2}$$

$$A_H = \left[1 + \left(\frac{\omega^2}{K_H}\right)^2\right]^{-1/2}$$

in which the transfer functions for each system are given by

$$G_L(s) = \frac{K_L}{s^2 + \sqrt{2K_L}s + K_L}$$

$$G_H(s) = \frac{s^2 + \sqrt{2K_H}s + K_H}{K_H}$$

Example 9.2.2 Rework Example 9.2.1 with second-order ($\zeta = 0.707$) filters.

Solution The first break occurs at $\omega = 20$ rad/s so that $K_1 = 20^2 = 400$. The curve is flattened out at 24 rad/s so that the second element requires $K_2 = (24)^2 = 576$. The two elements which return the Bode graph to the $A = 0$ axis are given by $K_3 = 26^2 = 676$ and $K_4 = 30^2 = 900$. The amplitude ratio of the filter is now given by

$$A = \frac{[1 + (\omega^2/576)^2]^{1/2}[1 + (\omega^2/676)^2]^{1/2}}{[1 + (\omega^2/400)^2]^{1/2}[1 + (\omega^2/900)^2]^{1/2}}$$

The system transfer function is given by

$$G(s) = 0.927\frac{(s^2 + 33.9s + 576)(s^2 + 36.8s + 676)}{(s^2 + 28.3s + 400)(s^2 + 42.4s + 900)}$$

9.2.4 Combinations of first- and second-order systems can be used to approximate arbitrary Bode plots. Slopes of even multiples of 20 dB/dec can be built up with cascades of second-order systems, whereas odd multiples can be generated by cascades of first-order systems or by combinations of second-order elements with first-order elements.

9.2.5 Once the approximate Bode graph has been synthesized, the general form of the performance equation is given. Beginning with the first approximation the coefficients of the filter performance equation can be varied by small increments on both sides of the value of the first approximation. Often, by trial and error maneuvering, a prescribed filter response diagram can be found. Analogue computer simulation is often the best technique. Hybrid analogue-digital computers can be programmed to print out Bode diagrams. Families of such diagrams can be printed as functions of design parameters and the best response selected from the family. For most instrument designs such refinement is not necessary; the first approximation does well enough.

9.2.6 The design of electrical or mechanical filters is a highly specialized profession which has spawned an extensive subliterature.[2] Once the transfer function has been found, a system has to be built to "realize" the performance equation.[3] That physical realization is the most difficult portion of filter design is well known in the trade—but too much has been made of it. The average instrument designer can find a commercial filter or a published filter design to meet most of his needs.[4] Almost any linear performance equation can be synthesized with operational amplifiers and simple electrical components.[5] The problem of separating the noise and the signal is solved as long as the noise frequency band and the useful signal frequency band are sufficiently different from each other to allow elimination of the noise by suppression of all signals in the noise band and amplification of all signals in the useful signal band.

9.3 Optimum Filters

In current practice it has become customary to design filters for high noise-to-signal ratios by statistical procedures. Usually the expression for the root mean square output error is written, and the available design parameters are then varied to obtain a minimum error. Since the procedure involves the use of the optimizing principles of variational calculus, such filters are called *optimum filters.*

9.3.1 If the noise input is random and is not presented to us in a well-defined band, filtering becomes more difficult to accomplish. Knowledge of certain statistics of the useful signal and the noise can be used to find the performance equation of filters which tend to assign a higher probability of transmission to useful signals than to noise. Such filters are called *optimum* filters.

How optimality is to be defined depends on the results desired and on the nature of the signal and noise. Commonly one would wish to minimize the error so as to make the instrument more precise (chap. 1). Unfortunately, not enough information is available about signal or noise, and one is reduced to the minimization of the mean square error or of the variance (Section 6.2). As a first approach to design it is usually assumed that the signal is statistically "stationary" (Section 6.5) and that the noise has a simple, well-defined spectrum such as "white" noise or "gaussian" noise. White noise is usually employed to make the computation simple, even though the methods discussed are applicable to any noise for which a mathematical model is available.

9.3.2 Let the useful input signal be $y(t)$ and the noise signal be $n(t)$. At the input there will appear a signal $x(t) = y(t) + n(t)$. It is assumed that both $y(t)$ and $n(t)$ are time-varying random functions and statistically stationary. The filter system is given by the transfer function $G(s)$ or

$$\mathfrak{L}z(t) = G(s)\mathfrak{L}x(t)$$

The problem becomes that of selecting $G(s)$ in such a fashion that the effects of $n(t)$ are minimized and such that the filter output approaches as closely as possible the ideal response $\bar{z}(t)$ defined by

$$\mathfrak{L}\bar{z}(t) = G(s)\mathfrak{L}y(t)$$

The signal $y(t)$ can now be recovered from $\bar{z}(t)$ by fairly straightforward methods.*

In effect, we are required to find the transfer function $G(s)$ or the

---

*Ideally it is wished to solve the equation

$$\mathfrak{L}y(t) = G(s)[\mathfrak{L}y(t) + \mathfrak{L}n(t)]$$

for $G(s)$. Such a procedure is impossible.

impulsive response $I(t) = \mathcal{L}^{-1}G(s)$ which will minimize the variance

$$\sigma^2 = \lim_{T\to\infty} \frac{1}{2T} \int_{-T}^{+T} [\bar{z}(t) - z(t)]^2\, dt \tag{9.3.1}$$

9.3.3 The impulsive response $I(t)$ was defined as the response of a system to the impulse $\delta(t)$ so that the output of the filter could be described by

$$z(t) = \int_{0^+}^{t} x(t-\tau)I(\tau)\, d\tau$$

It is required that $I(\tau) = 0$ for $\tau \leq 0$ and that $x(t-\tau) = 0$ for $\tau \geq t$ so that we can write

$$z(t) = \int_{-\infty}^{\infty} x(t-\tau)I(\tau)\, d\tau \tag{9.3.2}$$

Introduction of Equation (9.3.2) into (9.3.1) yields

$$\sigma^2 = \lim_{T\to\infty} \frac{1}{2T} \int_{-T}^{+T} \left[\bar{z}(t) - \int_{-\infty}^{+\infty} x(t-\tau)I(\tau)\, d\tau\right]^2 dt$$

which can be expanded into the limit of the sum of iterated integrals.[6] We introduce the correlation function (Section 6.6)

$$R_{\bar{z}\bar{z}} = \lim_{T\to\infty} \frac{1}{2T} \int_{-T}^{+T} \bar{z}(t-\tau)\bar{z}(\tau)\, d\tau$$

$$R_{xx} = \lim_{T\to\infty} \frac{1}{2T} \int_{-T}^{+T} x(t-\tau)x(\tau)\, d\tau$$

$$R_{\bar{z}x} = \lim_{T\to\infty} \frac{1}{2T} \int_{-T}^{+T} \bar{z}(t-\tau)x(\tau)\, d\tau$$

and obtain, after considerable reduction,

$$\sigma^2 = R_{zz}(0) - 2\int_0^\infty I(\tau)R_{zx}(\tau)\, d\tau + \int_0^\infty I(\mu)R_{xx}(\tau-\mu)\, d\mu \tag{9.3.3}$$

9.3.4 The minimization of σ^2 can be accomplished in several ways. The usual procedure is to use the calculus of variations. The variable is replaced by $\tau + \lambda$, in which λ is a "small variation." Expansion of the impulsive response by the Taylor series leads to

$$I(\tau+\lambda) = I(\tau) + \lambda\left[\frac{\partial I(\tau+\lambda)}{\partial \lambda}\right]_{\lambda=0} + \cdots \tag{9.3.4}$$

It is noted that the higher-order terms become negligible if λ is made sufficiently small. The variance σ^2 attains the form $\Delta(\sigma^2) + \sigma^2$, in which $\Delta(\sigma^2)$ is the variation of σ^2 for a variation of τ. It is known[7] that

$$\lim_{\lambda\to 0} \frac{\partial}{\partial \lambda}[\Delta(\sigma^2)] = 0 \tag{9.3.5}$$

is a necessary condition for the existence of an extremum of σ^2 (hopefully a minimum). The algebraic details are quite simple and it can be seen that application of Equations (9.3.5) and (9.3.4) to (9.3.3) leads to the necessary condition

$$R_{zx}(\tau) = \int_{0^+}^{\infty} R_{xx}(\tau - t)I(t)\,dt \qquad \textbf{(9.3.6)}$$

With Equation (9.3.6) we now obtain an equation for the variance

$$\sigma^2 = R_{\bar{z}\bar{z}}(0) + R_{zx}(\tau) - 2\int_{0^+}^{\infty} I(\tau)R_{zx}(\tau)\,d\tau \qquad \textbf{(9.3.7)}$$

It can be shown that (9.3.6) is also a sufficient condition for the existence of an optimum (minimum variance).

9.3.5 Solution of the integral Equation (9.3.6) has never been achieved in a satisfactory manner. The usual technique is to assume the form of $I(t)$, with some undetermined parameters, and then to use (9.3.6) and such additional conditions as may be needed to compute the undetermined quantities.[8]

We shall restrict ourselves to filters describable by linear ordinary differential equations with constant coefficients in which it is known (see chapter 2) that

$$I(t) = \sum_{i=1}^{m} A_i e^{-r_i t}$$

where the quantities r_i are complex numbers with positive real parts, m is the order of the highest-order derivatives in the performance equation of the filter and the A_i are functions of the r_i. It is required, therefore, to find the complex numbers r_i which satisfy the equation

$$R_{zx} = \sum_{i=1} A_i \int_0^{\infty} R_{xx} e^{-r_i t}\,dt$$

The quantities A_i are fully determined by the relationships

$$\sum_{i=1}^{m} \frac{d^n(A_i e^{-r_i t}}{dt^n}\bigg|_{t=0} = 0 \qquad n = 0, 1, 2, \ldots, m-2$$

$$\sum_{i=1}^{m} \frac{d^{m-1}(A_i e^{-r_i t})}{dt^{m-1}}\bigg|_{t=0} = K$$

which represent the initial state of the solution of

$$\frac{d^m I(t)}{dt^m} + a_{m-1}\frac{d^{m-1}I(t)}{dt^{m-1}} + \cdots + a_0 I(t) = 0 \qquad \textbf{(9.3.8)}$$

or, equivalently, the solution of

$$\frac{d^m z}{dt^m} + a_{m-1}\frac{d^{m-1}z}{dt^{m-1}} + \cdots + a_0 z = K\delta(t)$$

It follows immediately that

$$\sum_{i=1}^{m} A_i = 0$$

$$\sum_{i=1}^{m} r_i A_i = 0$$

$$\cdots$$

$$(-1)^{m-1} \sum_{i=1}^{m} r_i A_i = K$$

If we define the determinant

$$\mathcal{R} = \begin{vmatrix} 1 & 1 & \cdots & 1 \\ r_1 & r_2 & \cdots & r_m \\ r_1^2 & r_2^2 & \cdots & r_m^2 \\ \cdots & \cdots & \cdots & \cdots \\ r_1^{m-1} & r_2^{m-1} & \cdots & r_m^{n-1} \end{vmatrix}$$

and let $\mathcal{R}_{im}$ be the minor of the element in the ith column and the mth row (i.e., the bottom row) we can write

$$A_i = (-1)^{i-1} K \frac{\mathcal{R}_{im}}{\mathcal{R}}$$

(See chapters 2, 3, 4 and Appendix B. Some of the quantities A_i are given in Table 4.3.)

9.4 Constant Input with White Noise

It is well known that "white" noise, that is, noise with constant spectral density, is quite common in measurement systems. Such a noise signal is represented by $R_{nn} = N\delta(t)$. If a constant signal c is added to a white noise n we obtain

$$x = c + n$$

If signal and noise are uncorrelated (a not unnatural assumption if the noise comes from an external source), we obtain

$$R_{xx} = c^2 + R_{nn}(\tau) = c^2 + N\delta(\tau)$$

Since $z = c$ and $R_{cn} = 0$ we obtain $R_{\hat{z}x} = c^2$, and thus Equation (9.3.7) becomes

$$c^2 = \sum_{i=1}^{m} A_i\left(\frac{c^2}{r_i} + Ne^{-r_i\tau}\right) \qquad \textbf{(9.4.1)}$$

The first term of Equation (9.4.1) transforms rather easily if we use the definition of A_i

$$\sum_{i=1}^{m} \frac{A_i}{r_i} = \frac{K}{r_1 r_2, \ldots, r_m}$$

It is customary in systems represented by Equation (9.3.8) to set $a_0 = K$. But $a_0 = r_1 r_2 r_3, \ldots, r_m$ and thus $\sum A_i/r_i = 1$ and Equation (9.4.1) reduces to

$$\frac{N r_1 r_2, \ldots, r_m}{\mathcal{R}} \sum_{i=1}^{m} (-1)^{i-1} \mathcal{R}_{im} e^{-r_i\tau} = 0 \qquad \textbf{(9.4.2)}$$

The last result is impossible to achieve for all values of τ with a finite number of terms (m). The difficulty is easily explained by introduction of the results into Equation (9.3.7). We note that $R_{\hat{z}\hat{z}}(0) = c^2$ so that the equation reduces to

$$\sigma^2 = 2c^2\left(1 - \int_0^\infty I(t)\,dt\right)$$

But the integral was shown before to be equal to 1 and hence $\sigma^2 = 0$ which we know to be impossible to achieve. Therefore, we shall have to decide in advance what variance (rms error) we can tolerate and set

$$\frac{\sigma^2}{2c^2} = 1 - \sum \frac{A_i}{r_i}$$

Instead of letting $\sum (A_i/r_i) = 1$, we set $\sum (A_i/r_i) = 1 - \sigma_0^2$ so that $\sigma^2 = 2c^2\sigma_0^2$. For a given c^2 and a specified σ^2 we can see that σ_0 represents an estimate of the allowable error of the filter. Equation (9.4.2) is now replaced by

$$\frac{r_1 r_2, \ldots, r_m}{\mathcal{R}} \sum_{i=1}^{m} (-1)^{i-1} \mathcal{R}_{im} e^{-r_i \tau} \leq \frac{c^2\sigma_0^2}{N}$$

or alternately

$$I(\tau) \leq \frac{c^2\sigma_0^2}{N} \tag{9.4.3}$$

Any impulsive response function which satisfies (9.4.3) is a modified optimum filter within the limits of our new definition of optimum.

Usually $I(\tau)$ reaches a peak at some value of τ so that (9.4.3) can be replaced by the equality

$$I(\tau_p) = \frac{c^2\sigma_0^2}{N} \tag{9.4.4}$$

in which the peak value τ_p is found by solving the expression

$$\frac{dI}{d\tau} = 0 \tag{9.4.5}$$

for τ_p. We wish the filter to suppress the white noise with a maximum variance $2(c\sigma_0)^2$ which is to occur at time $t = \tau_p$.

In a first-order system $\tau_p = 0$ and thus since $I = K \exp(-a_0\tau)$ we obtain

$$K = \frac{c^2\sigma_0^2}{N}$$

We have

$$\frac{c^2\sigma_0^2}{Na_0} = 1 - \sigma_0^2$$

and thus

$$a_0 = \frac{c^2\sigma_0^2}{N(1 - \sigma_0^2)}$$

and finally

$$\frac{dz}{dt} + \frac{c^2\sigma_0^2}{N(1 - \sigma_0^2)} z = \frac{c^2\sigma_0^2}{N} y \tag{9.4.6}$$

Example 9.4.1 A spaceship radio transmits a pulse $c = 1$ volt and it is known that $c^2 = 10N$. How small must σ_0 be to obtain 1% rms error?

Solution $\sigma = 0.01$ and thus

$$\sigma_0^2 = \frac{\sigma^2}{2c^2} = 5 \times 10^{-5}$$

Obviously this is a very sluggish filter, since the time lag is $\tau_0 = 2{,}000$ s (33.3 minutes). If we were willing to accept a 10% rms error, on the other hand, we would have a time lag of only 20 seconds. Such a delay would not be intolerable in some instances.

9.5 The Simplest Optimum Filter

We do not have any a priori reasons for selecting a particular order filter from Equation (9.4.3). However, it can be seen that if $m > 2$ the system possesses $m - 2$ indeterminacies. We showed that two design criteria are available: the time τ_p and the error σ_0. The two equations (9.4.4) and (9.4.5) can be solved simultaneously for the r_i. It is clear that the r_i are uniquely determined if $m = 2$ and that for $m > 2$ some roots have to be given arbitrarily. The simplest optimum filter, therefore, is a second-order filter.

We obtain the following result by setting $m = 2$:

$$I(t) = \frac{r_1 r_2}{r_2 - r_1}(e^{-r_1 t} - e^{-r_2 t})$$

so that (9.4.4) and (9.4.5) become

$$\frac{r_1 r_2}{r_2 - r_1}(e^{-r_1 \tau_p} - e^{-r_2 t}) = \frac{c^2 \sigma_0^2}{N}$$

$$r_1 e^{-r_1 \tau_p} - e^{-r_2 \tau_p} = 0$$

The two equations reduce to the single expression

$$r_1 e^{-r_1 \tau_p} = r_2 e^{-r_2 \tau_p} = \frac{c^2 \sigma_0^2}{N} \tag{9.5.1}$$

Equation (9.5.1) is more meaningful if the roots are expressed in complex form. The impulsive response can be written as

$$I(t) = \frac{\omega e^{-\zeta \omega t}}{\sqrt{1 - \zeta^2}} \sin \sqrt{1 - \zeta^2}\,\omega t$$

so that application of the first derivative criterion for maximum immediately yields a pair of equations for τ_p.

$$\omega \exp(-\zeta \omega \tau_p) = \frac{c^2 \sigma_0^2}{N} \tag{9.5.2}$$

$$\begin{aligned} \sin(\sqrt{1 - \zeta^2}\,\omega \tau_p) &= \sqrt{1 - \zeta^2} \qquad |\zeta| \leq 1 \\ \sinh(\sqrt{\zeta^2 - 1}\,\omega \tau_p) &= \sqrt{\zeta^2 - 1} \qquad |\zeta| \geq 1 \end{aligned} \tag{9.5.3}$$

Equation (9.5.2) represents a family of solutions which, with the aid of (9.5.3), gives a family of relationship between ζ and ω.

$$\exp\left(\frac{-\omega \sin^{-1} \sqrt{1 - \zeta^2}}{\sqrt{1 - \zeta^2}}\right) = \frac{c^2 \sigma_0^2}{N}$$

As a practical method of design computation the relation $f = \omega \exp(-\zeta \omega \tau_p)$ has been plotted in Figure 9.5 as a function of ω. The

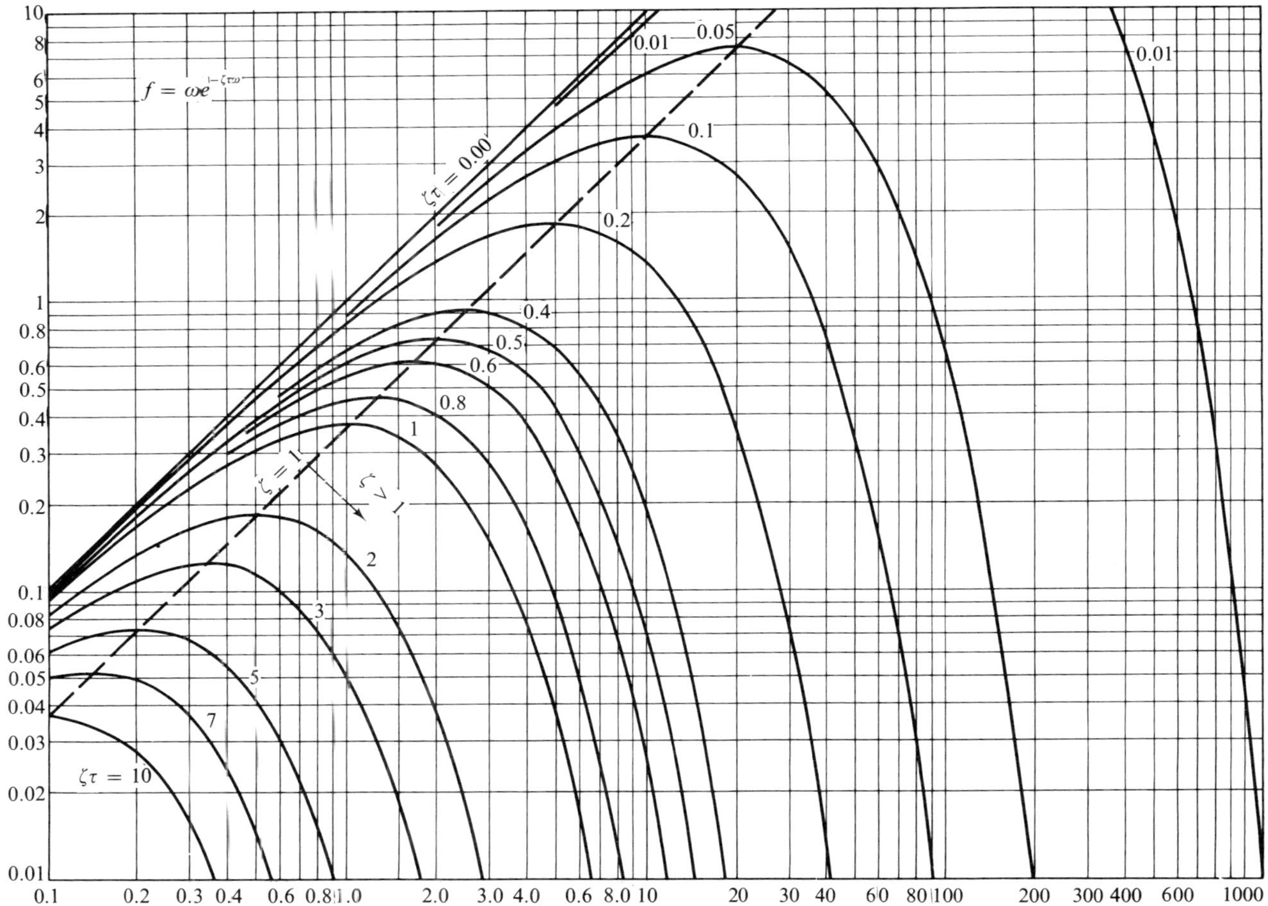

FIGURE 9.5 Design graph for optimum impulsive response

relationship between $\zeta\omega\tau_p$ and ζ has been computed from Equation (9.5.3) and plotted in Figure 9.6. To design an optimum filter, a line $f = c^2\sigma_0^2/N$ is drawn on Figure 9.5 and a table connecting ω to ζ is then constructed from Figure 9.6. A suitable combination of ζ and ω is then selected from the permissible values.

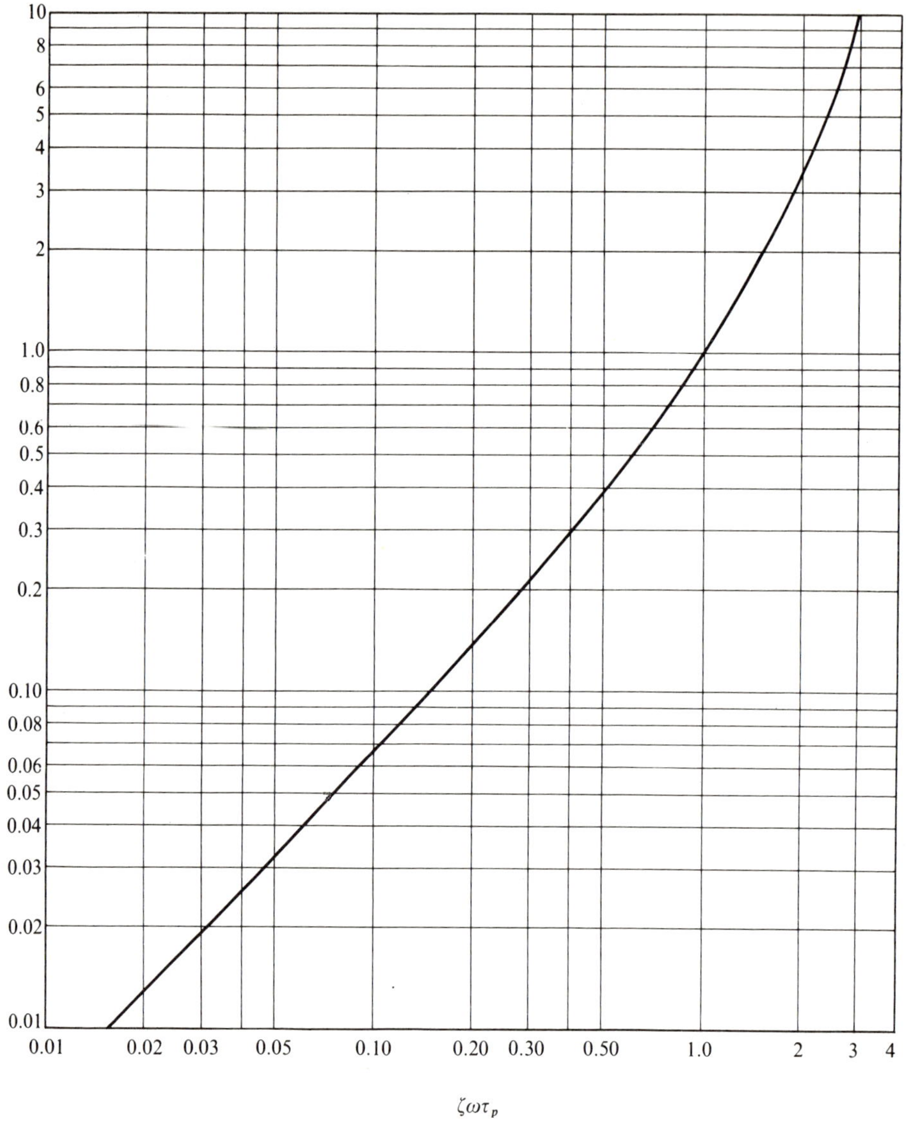

FIGURE 9.6 ζ *as a function of* $\zeta\omega\tau_p$

Example 9.5.1 A noisy input is known to be masked by white noise with a signal-to-noise ratio (rms) of 1.73. The allowable deviation variance is $\sigma_0^2 = 0.10$.

Design a second-order nearly optimum filter with a damping coefficient equal to 0.707.

Solution

$$\frac{c^2}{N}\sigma_0^2 = (1.73)^2 0.10 = 0.30$$

Table 9.1 can be obtained from Figures 9.5 and 9.6.

TABLE 9.1

800	0.01	8.00	
120	0.05	6.00	
52.0	0.10	5.20	
21.1	0.20	4.22	
8.40	0.40	3.36	
6.00	0.50	3.00	
4.60	0.60	2.76	
2.76	0.80	2.21	
1.76	1.00	1.76	2.69
0.50	1.00	0.50	0.395
0.40	0.80	0.32	0.230
0.36	0.50	0.18	0.125
0.32	0.20	0.06	0.040
0.30	0.00	0.00	0.000

Interpolation between 0.395 and 2.69 (or plotting of ζ vs. ω) yields the design point $\omega = 0.66$ and $\zeta = 0.707$.

PROBLEMS

9.1 Select the parameters for a first-order low-pass filter flat (within 1%) to 150 Hz.

9.2 Select the parameters for a first-order low-pass filter in which the output is twice the input (within 10%) up to 20 Hz.

9.3 Select the parameters for a high-pass filter flat above 50 Hz within 1%. Introduce a gain so that the amplitude ratio is equal to 1 (zero dB) in the high range and less than 1 in the low range.

9.4 The output of a high-pass filter is connected to a system which suppresses input signals of amplitude 10^{-2}. Design a filter which eliminates all signals below 50 Hz in frequency and which passes high-frequency signals with an output amplitude of 2, given that the input amplitude is 1.

9.5 A band-pass filter passes (within 1%) signals below 100 Hz and above 200 Hz. Signals between 100 and 200 Hz are reduced somewhat with 150-Hz signals being down -12 dB if the input-to-output ratio is equal to 0 dB near 0 Hz input frequency. Obtain the performance equation.

9.6 Work problem 9.5 with second-order filters with $\zeta = 0.707$.

9.7 Use the results of either problem 9.5 or problem 9.6 to obtain the output which corresponds to the input

$$z = 3.12 \sin 10t + 7.11 \sin 37.1t + 9.34 \sin 58.3t + \sin 277t$$

9.8 What are the filtering characteristics of a system with the performance equation

$$\frac{d^4z}{dt^4} + 9.11\frac{d^3z}{dt^3} + 10.13\frac{d^2z}{dt^2} + 17.4\frac{dz}{dt} + 40z = 40y$$

9.9 A signal has the form

$$x = 7.00 + 0.10 \sin 73t$$

Compute the correlation functions if $I(\tau) = \delta(0)$, i.e., if $x(t) = z(t)$ and if $y = 7$. What is the variance of the signal?

9.10 Let $I(\tau) = 2.11 \exp(-3.71\tau)$ and let $x = 9.32 + 0.25 \sin 3.15t$ where $y = 9.32$. Compute z, R_{zz} and σ^2.

9.11 Prove the equation for A_i in Section 9.3.5.

9.12 A constant signal $c = 1.29$ volts is subjected to white noise $N = 0.10$. How small must σ_0 be to obtain 1% rms error with a first-order linear filter?

9.13 A noisy constant input is subjected to white noise so that $c^2/N = 1$. The allowable value of σ_0^2 is 0.10. Design a near optimum second-order filter with $\zeta = 1.00$.

9.14 Let $c^2/N = 1$ and $\sigma_0^2 = 0.10$. Plot the possible characteristic times $(\zeta\omega)^{-1}$ as a function of τ_p for a near optimum second-order filter.

NOTES

[1]See, for example, Ref. 16, 19, 20, 29, 71.

[2]In addition to Ref. 16, 19, 20, 29, 71, the reader is referred to practically any textbook in circuit design such as Ref. 50. Mechanical isolation (filtering) is discussed in Ref. 56. Electrical filters of unusual response can be constructed from analogue computer components. See Ref. 67.

[3]The use of the word "realize" in the sense of "to make real" is sanctioned by dictionaries only insofar as "to make real" implies "to accomplish." The current use of "to realize" to mean "to make physically real" is curious if not barbarous.

[4]Ref. 67, 103.

[5]In addition to the material noted above in Note 2, the reader is referred to the book and periodical literature on high-fidelity sound-reproduction equipment.

[6]The operation is trivial but requires some skill. See Ref. 60.

[7]Ref. 60.

[8]Ref. 60, 84.

APPENDIX A

BIBLIOGRAPHY

The bibliographic material is divided into two sections, a numbered list of references keyed to the text, and a design bibliography arranged by measurand. The list is extensive but not exhaustive. The reader needs only to read a few references on a subject to acquire the vocabulary needed to pursue the research in the current periodical literature. It is suggested that the designer begin with one of the general volumes in the keyed reference list, proceed to more specialized material in the Design Bibliography and then use the vocabulary acquired in the reading to discover recent information in technical abstracting journals. Examination of a few current articles can generate a modern specialized bibliography on the specific measurand.

A.1 Keyed References

1. Draper, G. S.; McKay, E.; and Lees, S. 1952. *Instrument engineering*. New York, N.Y.: McGraw.
2. Lion, K. S. 1959. *Instrumentation in scientific research.* New York, N.Y.: McGraw.

3. Rhodes, T. J. 1941. *Industrial instruments for measurement and control.* New York, N.Y.: McGraw.
4. Eckman, Donald P. 1950. *Industrial instrumentation.* New York, N.Y.: Wiley.
5. Beckwith, T. G.; and Buck, M. L. 1961. *Mechanical measurements.* Reading, Mass.: A-W.
6. Considine, D. M. 1957. *Process instruments and controls handbook.* New York, N.Y.: McGraw.
7. Dove, R. D.; and Adams, P. H. 1964. *Experimental stress analysis and motion measurement.* Columbus, Ohio: Merrill.
8. ASME. 1959. *Fluid meters and their application.* New York, N.Y.
9. Pollermann, M. 1955. *Bauelemente der physikalischen technic.* Berlin, Germany: Springer Verlag.
10. Thomson, J., ed. 1963. *A guide to instrument design.* London, England: Taylor & Francis, Ltd.
11. Stein, P. K. 1964. *Measurement engineering.* Phoenix, Ariz.: Stein Engineering Services, Inc.
12. Pearlstein, J. 1963. *Searching the literature for transducer information, report PB* 161–320. Washington, D. C.: Office of Technical Services.
13. Minnar, E. J., ed. 1972. *ISA transducer compendium.* New York, N.Y.: Plenum Pr.
14. Minorski, N. 1970. *Nonlinear oscillations.* Princeton, N.J.: VanN-Rein.
15. Whitehead, T. N. 1954. *The design and use of instruments and accurate mechanisms.* New York, N.Y.: Dover.
16. Savant, C. J. 1964. *Control system design.* 2nd ed. New York, N.Y.: McGraw.
17. Hoel, P.G. 1954. *Introduction to mathematical statistics.* 2nd ed. New York, N.Y.: Wiley.
18. Cramer, H. 1946. *Mathematical methods of statistics.* Princeton, N.J.: Princeton U Pr.
19. Shearer, D.L. et al. 1964. *Dynamic systems.* Reading, Mass.: A-W.
20. Lynch, W. A. 1967. *Signals and systems in electrical engineering.* New York, N.Y.: McGraw.
21. Hix C. F.; and Alley, R. P. 1958. *Physical laws and their effects.* New York, N.Y.: Wiley.
22. Goode, H. H.; and Machol, R. E. 1957. *System engineering.* New York, N.Y.: McGraw.
23. Kaplan, W. 1958. *Ordinary differential equations.* Reading, Mass.: A-W.
24. Carslaw, H. S. 1930. *Introduction to the theory of Fourier series and integrals.* 3rd ed. New York, N.Y.: Dover

25. Knopp, K. 1947. *Theory of functions.* New York, N.Y.: Dover
26. Copson, E. T. 1935. *An introduction to the theory of functions of a complex variable.* New York: Oxford U Pr.
27. Churchill, V. 1948. *Introduction to complex variable and applications.* New York, N.Y.: McGraw.
28. Hildebrand, F. B. 1949. *Advanced calculus for engineers.* New York, N.Y.: P-H.
29. Murphy, G. J. 1966. *Basic automatic control theory.* Princeton, N.J.: D. VanN-Rein.
30. Hart, W. L. 1955. *Calculus.* Boston, Mass.: Heath.
31. Sobolev, S.L. 1963. *Applications of functional analysis in mathematical physics.* Providence, Rhode Island: American Mathematical Society.
32. Hobson, E.W. 1927. *The theory of functions of a real variable,* Vol. I, 3rd ed. Cambridge, Mass.: Cambridge U Pr. Reprinted by Dover, 1957.
33. McAuley, V. A. 1962. Extraction of roots. *JSIAM* 10: No. 4, 657.
34. Uspensky, J. V. 1948. *Theory of equations.* New York, N.Y.: McGraw.
35. Selby, D. 1954. *CRC standard mathematical tables.* 14th ed. Cleveland, Ohio: Chemical Rubber.
36. Units of weight and measure. *NBS Miscel.* Pub. 214. 1955.
37. *NBS Tech. News Bull.* 43, No. 1 (1959). Report on 1958 meeting of the international committee on weights and measures.
38. Tomovic, Rajko. 1963. *Sensitivity analysis of dynamic systems.* New York, N.Y.: McGraw.
39. Le Gette, M. A. *The theory of recording galvanometers.* Pasadena, Calif.: Consolidated Electrodynamics Corp (No date).
40. Ampex Corporation. 1957. *Magnetic tape instrumentation.* Redwood City, Calif.
41. Tech. Publishers, Inc. (Issued Annually). *Electronic engineer's master.* New York, N.Y.
42. F. W. Dodge Corporation (Issued Annually). *Sweet's catalog service.* New York, N.Y.
43. Hall, Bruce G. X-Y recorders. *Machine Design* 35: No. 21 (1963), 161.
44. Booker, H. G. 1959. *An approach to electrical science.* New York, N.Y.: McGraw.
45. Synge, J. L;. and Griffith, B. A. 1949. *Principles of mechanics.* New York, N.Y.: McGraw.
46. Stout, M. B. 1960. *Basic electrical measurements.* Englewood-Cliffs, N.J.: P-H.
47. Etkin, B. 1959. *Dynamics of flight.* New York, N.Y.: Wiley.

48. Consolidated Electrodynamic Corporation. 1962. *Galvanometers users handbook*. Monrovia, Calif.

49. Stott, V. An investigation of problems relating to the use of pivots and jewels in instruments and meters. *National Physics Laboratory Collected Researches* 24: No. 1 (1931), 1.

50. Brenner, E.; and Javid, M. 1960. *Analysis of Electric Circuits*. New York, N.Y.: McGraw.

51. Tyson, F. C. 1961. *Industrial Instrumentation*. Englewood-Cliffs, N.J.: P-H.

52. Byars, E. F.; and Synder, R. D. 1964. *Engineering mechanics of deformable bodies*. Scranton, Pa.: Intext.

53. Gray, P. E. 1960. *The dynamic behavior of thermoelectric devices*. New York, N.Y.: Wiley.

54. Ioffe, A. F. 1957. *Semiconductor thermoelements and thermoelectric cooling*. London, Eng.: Infosearch, Ltd.

55. Borrego, J. M. et al. 1958. The efficiency of thermoelectric generators. *WADC Technical Note* 58–200 *Project* 6058 *MIT DSR* 7672. Columbus, Ohio: WADC.

56. MacDuff, J. N.; and Curreri, John R. 1958. *Vibration control*. New York, N.Y.: McGraw.

57. Gilbert, E. G. Controllability and observability in multivariable control systems. *J. SIAM Control* AL No. 2 (1963), 128.

58. Coddington, E.A.; and Levinson, N. 1955. *Theory of ordinary differential equations*. New York, N.Y.: McGraw.

59. De Russo, P. M. et. al. 1965. *State variables for engineers*. New York, N.Y.: Wiley.

60. Solodonikov, V. V. 1960. *Statistical dynamics of linear automatic control systems*. Princeton, N.J.: VanN-Rein.

61. Stoker, J. J. 1957. *Non-linear vibrations*. New York, N.Y.: Interscience.

62. La Salle, J.; and Lefschetz, S. 1965. *Stability by Liapunov's direct methods*. New York, N.Y.: Acad Pr.

63. Timoshenko, S.; and Young, D.H. 1955. *Vibration problems in engineering*. Princeton, N.J.: VanN-Rein.

64. Dinter, H. A. Sensor threshhold definition. *Control Engineering* 13, No. 5 (1966), 93.

65. McClure, C. L. 1960. *Theory of inertial guidance*. Englewood-Cliffs, N.J.: P-H.

66. Dressler, Robert. Elastic shell-theory formulation for bourdon tubes. *JBE* 87, No. 4 (1965), 1072.

67. Korn, G. A.; and Korn, T. M. 1964. *Electronic analogue computers and hybrid computers*. New York, N.Y.: McGraw.

68. Routh, E. J. 1955. *Dynamics of a system of rigid bodies.* New York, N.Y.: Dover.
69. Bottaccini, M. R. et. al. Calibration of high-frequency manometers with the shock tube. *J. of Appl. Physiol.* 22, No. 4 (1967), 832.
70. Burton, D.H. 1965. *Dynamic calibration of intracardiac pressure transducer systems.* M. S. Thesis, University of Arizona, Tucson, Arizona.
71. Gille, J. C.; Pelegrim, M. J.; and Decaulne, P. 1959. *Feedback control systems.* New York, N.Y.: McGraw.
72. Singer, F. R. 1951. *Strength materials.* New York, N.Y.: Har-Row.
73. Leonhard, A. 1956. Determination of transient response from frequency response. *Frequency Response*, R. Oldenburger, ed. New York, N.Y.: Macmillan.
74. Brown, G. S.; and Campbell, D.P. 1948. *Principles of servomechanisms.* New York, N.Y.: Wiley.
75. Attinger, E.O. Pressure transmission in pulmonary arteries related to frequency and geometry. *Circulation Research* 12: No. 6 (1963), 623.
76. Evans, W. R. Control system synthesis by the root locus method. *Trans. AIEE* 69: (1950), 66.
77. Bartee, E. M. 1966. *Statistical methods in engineering experiments.* Columbus, Ohio: Merrill.
78. Pugh, E. M.; and Winslow, G. H. 1966. *The analysis of physical measurements.* Reading, Mass.: A-W.
79. Mood, A. M. 1950. *Introduction to the theory of statistics.* New York, N.Y.: McGraw.
80. Young, H. D. 1962. *Statistical treatment of experimental data.* New York, N.Y.: McGraw.
81. Kreyszig, E. *Advanced engineering mathematics.* 2nd ed. New York, N.Y.: Wiley.
82. Windred, G. 1940. *Electrical contacts.* London, Eng.: Macmillan.
83. Kaith, I. 1960. *The design and evaluation of a flight simulator apparatus*, M. S. Thesis, University of Arizona, Tucson, Arizona.
84. Peterson, E. L. 1961. *Statistical analysis and optimization of systems.* New York, N.Y.: Wiley.
85. Wilson, M. 1954. *American science and invention.* New York, N.Y.: Bonanger.
86. Blackwell, C. C. 1960. *A free piston apparatus for the determination of the compressibility characteristics of hydraulic fluids*, PhD. Dissertation, University of Arizona, Tucson, Arizona.
87. Downs, H. B. 1960. *The static bulk modulus of hydraulic fluids at elevated temperatures*, M.S. Thesis, University of Arizona, Tucson, Arizona.

88. Bottaccini, M. R.; and Timor, T. 1958. *Research notes.* Seattle, Wash.: Boeing Aircraft Co.

89. Cusick, C. F. 1961. *Flow meter engineering handbook.* Philadelphia, Pennsylvania: Minneapolis-Honeywell Regulator Co.

90. Lindahl, E. J. et. al. 1961. *Flowmeter computation handbook.* New York, N.Y.: ASME.

91. ASME. 1959. *Flow measurement.* New York, N.Y.

92. Bendict, R. P. Fluid flow measurement. *Electro-Technology* 55: Jan., 1966.

93. Bean, H. S. et. al. 1959. *Fluid meters.* New York, N.Y.

94. Mackenzie, D. D. Mass and volume flow measurement. *Intrumentation Technology* 14: No. 3 (1967), 23.

95. Lay, J. E. 1963. *Thermodynamics.* Columbus, Ohio: Merrill.

96. Sibley, F.S. Aging of type K thermocouples. *Instrumentation Technology* 15: No. 7 (1968), 53.

97. Carnes, F. C. Selecting paper for recorder charts. *ISA J.* 7: No. 12 (1960), 55.

98. Hornbaker, D. R.; and Roll, D. L. Thermal peturbation caused by heat flux transducers and their effect on the accuracy of heating rate measurements. *ISA Proc.* 18: (1963), paper 8.3.63.

99. Jakob, M. 1949. *Heat transfer.* New York, N.Y.: Wiley.

100. Brookley, C. E. Measurement of heat flux in solid propellant rocketry. *ISA Proc.* 18: (1963), Paper 8.43.63–1.

101. *Product Engineering.* April 1956. 162. "Calculating Damping Factors for Dashpot Dampers."

102. Streeter, V. L. 1962. *Fluid mechanics.* 3rd ed. New York, N.Y.: McGraw.

103. Philbric, G. 1966. *Applications manual for computing amplifiers.* Dedham, Mass.: Philbric Researches, Inc.

104. Freymuth, P. Feedback control theory for constant-temperature hot wire anemometer. *Rev. Sc. Instr.* 38: No. 5 (1967), 677.

105. Freymuth P. Noise in hot wire anemometers. *Rev. Sc. Instr.* 39, No. 4 (1968), 550.

106. Eckert, E. R. G.; and Drake, R.M. 1959. *Heat and mass transfer.* New York, N.Y.: McGraw.

107. Garrison, J. B; and Lawson, A. W. An absolute noise thermometer for high temperatures and high pressures. *Rev. of Sc. Instr.* 20: No. 11 (1949), 785.

108. Johnson, J. B.; and Llewellyn, F.B. Limits to amplification. *Electrical Engin.* 24: No. 11 (1934), 1449.

109. Budzilovich, P. N. Electrical noise, its nature, causes, solutions. *Control Engin.* 16: No. 5 (1969), 74

A.2 Design Bibliography

This bibliography is divided alphabetically according to the following subject list:

A	Acceleration
C	Calorimetry, Capacitance, Conductivity, Consistency, Current (Electrical)
D	Density, Displacement
E	Error
F	Flow (Mass), Flow (Volumetric), Force
G	Galvanometers
H	Heat Flux, Humidity
I	Inductance, Infrared
L	Light, Liquid Level
M	Magnetism, Moisture, Motion
P	pH, Pressure
R	Resistance
S	Shock, Specific Volume, Specific Weight, Strain, Stress, Surface Finish
T	Temperature, Thickness, Torque
U	Ultraviolet
V	Vacuum, Vibration, Viscosity, Voltage
W	Weight
X	X-Rays

ACCELERATION

(*See also* "Shock and Vibration," also Ref. 1, 2, 5, 7, 11, 13, 56, 63, 65.)

Baker, W. E.; and Dove, R. C. Transient calibration of piezo-electric accelerometers. *The Journal of Environmental Sciences* 5: No. 5 (1962).

Bradley, W. Performance of three 500° F crystal accelerometers. Paper presented at the National Telemetering Conference, El Paso, Texas (May 1957).

Conrad, R. W.; and Vigness, I. Calibration of accelerometers by impact techniques. *Proc of the Instr. Soc. of America* 8: (1953), 166.

Feder, E. I.; and Gillen, A. M. High-frequency, high-G calibration. *IRE Trans. Instr* 6: No. 2 (1957), 74.

Fitzgerald, J. V.; and Matuskik, F. J. Rotating-reed accelerometers. *Instr. Control Systems* 36: No. 12 (1963), 89.

Johansson, J. Accelerometer calibration. *Instr. Control Systems* 36: No. 12 (1963), 79.

Kearns, R. W. Velocities and displacements associated with the transient response of accelerometers. *Soc. for Exper. Stress Anal.* Paper No. 534, May 1961.

Keast, D. M. Calibration of accelerometers in a simulated space environment. *J. of the Acoust. Soc. Amer.* 31: Part 1 (1959), 584.

Palmer, D. F. An accurate shock calibrator for accelerometers. *SCTM* 96–91. 73 Sandia Corp., Albuquerque, New Mexico (1961).

Pennington, D. In-place calibration of piezo-electric crystal accelerometer amplifier systems. *Sixth ISA National Flight Test Instrumentation Symposium.* San Diego, California (May 1960).

Perls, T. A. Tests of accelerometer linearity at high accelerations. *National Bureau of Standards Report.* No. 2624 (1953), 133.

Perls, T. A.; and Kissinger, C.W. High G accelerometer calibrations by impact methods with ballistic pendulum, air gun and inclined through. *Proc. of the Instr. Soc. of Amer.* 9: Part 5 (1954).

Rosenberg, J. D. A flat vibration generator for accelerometer calibration. *Natl. Bur. Std. Rept.* No. 2524 (1953), 85.

Stedman, C. K. Some characteristics of gas damped accelerometers. *Instrument Notes* No. 33 (Oct. 1958): Statham Instruments, Inc.

Sulzer, P. G.; Smith, E. R.; and Edelman, S. High-speed stroboscope for accelerometer calibration. *Review of Scientific Instruments* 25: No. 8 (1954), 837.

Calorimetry

(*See* "Heat Flux.")

Capacitance

Balsbaugh, J. C. et al. Generalized bridge network for dielectric measurements. *Trans. A.I.E.E.* 59: No. 12 (1940), 950.

Behr, L.; and Williams, A. J. The campbell-shackleton shielded ratio box. *Proc. IRE* 20: No. 6 (1932), 969.

Hague, B. 1946. *Alternating current bridge methods.* London, England: Sir Isaak Pitman & Sons, Ltd.

Jones, A. R. Nonferrous conductivity standards. *ISA Transactions* 8: No. 1 (1969), 79.

Kowenhoven, W. B.; and Baros, T. A high sensitivity power factor bridge. *Trans. AIEE* 51: No. 1 (1932), 40.

Kowenhoven, W. B.; and Berberich, L. A. A standard of low power factor. *Trans. AIEE* 52: No. 1 (1933), 48.

McGregor, M. C. et al. New apparatus at national bureau of standards for absolute capacitance measurement. *Trans. IRE* I-7: (1958), 255.

Thompson, A. M. The precise measurements of small capacitances. *Trans. IRE* I-7: (1958), 245.

CONDUCTIVITY

(*See* "Resistance.")

CONSISTENCY

(*See* "Density.")

CURRENT (ELECTRIC)

(*See also* "Galvanometers.")

Curtis, H. L.; and Curtis, R.W. An absolute determination of the ampere. *J. Research NBS* 12: R. P. 685 (1934), 665.

Vignone, E. J. RF current standard for the calibration of current probes. *ISA Trans.* 8: No. 2 (1969), 145.

DENSITY, SPECIFIC WEIGHT, SPECIFIC VOLUME AND CONSISTENCY

Consistency Control and Metering Systems. *Pulp Paper Can.* 43: No. 13, 974.

Definitions relating to density and specific gravity of solids, liquids and gases. *ASTM Standards* Parts 4–11: (1961).

Density test of liquids by bingham pyconometer. *ASTM Standards* Part 5: (1952), 618.

Specific gravity test of gaseous fuels. *ASTM Standards* Part 5: (1952), 957.

Technical publication on bulk quantity measurement. *ASTM Standards.* STP 249, (1959).

United Trade Press, Ltd. 1953. The instrument manual. Sec. X. *Measurement and Control of the Specific Gravity of Liquids.* London, Eng.

Hopkins, V.; Wilson, D. R.; and Bolze, C. Isothermal bulk modulus of selected fluids to 700°F and 10,000 psig. Reprint Paper No. 63-WA-111: ASME Winter Annual Meeting, 1963

DISPLACEMENT

(*See also* "Strain," "Surface Finish," also Ref. 1, 2, 5, 7, 11, 13.)

Campbell, D. P. 1958. *Process dynamics.* New York, N.Y.: Wiley.

Englehard, E. Metrology of gage blocks. Symposium in Washington, NBS, Aug. 1955: *Natl. Bur. Std. Circ.* 581, U.S. Dept. of Commerce, 1957, 1.

Fastle, D. L. A film-plane, image-motion synchronization (FIS) camera. *SCTM* 116-61 (72) Sandia Corp.: Albuquerque, New Mexico, Nov. 1961.

Kinder, W. Comparator measures up to 40 inches with accuracy of 0.4 microinch. *Instr. Control Systems* 36: No. 4

Mckinnon, M. G. et al. A fluidic digital position sensor. *ISA Transactions* 8: No. 4 (1969), 198.

Schaevitz, H. The linear variable differential transformer. *Pro. of the Soc. for Exper. Stress Anal.* 4: No. 2, 79.

Watson, R. C. et al. Instruments for motion measurement using laser doppler heterodyning techniques. *ISA Transactions* 8: No. 1 (1969), 20.

Errors in Measurement

Acton, F. S. 1959. *Analysis of straight-line data.* New York, N.Y.: Wiley.

Baird, D. C. 1962. Experimentation. *An Introduction to Measurement Theory and Experimentation Design.* Englewood Cliffs, N.J.: P-H.

Beers, Y. 1957. *Introduction to the theory of error.* 2nd ed. Reading, Mass.: A-W.

Deekshatulu, B. L. Graphic solution of higher-order equations. *Instr. Control Systems* 37: No. 1 (1964), 130.

Dixon, W. J.; and Massey, F.J. 1957. *An introduction to statistical analysis.* 2nd ed. New York, N.Y.: McGraw.

Hicks, C. R. 1964. *Fundamental concepts in the design of experiments.* New York, N.Y.: HR & W.

Jarve, P. A. Approximating polynomials by polynomials of lower degree. *Instr. Control Systems* 37: No. 4 (1964), 140.

Lion, K. S. 1959. *Instrumentation in scientific research.* New York, N.Y.: McGraw.

Mandel, J.; and Linning, F.J. Study of accuracy in chemical analysis using linear calibration curves. *Anal. Chem.* 29: No. 5 (1957), 743.

Todd, J. 1962. *Survey of numerical analysis.* New York, N.Y.: McGraw, 131.

Turner, R. M.; and Anderson, T. G. Approximation polynomials by relaxation. *Instr. Control Systems* 36: No. 9 (1963), 161.

Wilson, W. A. Design of power plants to insure reliability of results. *Trans. ASME* 77: (1955), 405.

Flow (Mass)

Mass flow by temperature measurement. *Instr. Control Systems* 37: No. 6 (1964), 35.

Ellis, P. F. Advancements in magnetic flowmetering. *ISA Trans.* 7: No. 4 (1968), 332.

Gray, W. C.; and Astley, E.R. Liquid metal magnetic flow meters. *ISA J.* 1: No. 6 (1954), 15.

Mortenson, L. N. Mass flowmeter calibration. *Instr. Control Systems* 37: No. 3 (1964), 133.

Stoll, H. W. The measurement of flow with density compensation as applied to liquids and gases. *ISA J.* 2: No. 5 (1955), 159.

FLOW (VOLUMETRIC)

[*See also* "Flow (Mass)."]

ASME. 1961. *Flowmeter computation handbook.* New York, N.Y.

ASME. 1959. *Fluid meters, their theory and application.* 5th ed. New York, N.Y.

Orifice metering of natural gas. Gas Measurement Committee Report No. 3, American Gas Association: New York, N.Y., 1956.

ASME. 1954. *Symposium on measurement in unsteady flow.* H41. New York, N.Y.

Baird, R. C. Dynamics of pulsative flow through orifices. *Instruments.* No. 4: (1952), 481.

Beitler, S. R.; Lindahl, E. J.; and McNichols, H. B. Developments in the measuring of pulsating flows with inferential meters. *Trans. ASME* 75: No. 5 (1953).

Crawford, W. A. 1959. Flow and its measurement. *Variable Area Flow Meters.* ISA Symposium.

Fleming, F. W.; and Binder, R. C. 1950. Study of linear resistance flow meters. ASME paper 50-A-37.

Galley, R. 1954. *Handbook of measurement and control.* Chap. 5. Pittsburgh, Pa.: Instru Soc.

Hawley, R. H. Consistency of coefficients in small venturi nozzles. *Instr. Control Systems* 37: No. 3 (1964), 121.

Johnson, R. C. Calculations of real-gas effect in flow through critical-flow nozzles. Reprint paper No. 63-WA-71: ASME Annual Winter Meeting, 1963.

Jorrison, A. L. On the evaluation of the accuracy of the coefficient of discharge in the flow-measurement equation. *Trans. ASME* 75: (1953), 1323.

Kemp, L. J. Today's flow measurement. *Instr. Control Systems* 37: No. 4 (1964), 147.

King, R. W. 1954. *Handbook of hydraulics.* 4th ed. New York, N.Y.: McGraw.

Oppenheim, A. K.; and Chilton, E.G. Pulsating flow measurements– a literature survey. ASME Paper 53-A-a57: Dec., 1953.

Shafer, M. R. Performance characteristics of turbine flowmeters. *J. Basic Eng.* 84: No. 4 (1962), 471.

Shafer, M. R.; and Ruegg, F.W. Liquid-flowmeter calibration techniques. *Trans. ASME* 79: No. 10 (1958), 1369.

Souers, R. C.; and Binder, R. C. Linear resistance meters for liquid flow. ASME Paper 51-A-57.

Walker, J. M. 1960. *Brooks rotameter handbook.* Langsdale, Pa.: Brooks Rotameter Co.

Wood, R. D. Steam measurement by orifice meter. *Instr. Control Systems* 36: No. 4 (1963), 135.

Yard, J. S. Characteristics and uses of turbine flowmeters. *ISA J.* 6: No. 5 (1959), 54.

FORCE

(*See also* "Acceleration," Torque," "Shock and Vibration," "Weight.")

Mason, W. P.; and Thurston, R. N. Piezoresistive materials in measuring displacement, force and torque. *J. Acoust. Soc. Amer.* 29: No. 10 (1957), 1096.

GALVANOMETERS

(*See also* Ref. 46, 48.)

Hill, A. W. The theory of moving-coil galvanometers. *J. Sci. Instruments* 11: (1934), 309.

Judkins, J. R.; and Kessler, E. Speed servo recorder has 1/8-second response. *Instr. Control Systems* 36: No. 8 (1963), 100.

Miller, J. H. Instruments, electrical measuring. *Encyclopedia Britannica* 12: (1967), 325.

Spreadbury, F. G. 1956. *Electronic measurements and measuring instruments,* London, Eng.: Constable, 31–35.

Tengwell, R. Driving galvanometers. Endevco Corp.: Tech. Data Report (May 25, 1961).

Whipple, R. S. The evolution of the galvanometer. *J. Sci. Instruments* 3: No. 11 (1934), 37.

HEAT FLUX AND CALORIMETRY

(*See also* Ref. 100.)

Burnstein, F. V. et al. The development and evaluation of a radiative heat source for the calibration of transient and steady-state heat flux sensors. *ISA Trans.* 8: No. 2 (1969), 110.

Drummond A. J. The absolute calibration of thermal radiation detectors. *ISA Trans.* 7: No. 3 (1968), 194.

Hager, N. E., Jr. Miniature thin-heater thermal conductivity apparatus. *ISA Trans.* 8: No. 2 (1969), 104.

Malone, E. W. Design and calibration of thin-foil heat flux sensors. *ISA Trans.* 7: No. 3, (1968), 175.

Starner, K. E. Use of thin-skinned calorimeters for high heat flux arc-jet measurements. *ISA Trans.* 7: No. 3 (1968), 181.

HUMIDITY

Beck, R. E. Measurement of water vapor in catalytic reformer gases. *ISA Trans.* 7: No. 4 (1968), 367.

Cole, K. M. et al. Precision humidity analysis. *ISA Trans.* 9: No. 1 (1970), 68.

Jennings, B. H.; and Torloni, A. Psychrometric charts for use at altitudes above sea level. *Refrigeration Engineer* 62: No. 6 (1954), 71.

Leisey, F. A. Determination of low moisture concentration in refinery gases. *ISA Trans.* 7: No. 4 (1968), 363.

Long, D. D. Low-humidity test measurements. *ISA Trans.* 8: No. 1 (1969), 71.

Wexler, A. Calibration of humidity-measuring instruments at the national bureau of standards. *ISA Trans.* 7: No. 4 (1968), 356.

Wexler, A. Electric hygrometers. *National Bureau of Standards Circular* 586: Sept. 3, 1957.

INDUCTANCE

Laws, F. A. 1938. *Electrical measurements.* New York, N.Y.: McGraw.

Stout, M. S. 1960. *Basic electrical measurements.* 2nd ed. Englewood Cliffs, N.J.: P-H.

INFRARED

(*See also* Ref. 2, 6, 21.)

Friedel, R. A.; Queiser, J. A. 1966. *Infrared and ultraviolet spectrometric techniques.* Washington, D. C.: Bureau of Mines.

Ginzburg, A. S. 1969. *Application of infrared radiation in food processing.* London, Eng.: Leonard Hill Books.

Hill, D. W. 1968. *Non dispersive gas analysis in science, medicine and industry.* New York, N.Y.: Plenum Pr. Plenum Pub.

Jamieson, J. A. 1963. *Infrared physics and engineering.* New York, N.Y.: McGraw.

Martin, A. E. 1966. *Infrared instrumentation and techniques.* New York, N.Y.: Elsevier.

Nakaniski, K. 1962. *Infrared absorption spectroscopy,* San Francisco, Calif.: Holden-Day.

Rao, C. N. R. 1963. *Chemical applications of infrared spectroscopy.* New York, N.Y.: Acad Pr.

Smith, R. A. 1957. *The detection and measurement of infrared radiation.* Oxford, Eng.: Clarendon Press.

Strobel, H. A. 1962. *Chemical instrumentation.* Reading, Mass.: A-W.

LIGHT

Boltz, D. F., ed. Colorimetric determination of nonmetals. *Chemical Analysis* 8: (1958).

Ditchburn R. W. 1953. *Light.* New York, N.Y.: Interscience.

Houston, R. A. 1958. *Physical Optics*, New York, N.Y.: Interscience.

Hume, D. N.; Marple, T. L. et al. Automatic photometric titrations. *Anal. Chem.* 28: No. 7 (1956), 1116.

Kortum, G. 1955. *Colorimetry photometry and spectrometry* (German text). 3rd ed., vol. II. Berlin, Germany: Springer.

Monk, G. S.; and McGorkle, W. H. 1954. *Optical instrumentation.* New York, N.Y.: McGraw.

Optical Society of America Comm. 1953. *Calorimetry: the science of color.* New York, N.Y.: Thomas Y. Crowell Co.

Robertson, J. K. 1954. *Introduction to optics, geometrical and physical.* 4th ed. New York, N.Y.: VanN-Rein.

Sommerfield, A. 1954. *Optics.* New York, N.Y.: Acad.

Thodurn, J. M.; Jankowski, C. M.; and Reynolds, M. S. Development of commercial automatic calorimetric titrator. *Anal. Chem.* 31: No. 1 (1959), 124.

Van der Hulst, H. C. 1957. *Light scattering by small particles.* New York, N.Y.: Wiley.

Wahlstrom, E. E. 1957. *Optical crystallography.* 2nd ed. New York, N.Y.: Wiley.

Wallach, D. F.; and Surgenor, D. M. Simple microspectrophotometers. *Anal. Chem.* 30: No. 11 (1958), 1879.

Liquid Level

Elfers, P. A. Liquid level control and controllers. *Petroleum Engr.* 2: No. 11 (1951), 124.

Ross, S. D.; and Drake, W. W. Liquid level measurement and control. *Petroleum Engr.* 25: No. 8 (1953), C7.

Magnetism

(*See also* "Galvanometers" and Ref. 46.)

Standard methods of testing for alternating current core loss and permeability of magnetic materials. *ASTM Stand.* Part 1: (1954), 485.

Standard methods of testing for alternating current magnetic properties of laminated core samples. *ASTM Stand.* Part 3: (1958), 416.

Standard methods of testing for normal induction and hysteresis of magnetic materials. *ATSM Stand.* Part 1: (1955), 1442.

Standard methods of testing magnetic materials. *ASTM Stand.* Part 1: (1953), 315.

Bozorth, R. M. Magnetism. *Rev. Mod. Phys.* 19: No. 1 (1947), 29.

Golding, E. W. 1948. *Electrical measurements and measuring instruments.* London, Eng.: Sir Isaak Pitman and Sons, Ltd.

Pearson, G. L. Magnetic field-strength meter employing the hall effect in germanium. *Rev. Sci. Instr.* 19: No. 4 (1948), 263.

Smith, George S. A new magnetic flux meter. *Trans. AIEE* 56: No. 4 (1937), 441.

Smith, G. S. Use of bismuth-bridge fluxmeter for AC fields. *Trans. AIEE* 58: No. 2 (1939), 52.

MOISTURE

(*See* "Humidity.")

MOTION

(*See* "Acceleration," "Displacement," "Surface Finish," "Strain," "Velocity.")

pH

Bates, R. G. 1954. *Electrometric pH determinations, theory & practice.* New York, N.Y.: Wiley.

Clark, W. H. 1957. *The determination of hydrogen ions.* Baltimore, Md.: William & Wilkins.

Keyhoe, T. J. pH control in industrial waste treatment. *Proc. ISA* 10 Part 3: paper no. 55-15-1 (1955).

Sims, R. W. et al. Continuous, process parts-per-million pH measurements.: *ISA Trans.* 7: No. 4 (1968), 299.

PRESSURE

Ablidge, D. B. How to measure vacuum. *Instr. Control Systems* 36: (1963), 109.

Capacitive Pressure Sensors. *Instr. Control Systems* 35: No. 5 (1962), 119.

ASME. 1962. *High-pressure measurement.* New York, N.Y.

Metal Bellow and Bellows Devices. 1959. Bridgeport, Conn.: Robertshaw-Fulton Controls Co., Bridgeport Thermostat Division.

Baird, R. C.; Solnich, R. L.; and Amiss, J. R. Calibrator for dynamic pressure transducer. *Instr. Automation* 27: No. 7 (1954), 1074.

Blair, E. J. 1962. *Introduction to chemical instrumentation.* New York, N.Y.: McGraw.

Bowersox, E. Calibration of high frequency response pressure transducers. *ISA J.* 5: No. 11 (1958), 98.

Davis, W. R. Measuring high pressure transients. *Autom. Control* 4: No. 1 (1956), 24.

Hall, H. T. Some high-pressure, high temperature apparatus considerations: equipment for use at 100,000 atmospheres and 1000°C. *Rev. Sci. Instr.* 29: No. 4 (1958), 267.

Haringx, J. A. Design of corrugated diaphragms. *ASME.* Paper 55-A-112, (1955).

Haringx, J. A. The rigidity of corrugated diaphragms. *Appl. Sci. Research,* VA2-1950, 299.

Hylkema, E. G.; and Bowersox, R.B. Experimental and mathematical techniques for determining the dynamic response of pressure gages. *ISA Proc.* 8: (1953), 115; *ISA J.* 12: (1954), 27.

Kankel, A. V.; and Whitten D. C. Corrugated metal diaphragm performance. *ASME.* Paper 55-A-180.

Kreismean, W.S. 1960. *Transactions* 1960 *vacuum symposium.* London, Eng.: Pergamon.

Leck, J. H.; and Martin, G. S. Feedback controlled heat conductivity gage measuring pressure in vacuum systems. *Rev. Sci. Instr.* 28: No. 2 (1957), 119.

Lee, G. H.; and Van der Pyl, L. M. A bibliography of diaphragms and aneroids. *ASME. Paper* 55-A-180, (1955).

Legat, J. W.; Rassweiller, G. M.; and Sihvonen, Y. T. Engine pressure indicators application of a capacitor type. *ISA J.* 2: No. 1 (1955), 9.

McKay, B. P. A mechanical-electronic transducer for low range pressure and force. *ISA J.* 3: No. 6 (1956), 194.

Norton, H. N. Specification characteristics of pressure transducers. *Instr. Control Systems* 36: No. 12 (1963), 91.

Philips, K. Some experiments with a cold cathode vacuum gauge. *J. Sci. Instr.* 31: No. 3 (1954), 110.

Rassweiller, G. M. et al. Recent improvements in a capacitor type pressure transducer. *ISA J.* No. 11 (1955), 497.

Rogers, E. J. Semiconductor pressure transducer features mechanical comparison. *Inst. Control Systems* 36: No. 4 (1963), 135.

Sibley, C. B.; and Roehrigh, J.R. Wide range vacuum gage. *Electronics* 26: No. 11 (1953), 176.

Van der Pyl, L. M. Bibliography on bourdon tubes and bourdon-tube gauges. *Mech Eng.* 75: No. 10 (1953), 915.

Wahl, A. M. Recent researches in flat diaphragm and circular plates with particular reference to instrument applications. *ASME.* Paper 55-A-116, (1955).

Wildhack, W. A.; Dressler, R. F.; and Lloyd, E. C. Investigation of the properties of corrugated diaphragms. *ASME.* Paper 55-A-181, (1955).

Curtis, Harvey L. 1937. *Electrical measurements.* New York, N.Y.: McGraw.

Stack, S. S. Resistance-thermometer bridge. *Gen. Elec. Rev.* 51: No. 7 (1948), 17.

Thomas, James L. A new design of precision resistance standard. *NBS J. Research* 5: No. 2 (1930), 295

Thomas, James L. Precision resistors and their measurement. *NBS Circular* 470: Oct. 8, 1948, 32.

Wenner, F. Methods, apparatus and procedures for comparison of precision standard resistors. NBS Research Paper (1940), RP 1323.

SHOCK AND VIBRATION

(*See also* "Acceleration.")

Amer. Stand. Assoc., Inc. 1965. American standard methods for the calibration of shock and vibration pickups. New York, N.Y.

Bouche, R. R. The absolute calibration of pickups on a drop-ball shock machine of the ballistic type. *Proc., Inst. Environmental Sci.* (1961), 115.

Bouche, R. R. Improved standard for the calibration of vibration pickups. *Experimental Mechanics.* Apr., 1961, 116.

Brooks, R. O. et al. Distortions of controlled transients by shock-monitoring systems. *ISA Trans.* 7: No. 3 (1968), 207.

Church, A. H. 1963. *Mechanical vibrations.* 2nd ed. New York, N.Y.: Wiley.

Edelman, S.; Jones, E.; Smith, E. A. Some developments in vibration measurements. *J. Acoustical Soc. of Amer.* 27: No. 4 (1955), 728.

Fibikar, R. J. Touch and vibration sensitivity. *Prod. Eng.* 27: (1956), 177.

Harris, C. M.; and Crede, C. E. Basic theory & measurement, *Shock & Vibration Handbook* 1 1961. New York, N.Y.: McGraw.

Holmgren, W. A. et al. Shock-pressure instrumentation for severe-environment applications. *ISA Trans.* 7: No. 3 (1968), 217.

Lawrence, A. F. 1956. Crystal accelerometer response to mechanical shock impulses. *SCTM* 168-56-16. Sandia Corp., Albuquerque, New Mexico.

Levy, S.; and Bouche, R. R. Calibration of vibration pickups by the reciprocity method. *J. Research of Nat. Bur. of Stand.* 57: No. 4 (1956), 227.

Smith, E. R. et al. Stroboscopic interferometer for vibration measurement. *J. Acous. Soc. of Amer.* 39: Part 2. (1958), 867.

Vigness, I. 1960. Measurement of shock. Shock and Structural Response. *ASME.*

Williams, E. O. An etched manganin gauge system for shock-pressure measurement in a high-noise environment. *ISA Trans.* 7: No. 3 (1968), 223.

Wilson, R. A. Understanding torsional vibrations. *Gen. Motors Eng. J.* 9: No. 1 (1962), 8.

Yates, W. A.; and Davidson, M. Wide range calibrator for vibration pickups. *Electronics* 26. Part 2. (1953), 183.

Ziegler, C. A. Electromechanical pickup calibration by the interferometer method. *J. Acous. Soc. of Amer.* 25: No. 1 (1953), 135.

Specific Volume

(See "Density.")

Specific Weight

(*See* "Density.")

Strain

Boiten, R. G. Characteristics and applications of resistance strain gauges. *Natl. Bur. Std. Circ.* 528, (1954).

Dean, M.; and Douglas, P. 1962. *Semiconductor and conventional strain gages.* New York, N.Y.: Acad Pr.

Dorsey, J.; and Reiters, L. O. Bakelite backed semiconductor strain gage. *Instr. Soc. of Amer.* 8: (1961), 166.

Gillette, O. L. et al. Development of high-temperature capacitance strain gages. *ISA Trans.* 8: No. 1 (1969), 52.

Gunn, J.; and Billinghurst, E. Magnetic fields affect strain gages. *Control Eng.* 4: No. 8 (1957), 109.

Hampton, D. et al. Stress and strain gauges for use in soil dynamics research. *ISA Trans.* 7: No. 3 (1968), 242.

McDonald, V. J.; and Roberts, H. C. Practical strain gage circuit assembly. *ISA.* Paper 54-23-1.

Miller, R. L.; and Wannamaker, W. H. Strain gage measurement. A private publication. No. B-15-17, Brown Instruments Division Minneapolis-Honeywell Regulator Co., Philadelphia, Pa.

Millis, D.; and Douglas, R. D. 1962. *Semiconductor and conventional strain gages.* New York, N. Y.: Acad Pr.

Murray, W. W.; and Stein, P. K. 1960. *Strain gage techniques.* Cambridge, Mass.: MIT Pr.

Perry, C. C.; and Lissner, H. R. 1955. *The strain gage primer.* New York, N.Y.: McGraw.

Statham, L. Bonded vs. unbonded strain gage transducers. *Instr. Control Systems* 35: No. 9 (1962), 123.

Stein, P. K. A survey of commercially available semiconductor strain gages. *Strain Gage Readings* 3: No. 5 (1960–61), 17.

Stress

(*See* "Strain.")

SURFACE FINISH

Hagan, J. F.; and Lindberg, E. 1953. Specifications and measurement of surface finish. Mechanical Development Dept., Research Laboratories Division, General Motors Corp.

Reason, R. E. Surface finish. *Prod. Eng.* 28: No. 10 (1957), 77.

Wite, P. J. The high cost of overfinishing. *Am. Machinist* 107: No. 3 (1963), 53.

Williamson, D. E. Tracer-point sharpness as affecting roughness measurements. *Trans. ASME* 69: No. 5 (1947), 319.

TEMPERATURE (LIQUID- AND GAS-FILLED THERMOMETERS)

Natl. Bur. Stds. Circ. Clinical Thermometers. CS a-53.

Construction of filled thermal systems. Can. Chem. Processing 37; No. 7 (1953), 42, 64.

Industrial filled system thermometers. *Mech. World* 134: No. 3420 (1954), 306.

Standard specifications for AST thermometers. *Report of ASTM-El-*53. (1953)

Muller and Girard. Dynamics of filled temperature-measuring systems. *ASME. Paper* 54-SA-29, (1954).

Murdock, J. W. et al. Practical method of determining response of thermometers in liquid baths. *ASME.* 61-WA-32, (1961).

Robertson, T. L. Modern clinical temperature measurement. *ISA Trans.* 7: No. 2 (1968), 112.

Thompson, R. D. Liquid in-glass thermometers: principles and practices. *ISA Trans.*, 7, No. 2 (1968), 87.

Ween, S. Care and use of liquid in-glass laboratory thermometers. *ISA Transactions* 7: No. 2 (1968), 93.

TEMPERATURE (MISC.)

Blakemore, J. S. Germanium for low-temp resistance thermometry. *Instr. Control Systems* 35: No. 5 (1962), 94.

Dengler, C. O. Mercury freezing-point cell. *Instr. Control Systems* 36: No. 12 (1963), 119.

Friberg, S. A. 1955. *Temperature, its measurement and control in science and industry.* New York, N.Y.: VanN-Rein.

Giedt, W. H. et al. Determination of high gas-stream temperatures from short-exposure probe responses. *ISA Trans.* 8: No. 1 (1969), 43.

Harrison, T. R. 1960. *Radiation pyrometry and its underlying principles of radiant heat transfer.* New York, N.Y.: Wiley.

Kostkowski, H.J.; and Lee, R.D. Theory and method of optical pyrometry. *Natl. Bur. Std. Monograph* 41: (1962).

TEMPERATURE (THERMISTORS AND RESISTANCE THERMOMETRY)

Positive temperature coefficient thermistors. *Electronic Design* 4: No. 13 (1956), 5.

Ayers, D. L. et al. A comparison of thermocouples and thermistors for surface-temperature measurement of low conductivity materials. *ISA Trans.* 8: No. 2 (1969), 117.

Bjork, N.; and Davidson, R. 1955. Small signal behavior of directly heated thermistors. *Trans. Chalmers Univ. Tech. Bull.* No. 169.

Daneman, H. L. Cryogenic temperature sensor calibration with automated data readout. *ISA Trans.* 8: No. 2 (1969), 151.

Drog, J. Thermistor hypodermic needle for subcutaneous temperature measurement *Rev. Sci. Instr.* 27: No. 6 (1956), 408.

Ekelog, S.; and Kihlberg, G. Theory of the thermistors as an electric circuit element. *Trans. Chalmers Univ. of Technol.* No. 142, (1954).

Hill, J. J. Platinum-resistance thermometry using an AC inductive ratio bridge. *ISA Trans.* 7: No. 2 (1968), 101.

Johnson, D. E.; and Schnelle, P. D. A precision recording refractometer. *Rev. Sci. Instr.* 24: No. 1 (1953), 26.

Loerner, L. F. Oven control with thermistors. *Bell Labs Record.* 31: No. 3 (1953), 115.

Lortie, Y. Use of thermistors as vacuum gauges. *J. de Phys. et Radium* 16: (1953), 317.

Malmberg, P. R.; and Matland, C. G. Thermistor temperature control. *Rev. Sci. Instr.* 27: No. 3 (1956), 136.

Roberts, M. L. Potentiometric bridge for resistance thermometry. *ISA Trans.* 8: No. 2 (1969), 157.

Schurr, R. A. Thermopiles aid in measuring the rate of heat rejection during automotive engine tests. *Gen. Motors Eng. J.* 10: (1963), 8.

Smith, R. W. An evaluation of thermistors: thermally sensitive semiconductors. *Gen. Motors Eng.* J. 7: (1960), 14.

TEMPERATURE (THERMOCOUPLES)

Baker, H. D. et al. 1953. *Temperature measurement in engineering* 1. New York, N. Y.: Wiley.

Benedict, R. P.; and Ashby, J. F. 1961. Empirical determination of thermocouple characteristics. *ASME* 61-WA-19.

Moffat, R. J. How to specify thermocouple response. *ISA J.* 4: No. 6 (1957), 219.

Nanigian, J. Thermal properties of thermocouples. *Instr. Control Systems* 36: No. 10 (1963), 86.

Parker, R. Response of thermocouples to a rapid temperature rise. *ISA Trans.* 8: No. 2 (1969), 122.

Roeser, W. F.; and Lonberger, S. T. Methods of testing thermocouples and thermocouple materials. *Natl. Bur. Std. Circ.* 590, 1958.

THICKNESS

(*See* "Displacement.")

TORQUE

(*See also* "Acceleration," "Force.")

Amsler, A. J. *Amsler torsion dynamometers.* Schaffhausen, Switzerland: Technical Industrial Literature by Alfred J. Amsler & Co.

Measurement of shaft horsepower. *Instr. Control Systems* 35: No. 9 (1962), 111.

Torque pickups. Baldwin-Lima-Hamilton Corp., Electronics Division. Bull. 4370A, 1962.

Cook, N. H., Loewen, E. G.; and Shaw, M. C. Machine-tool dynamometers. *Am. Machinist.* May 10, 1954.

Dillon, R. P. How to measure with dynamometers. *Power Eng.* 63: No. 7 (1959), 80.

Hartman, H. R. Micro torque measurement. *Instr. Control Systems* 35: No. 11 (1962), 105.

Knudsen, R. F. A discussion of present day dynamometers: their application, operation and control. *Gen. Motors Eng. J.* 4: No. 4 (1957), 18.

Lockery, H. E. Theory & applications of strain-gage torque measuring devices. *ASME.* Paper No. 59-A-314.

Rebeska, J. J. 1950. Investigations of an NACA high-speed gage torquemeter. *NACA Tech. Note* 2003.

ULTRAVIOLET

(*See also* Ref. 2, 6, 21.)

Glasser, L. G. Automatic photoelectric ultraviolet analyzer for continuous chemical analysis of process streams. *J. Opt. Soc. Amer.* 45: No. 7 (1955), 556.

Koller, Lewis, R. 1965. *Ultraviolet radiation.* 2nd ed. New York, N.Y.: Wiley.

Strobel, Howard A. 1962. *Chemical instrumentation.* Reading, Mass.: A-W.

Summer, W. 1962. *Ultraviolet and infrared engineering.* New York, N. Y.: Interscience.

Troy, D. J. Measurement of atmospheric pollution by ultraviolet photometry. *Anal. Chem.* 27: No. 8 (1955), 1217.

VACUUM

(*See also* "Pressure.")

Vacuum instrumentation. *Instr. Control Systems* 37: No. 9 (1964), 110.

Bryant, P. J. et al. *Extreme vacuum technology.* Washington, D. C.: NASA Cr84, 1964.

Dushman, S. 1949. *Scientific foundation of vacuum technique.* New York, N. Y.: Wiley.

Hnilicka, M. P. Extreme high vacuum. *Ind. Res.* 6, No. 9 (1964), 36.

Lafferty, J. M.; and Van der Slice, T. A. Vacuum measurement by ionization. *Instr. Control Systems* 36: No. 3 (1963), 90.

Melling, R. J. Ionization vacuum gage measures absolute pressure up to 1 mm hg. *Instr. Control Systems* 37: No. 9 (1964), 119.

Rohrig, J. R.; and Simons, J. C. Calibrating vacuum gages to 10^{-9} torr. *Inst. Control* 36: (1963), 107.

Vibration

(*See* "Shock and Vibration.")

Viscosity

ASTM standards on petroleum products methods D88-44, D341-43, D445-46T, D4646-39, D567,41, D666-44 & appendix V. *American Soc. of Testing Materials.* Phil., Pa. (1965).

Calculating viscosity index. *ASTM Special Tech. Publ.* 168 or 43B: *ASTM Method* D567-53. (1953)

Eakin, B. E.; and Ellington, R. T. Improved high pressure capillary tube viscometer. *Petrol. Trans.* 11: No. 4 (1959), 71.

Gilmont, R. A falling-ball viscometer. *Instr. Control Systems* 36: No. 9 (1963), 121.

Holcomb, L. A. Calibration of kinetic viscometers. *Instr. Control Systems* 37: No. 2 (1964), 109.

Nelson, R. C. Automatic force viscometers. *Instr. Control Systems* 36: No. 4 (1963), 94.

Roth, W.; and Rich, S. Principles of operation for the ultra-viscoson. *J. Applied Physics.* July, 1953.

Wiberly, J. S. *Basic calibration of master viscometers & viscosity oil standards.* ASTM Subcommittee on Viscosity, 1962.

Voltage

(*See also* "Galvanometers," "Potentiometers," "AC Bridges.")

Braudaway, D. W. Optimal control of the radioactive voltage standard. *ISA Trans.* 8: No. 3 (1969), 202.

Eppley, M. International standard of electromotive force. *Trans. AIEE* 50: No. 4 (1931), 1293.

Kennedy, V. C. Electronic weight determination as a tool for control and measurement procedures. *ASME Instruments and Regulators Conf.* Paper 53-IRD-8, 1953.

Kirwin, J. O.; and Demler, L. E. Continuous weighing meters and feeder. *ASME Instrument and Regulators Conf. Paper* 53-IRD-9, 1953.

Klein, E. Processing and proportioning materials by weight. *ASME Instrument and Regulators Conf.* Paper 53-IRD-10, 1953.

Lowe, R. P. Continuous gravimetric proportioning systems. *ASME Instrument and Regulators Conf.* Paper 53-IRD-7, 1953.

Maloney, N. G. High speed weighing. *ASME Instrument and Regulators Conf.* Paper 53-IRD-10, 1953.

Milroy, J. W.; and Mayor, G. C. Characteristics of Components in Pneumatic Weighting Systems. *ASME Instrument and Regulators Conf.* Paper 53-IRD-3, 1953.

Richardson, I. H. Trend toward automation in automatic weighing and bulk materials handling. *ASME Instruments and Regulators Conf.* Paper, 53-IRD-12, 1953.

O'Brian, L. Standard Cells. *Gen. Electric Rev.* 35: No. 11 (1932), 591.

Vinal, George W. 1950. *Primary batteries.* New York, N.Y.: Wiley.

Vincent, George D. The construction and characteristics of standard cells. *IRE Trans. on Instr.* I-7 (1958), 221.

Warwick, A. E. A method of maintaining a reference-standard of EMF using shippable saturated standard cells. *ISA Trans.* 8: No. 2 (1969), 138.

WEIGHT

Barnett, L. Electrical strain-cell weighing. *ASME, Instruments and Regulators Conf.* Paper 53-IRD-12, 1953.

Berg, R. H. Solids flow and level measurements by continuous weighing. *ASME Instruments & Regulators Conf.* Paper 53-IRD-14, 1953.

Carlcton, R. J., Jr. Electric weighing in steelmill processes. *ASME Instruments & Regulators Conf.* Paper 53-IRD-13, 1953.

Hicks, J. J. Remote weight measurement by means of hydraulic load, cells with electrical transmitters. *ASME Instruments and Regulators Conf.* Paper 53-IRD-4, 1953.

X-RAYS

Symposium on radiography and x-ray diffraction methods. 1937. New York, N.Y.: Am. Soc. for Testing Materials.

Berry, Leonard G.; and Thompson, R. M. 1962. *X-ray powder data for ore minerals.* New York, N.Y.: Geological Society of Amer.

Birks, L. 1959. *X-ray spectrochemical analysis.* New York, N.Y.: Interscience.

Brandenberger, E. 1945. *Roentgennogrphischanalytische chemie*. Basel, Switzerland: Birkhauser.

Clark, George L. 1955. *Applied x-rays*. 4th ed., New York, N.Y.: McGraw.

Isenburger, Herbert R. *Bibliography on x-ray stress analysis*. Califon, N.J.: St. John X-Ray Labs, 1953.

Kovacs, T. 1964. *Principles of x-ray metallurgy*. New York, N. Y.: Plenum Pr.

Krivoglaz, M. A. 1969. *Theory of x-ray and thermal-neutron scattering by real crystals*. New York, N.Y.: Plenum Pr.

Pattee, H. H. et al. 1963. *X-ray optics and x-ray microanalysis*. New York, N.Y.: Acad Pr.

APPENDIX B

LAPLACE TRANSFORMS

B.1 Introduction

Transform methods enjoy an undeserved reputation in engineering. The Laplace transform, for example, is taught as a method for the solution of ordinary differential equations with constant coefficients, in spite of the fact that a glance at current engineering literature will show that the $\mathcal{L}$ transforms are never used to solve differential equations directly.[1] Engineers and physicists use methods of design based on the transform calculus but do not actually solve the system equation. It must be admitted that the derivation of particular solutions appears more straightforward in the transform calculus than in the "classical" methodology, but even in this instance the advantage is imaginary. It is just as simple to compute particular solutions with the superposition integral and a good table of integrals as it is with tables of transform. It is with respect to the solution of the characteristic equations that both the transform procedure and the classical methods founder: there are no closed form solutions of polynomial equations of order higher than the fourth. It is just as difficult to compute the roots of the characteristic polynomial with $\mathcal{L}$-transformed equations as it is with classical methods.

If the solution of ordinary differential equations with constant coefficients were our sole aim, we would find it useless to burden ourselves with the apparatus of the transform calculus. Fortunately, the transformed equations can be used as design tools without regard to the actual solution of the equation. The calculus of block diagrams (chap. 2), and the frequency plane methods (chap. 5) are all based on transformed differential equations. It would be possible to rederive the procedures by classical methods, but since most modern authors use Laplace transforms, it is more convenient to follow the precedents.

B.2 Fourier Integral and the Fourier Transform

The mathematical foundation of the Laplace transform is found in the theory of the Fourier integral. It is not necessary, for the purposes of this book, to delve too deeply into the theory. What is needed is a clear understanding of the methodological limitations.[2]

Definition B.2.1 A FUNCTION $f(t)$ IS SAID TO SATISFY DIRICHLET'S CONDITION IN AN INTERVAL $a < t < b$ IF: (A) $-\infty < f(t) < \infty$ IN (a, b), THAT IS, THE FUNCTION IS BOUNDED EVERYWHERE IN THE INTERVAL; (B) (a, b) CAN BE DIVIDED INTO A FINITE SET OF SUBINTERVALS $(a, a_1)(a_1, a_2) \ldots (a_n, b)$ IN WHICH $f(x)$ IS MONOTONIC.

The set of functions which satisfy the Dirichlet condition as stated above can be broadened to include a class of functions which is unbounded.

Definition B.2.2 A FUNCTION $f(t)$ IS SAID TO SATISFY THE MODIFIED DIRICHLET CONDITION IN AN INTERVAL (a, b) IF: (A) $f(t) \longrightarrow \infty$ AT A FINITE NUMBER OF POINTS t_i $(i = 1, 2, \ldots, n)$ IN SUCH A FASHION THAT IF ARBITRARILY SMALL NEIGHBORHOODS $t_i \pm \epsilon_i$ $(\epsilon_i > 0)$ CAN BE DELETED FROM (a, b) THEN IN THE REMAINDER OF THE INTERVAL $f(t)$ IS BOUNDED.

(B) $\int_a^b f(t)\,dt$ IS ABSOLUTELY CONVERGENT.

Theorem B.2.1 LET $f(t)$ BE DEFINED IN THE INTERVAL $I = (-\infty, +\infty)$ AND LET $f(t)$ SATISFY A DIRICHLET CONDITION ON ALL FINITE SUBINTERVALS OF I NO MATTER HOW SELECTED. AND LET $\int_{-\infty}^{+\infty} |f(t)|\,dt$ EXIST (I.E., LET $f(t)$ BE ABSOLUTELY CONVERGENT). THEN

$$\lim_{\epsilon \to 0} \frac{1}{2}[f(t+\epsilon) + f(t-\epsilon)] = \frac{1}{2\pi}\int_{-\infty}^{+\infty} e^{j\omega t}\,d\omega \int_{-\infty}^{+\infty} f(\tau)e^{-j\omega\tau}\,d\tau \tag{B.2.1}$$

IF $f(t)$ IS CONTINUOUS AT A POINT t_k, THEN

$$\lim_{\epsilon \to 0} \frac{1}{2}[f(t_k+\epsilon) + f(t_k-\epsilon)] = f(t_k)$$

Thus if $f(t)$ is continuous at t_k, the Fourier integral [Equation (B.2.1)] represents the function at t_k. If, on the other hand, $f(t)$ has a jump discontinuity $f(t_k + \epsilon) \neq f(t_k - \epsilon)$, then the integral represents the *average* value of the function.

Equation (B.2.1) is the fundamental equation of the transform calculus. The inner integral in (B.2.1) can be written as

$$F_1(\omega) = \int_{-\infty}^{+\infty} f_1(\tau)e^{-j\omega\tau}\, d\tau \qquad \textbf{(B.2.2)}$$

The integration can be thought of as an operation which transforms the function $f_1(\tau)$ into the function $F_1(\omega)$. $F_1(\omega)$ is said to be the *Fourier transform* of $f_1(\tau)$. With this notation, Equation (B.2.1) becomes

$$f_1(t) = \frac{1}{2\pi}\int_{-\infty}^{+\infty} F_1(\omega)e^{j\omega t}\, d\omega \qquad \textbf{(B.2.3)}$$

Equation (B.2.3) is called the *inversion* integral since it is the operation which "inverts" the transform $F_1(\omega)$ back into $f_1(t)$.[3]

B.3 The Laplace Transform

Many functions of practical interest are not absolutely integrable and thus not Fourier transformable. For example, ϵ^{at} does not have a Fourier transform since $\epsilon^{at} \longrightarrow \infty$ as $t \longrightarrow \infty$ if $a > 0$. There is a modified transformation, however, which permits a substantial increase of the set of transformable functions. Let us substitute in Equation (B.2.1) the following function

$$f_1(t) = \begin{cases} f(t)e^{-\sigma t} & \text{if } t > 0 \\ 0 & \text{if } t \leq 0 \end{cases} \quad \sigma > 0 \qquad \textbf{(B.3.1)}$$

Then the absolute integrability condition is replaced by

$$\int_{-\infty}^{+\infty} |f(t)|\, e^{-\sigma t}\, dt < \infty$$

For most functions it is usually possible to find a σ sufficiently large to guarantee absolute integration.[4]

Substitution of (B.2.4) into (B.2.1) gives

$$e^{-\sigma t}f(t) = \frac{1}{2\pi}\int_{-\infty}^{+\infty} e^{j\omega t}\, d\omega \int_{-\infty}^{+\infty} f(t)e^{-\sigma\tau}e^{-j\tau}\, d\tau$$

We now substitute the complex variable $s = \sigma + j\omega$ and remember that $ds = jd\omega$. Then

$$f(t) = \frac{1}{2\pi j}\int_{\sigma - j\infty}^{\sigma + j\infty} e^{st}\, ds \int_{0^+}^{\infty} f(\tau)e^{-s\tau}\, d\tau \qquad \textbf{(B.3.2)}$$

The inner integral

$$F(s) = \int_{0^+}^{\infty} f(t)e^{-st}\, dt \qquad \textbf{(B.3.3)}$$

is called the *Laplace transform* of $f(t)$. Substitution of (B.3.3) into (B.2.3) gives the *inversion integral*

$$f(t) = \frac{1}{2\pi j} \int_{\sigma - j\infty}^{\sigma + j\infty} F(s) e^{st} \, ds \tag{B.3.4}$$

The transformation from the time domain (t) to the complex domain (s) is symbolized by $\mathcal{L}f(t) = F(s)$ and the inversion is symbolized by $\mathcal{L}^{-1}F(s) = f(t)$. Obviously $\mathcal{L}^{-1}\mathcal{L}f(t) = f(t)$. If a function $f(t)$ satisfies a Dirichlet condition and if $f(t)e^{-\sigma t}$ is absolutely integrable for some $\sigma > 0$, then $f(t)$ belongs to the set $\mathcal{L}$ of Laplace transformable functions. In short, we say that $f(t)$ is in $\mathcal{L}$. If a function $f(t)$ is in $\mathcal{L}$, it can be transformed to a function of a complex variable $F(s)$. By this we mean that there exists a set of values of $s = \sigma + j\omega$ upon which a unique $F(s)$ is defined for every s.

B.4 Functions of a Complex Variable

The Laplace transform is a function of a complex variable and its theory is the theory of complex functions. Although it is not necessary for an engineer to have a deep understanding of complex variables theory to use Laplace transforms, a casual acquaintance with a few basic concepts will help in the manipulations. A number of definitions and theorems will be presented in this section, mostly without proof. (An occasional proof is given when the understanding of the methodology is thereby enhanced.[5])

B.4.1 If on a plane Cartesian coordinate system the horizontal coordinate is labeled σ and the vertical coordinate is labeled ω, then it becomes possible to locate a unique point (σ, ω) on the plane as a representation of the complex number $s = \sigma + j\omega$. The length of the line from the origin (0, 0) to (σ, ω) shall be called the *absolute value* $|s|$, or *modulus* of s. The angle ϕ between the modulus and the σ axis is called *argument* of s (Fig. B.1). A number of operations are defined.

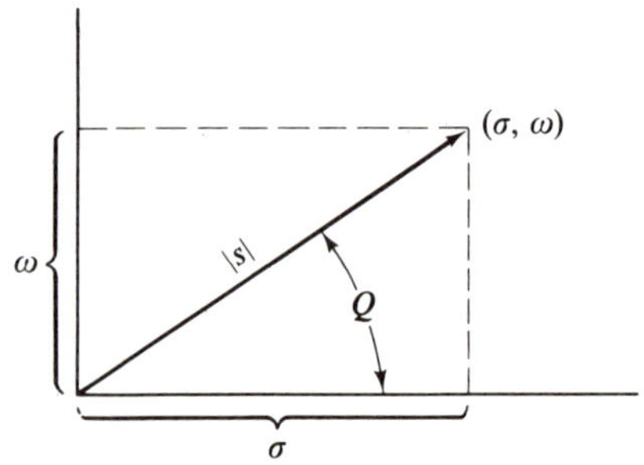

FIGURE B.1 The Argand diagram; graphical representation of a complex number

Definition Set B.4.1 LET $s_1 = \sigma_1 + j\omega_1$ AND $s_2 = \sigma_2 + j\omega_2$ BE TWO COMPLEX NUMBERS. THEN THE FOLLOWING OPERATIONS CAN BE DEFINED:

(A) ADDITION AND SUBTRACTION

$$s_1 \pm s_2 = (\sigma_1 \pm \sigma_2) + j(\omega_1 \pm \omega_2)$$

(B) MULTIPLICATION

$$\begin{aligned} s_1 s_2 &= (\sigma_1 + j\omega_1)(\sigma_2 + j\omega_2) \\ &= (\sigma_1\sigma_2 - \omega_1\omega_2) + j(\omega_1\sigma_2 + \omega_2\sigma_1) \end{aligned}$$

SPECIFICALLY LET $s_1 = s_2 = j$, THAT IS, $\sigma_1 = \sigma_2 = 0, \omega_1 = \omega_2 = 1$. THEN $jj = -1$.

LET c BE A REAL CONSTANT; THEN IT FOLLOWS THAT

$$cs = c\sigma + jc\omega$$

(C) $s = 0$ IMPLIES THAT $\sigma = 0, \omega = 0$.

(D)

$$\begin{aligned} s_1 + s_2 &= s_2 + s_1 \\ s_1 s_2 &= s_2 s_1 \\ s_3(s_1 + s_2) &= s_3 s_1 + s_3 s_2 = s_1 s_3 + s_2 s_3 = (s_1 + s_2)s_3 \end{aligned}$$

(E) IF $s_1 = \sigma_1 + j\omega_1$, THEN $\bar{s}_1 = \sigma_1 - j\omega_1$ IS CALLED THE COMPLEX CONJUGATE OF s_1. IT FOLLOWS FROM DEFINITION B THAT $s_1\bar{s}_1 = \sigma_1^2 + \omega_1^2$. FROM FIGURE B.1 AND FROM THE PYTHAGOREAN THEOREM IT FOLLOWS THAT $|s|^2 = s\bar{s} = \sigma^2 + \omega^2$.

B.4.2 The complex number s can be said to represent a vector[6] of length $|s|$ and direction ϕ (Fig. B.1). The sum of the two vectors s_1 and s_2, according to Definition B.4.1(a), is a new vector $(\sigma_1 + \sigma_2) + j(\omega_1 + \omega_2)$. Addition by components is standard vector addition and it can be shown graphically that the vector $s_1 + s_2$ is obtainable by parallelogram composition (Fig. B.2a). Subtraction of two vectors yields a vector between (σ_1, ω_1) and (σ_2, ω_2) as is to be expected (Fig. B.2b). The vector $s_2 - s_1$ is not tied to the origin. It is not difficult to show that even in this case all the rules of Definition set B.4.1 apply and that specifically

$$\begin{aligned} |s_1 - s_1|^2 &= (\sigma_2 - \sigma_1)^2 + (\omega_2 - \omega_1)^2 = (s_2 - s_1)(\overline{s_2} - \overline{s_1}) \\ &= |s_2|^2 + |s_1|^2 - (s_1\bar{s}_2 + s_2\bar{s}_1) \\ &= |s_2|^2 + |s_1|^2 \quad 2 \text{ Rcal Part } (\bar{s}_1 s_2) \end{aligned}$$

B.4.3 A complex number has two polar coordinate representations. If we let $|s| = R$ it can be seen from Figure B.1 that

$$s = \sigma + j\omega = R(\cos\phi + j\sin\phi) \qquad \textbf{(B.4.1)}$$

Let us consider now the well-known series expansion

$$\begin{aligned} e^{j\phi} &= 1 + (j\phi) + \frac{1}{2!}(j\phi)^2 + \frac{1}{3!}(j\phi)^3 + \frac{1}{4!}(j\phi)^4 + \cdots \\ &= \left(1 - \frac{1}{2!}\phi^2 + \frac{1}{4!}\phi^4 + \cdots\right) + j\left(\phi - \frac{1}{3!}\phi^3 + \frac{1}{5!}\phi^5 + \cdots\right) \end{aligned}$$

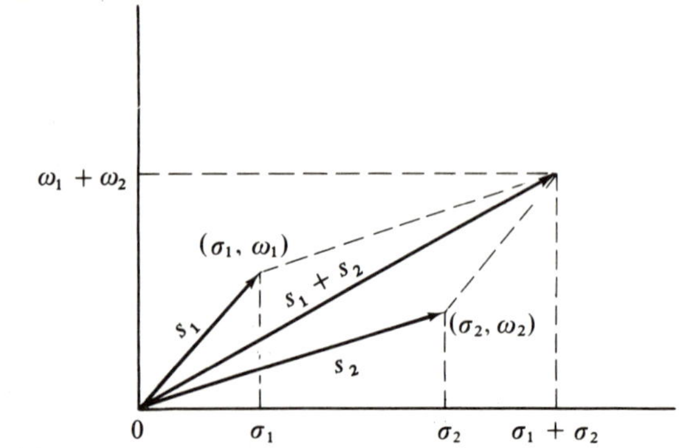

FIGURE B.2a Sum of two complex numbers

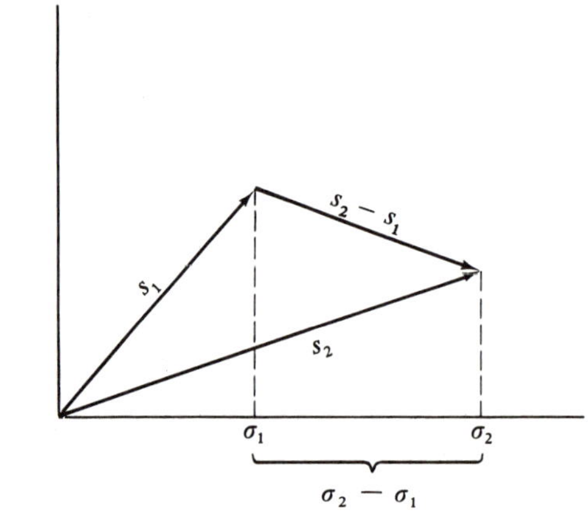

FIGURE B.2b Difference of two complex numbers

It is also well known that

$$\cos \phi = 1 - \frac{1}{2!}\phi^2 + \frac{1}{4!}\phi^4 + \cdots$$

$$\sin \phi = \phi - \frac{1}{3!}\phi^3 + \frac{1}{5!}\phi^5 + \cdots$$

and thus

$$s = R(\cos \phi + j \sin \phi) = Re^{j\phi} \qquad \textbf{(B.4.2)}$$

Example B.4.1 Relate the cosine of a real number to a hyperbolic cosine of an imaginary number.

Solution

$$e^{j\phi} = \cos \phi + j \sin \phi$$

$$e^{-j\phi} = \cos \phi - j \sin \phi$$

Thus

$$\cos \phi = \frac{1}{2}(e^{j\phi} - e^{-j\phi}) = \cosh (j\phi)$$

Example B.4.2 Prove the trigonometric identity

$$\sin 2\theta = 2 \sin \theta \cos \theta$$

Solution By the method used in the previous example it can be shown that

$$\sin \theta = \frac{1}{2j}(e^{j\theta} - e^{-j\theta})$$

Thus

$$\sin 2\theta = \frac{1}{2j}(e^{j2\theta} - e^{-j2\theta}) = \frac{1}{2j}(e^{j\theta} - e^{-j\theta})(e^{j\theta} + e^{-j\theta}) \qquad \textbf{(B.4.3)}$$
$$= 2 \sin \theta \cos \theta$$

Definition B.4.2 LET σ AND ω BE REAL NUMBERS AND LET $s = \sigma + j\omega$. LET THERE BE A CLASS OF FUNCTIONS $F(s)$ SUCH THAT

$$F(s) = u(\sigma, \omega) + jv(\sigma, \omega)$$

CAN BE OBTAINED BY LEGITIMATE MANIPULATIONS ON s ACCORDING TO DEFINTION SET B.4.1 AND u AND v ARE FUNCTIONS OF σ AND ω. THEN $F(s)$ IS CALLED A FUNCTION OF A COMPLEX VARIABLE.

Example B.4.3 Write s^2 in the form $u + jv$.

Solution

$$s^2 = (\sigma^2 - \omega^2) + j(2\sigma\omega)$$
$$u = \sigma^2 - \omega^2 \qquad v = 2\sigma\omega$$

Example B.4.4 Write $s^2 + s + 1 = F(s)$ in the form $u + jv$.

Solution

$$F(s) = (\sigma^2 - \omega^2 + \sigma + 1) + j(2\sigma\omega + \omega)$$

Example B.4.5 Separate $e^s = F(s)$ into the real part u and the imaginary part v.

Solution

$$e^s = e^{\sigma + j\omega} = e^{\sigma}(\cos \omega + j \sin \omega)$$
$$u = e^{\sigma} \cos \omega \qquad v = e^{\sigma} \sin \omega$$

Example B.4.6 Separate $\cos s$ into a real and an imaginary part.

Solution

$$\cos s = \cos (\sigma + j\omega) = \cos \sigma \cos j\omega - \sin \sigma \sin j\omega$$
$$= \cos \sigma \cosh \omega + j \sin \sigma \sinh \omega$$

since it can be shown that

$$\cos j\omega = \cosh \omega$$
$$-\sin j\omega = j \sinh \omega$$

Example B.4.7 Write $\sqrt{s}$ into the form $u + jv$.

Solution This problem involves a peculiar ambiguity. We note that $\cos \phi = \cos(\phi \pm 2n\pi)$, $\sin \phi = \sin(\phi \pm 2n\pi)$, $n = 0, 1, 2, \ldots$ and thus

$$s = Re^{j(\phi \pm 2n\pi)}$$

Hence

$$s^{1/2} = R^{1/2} e^{j(\phi/2 \pm n\pi)}$$

It is evident that $s^{1/2}$ can have two values. If we compute $s^{1/2}$ for various n we see that

$$n = 0 \qquad s^{1/2} = R^{1/2}\left(\cos\frac{\phi}{2} + j\sin\frac{\phi}{2}\right)$$

$$n = 1 \qquad s^{1/2} = R^{1/2}\left[\cos\left(\frac{\phi}{2} + \pi\right) + j\sin\left(\frac{\phi}{2} + \pi\right)\right]$$

$$= -R^{1/2}\left(\cos\frac{\phi}{2} + j\sin\frac{\phi}{2}\right)$$

For $n = 2$, $s^{1/2}$ is the same as $n = 0$. If we continue, we discover that for n even, $s^{1/2}$ is the same as for $n = 0$ and for n odd, $s^{1/2}$ is the same as for $n = 1$.

Example B.4.8 Separate $F(s) = \ln s$ into its real and imaginary parts.

Solution

$$\ln s = \ln\left[Re^{j(\phi \pm 2n\pi)}\right] \qquad n = 0, 1, 2, \ldots$$

$$= \ln R + j(\phi \pm 2n\pi)$$

$s^{1/2}$ and $\ln s$ are examples of multiple-valued functions. Whenever we use such functions, we have to restrict the variables in such a fashion that the function is single valued, since most of our theorems presuppose single-valued functions, that is, for each (σ, ω) there exist only one function $u(\sigma, \omega)$ and only one function $v(\sigma, \omega)$.

B.4.4 On the complex plane (σ, ω) let there be drawn a closed boundary C which separates the infinite plane into an interior region R and an exterior region $\bar{R}$ (Fig. B.3). Let a point $s_0 = \sigma_0 + j\omega_0$ be placed in R but not on C. Then any other point s in R is said to be in the neighborhood of s_0. If a function $w = F(s)$ has a value $w_0 = F(s_0)$ it is said to be defined at s_0; if it is defined for every point in R (excluding perhaps C and a finite number of points in R), it is said to be defined in the region R.

Definition B.4.3 A FUNCTION $F(s)$ IS SAID TO BE ANALYTIC AT A POINT s_0 IF AND ONLY IF $F(s)$ IS DIFFERENTIABLE THROUGHOUT SOME NEIGHBORHOOD OF s_0.

Definition B.4.4 IF IT IS POSSIBLE TO DEFINE A COMPLEX NUMBER Δs IN SUCH A FASHION THAT $(s_0 + \Delta s)$ IS IN SOME NEIGHBORHOOD OF s_0 AND IF IT IS POSSIBLE TO CARRY OUT THE OPERATION

$$F'(s_0) = \lim_{\Delta s \to 0}\left[\frac{F(s_0 + \Delta s) - F(s_0)}{\Delta s}\right]$$

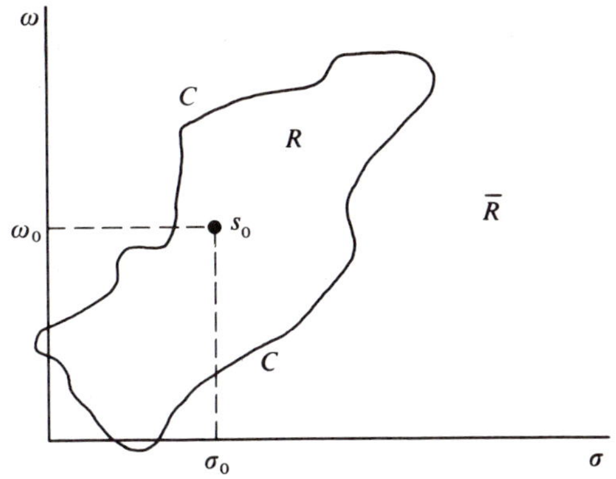

FIGURE B.3 A neighborhood of S_0

SO THAT $F'(s_0)$ IS UNIQUE, THEN $F'(s_0)$ IS CALLED THE DERIVATIVE OF $F(s)$ AT THE POINT s_0.

If the above definition is applicable it follows that the common formulas for the differentiation of continuous, differentiable real functions are usable directly for the differentiation of complex variables. Great care must be exercised in the application of ordinary differentiation formulas to complex functions. It must be shown that the function $F(s)$ is defined throughout a neighborhood of s_0 and that certain other conditions are satisfied.

We now proceed to the determination of differentiability conditions. Let the partial derivatives of u, v with respect to ω and σ exist. From the definition of the derivative it follows that

$$\begin{aligned}\frac{\Delta F}{\Delta s} &= \frac{F(s_0 + \Delta s) - F(s_0)}{\Delta s}\\ &- \left\{\left[\frac{u(\sigma_0 + \Delta\sigma, \omega_0 + \Delta\omega) - u(\sigma_0, \omega_0)}{\Delta\sigma + j\Delta\omega}\right]\right.\\ &\left. + j\left[\frac{v(\sigma_0 + \Delta\sigma, \omega_0 + \Delta\omega) - v(\sigma_0, \omega_0)}{\Delta\sigma + j\Delta\omega}\right]\right\}\end{aligned}$$

In the limit $\Delta s \longrightarrow 0$ implies $\Delta\sigma \longrightarrow 0$ and $\Delta\omega \longrightarrow 0$ independently. By this we mean that we can let $\Delta\sigma$ go to zero first and then $\Delta\omega$ or vice versa. If $\Delta\omega$ is first let go to zero we obtain

$$\begin{aligned}\left(\frac{dF}{ds}\right)_1 &= \lim_{\Delta\sigma\to 0}\left\{\left[\frac{u(\sigma_0 + \Delta\sigma, \omega_0) - u(\sigma_0, \omega_0)}{\Delta\sigma}\right]\right.\\ &\left. + j\left[\frac{v(\sigma_0 + \Delta\sigma, \omega_0) - v(\sigma_0, \omega_0)}{\Delta\sigma}\right]\right\}\\ &= \frac{\partial u}{\partial\sigma} + u\frac{\partial v}{\partial\sigma}\end{aligned}$$

If, on the other hand, we let $\Delta\sigma \longrightarrow 0$ first and then $\Delta\omega$ we obtain

$$\left(\frac{dF}{ds}\right)_2 = \frac{\partial v}{\partial \omega} - j\frac{\partial u}{\partial \omega}$$

In Definition B.4.4 it was required that the derivative be unique, and thus

$$\left(\frac{dF}{ds}\right)_1 = \left(\frac{dF}{ds}\right)_2$$

which, by equating real and imaginary parts, yields the Cauchy-Rienman equation (or conditions) of differentiability:

$$\frac{\partial u}{\partial \sigma} = \frac{\partial v}{\partial \omega} \qquad \frac{\partial v}{\partial \sigma} = -\frac{\partial u}{\partial \omega} \tag{B.4.4}$$

Theorem B.4.1 LET $F(s) = u + jv$ BE DEFINED IN A NEIGHBORHOOD R OF s_0 AND LET u AND v BE DIFFERENTIABLE WITH RESPECT TO σ AND ω. THEN $F(s)$ IS DIFFERENTIABLE AT s_0 IF AND ONLY IF THE CAUCHY-RIENMAN EQUATIONS ARE SATISFIED AT s_0.[7]

Example B.4.9 Show that $F(s) = s^2$ is everywhere differentiable.

Solution

$$s^2 = (\sigma^2 - \omega^2) + (2\sigma\omega)j$$

Then

$$\frac{\partial u}{\partial \sigma} = \frac{\partial v}{\partial \omega} = 2\sigma$$

$$\frac{\partial v}{\partial \sigma} = \frac{\partial u}{\partial \omega} = 2\omega$$

Example B.4.10 Show that $\bar{s}$ is nowhere differentiable.

Solution

$$\bar{s} = \sigma - j\omega$$

$$\frac{\partial u}{\partial \sigma} = 1 \neq \frac{\partial v}{\partial \omega} = -1$$

Example B.4.11 Is $F(s) = 1/s$ differentiable for every value of s?

Solution At $s = 0$, $\sigma = 0$ and $\omega = 0$ and thus

$$\lim_{s\to 0} \frac{1}{s} = \lim_{s\to 0} \left(\frac{\sigma - j\omega}{\sigma^2 + \omega^2}\right) \longrightarrow \infty + j\infty$$

i.e., both the real and the imaginary parts of $1/s$ fail to exist at $s = 0$. Thus $1/s$ is not differentiable at $s = 0$.

Example B.4.12 A function $F(s)$ satisfies the C-R equations and $\partial u/\partial \omega$, $\partial u/\partial \sigma$, $\partial v/\partial \omega$ and $\partial u/\partial \sigma$ are differentiable with respect to σ and ω in a region R.

Show that both u and v satisfy the Laplace equation

$$\nabla^2 u = \frac{\partial^2 u}{\partial \sigma^2} + \frac{\partial^2 u}{\partial \omega^2} = 0$$

Solution

$$\frac{\partial^2 u}{\partial \omega\, \partial \sigma} = \frac{\partial^2 v}{\partial^2 \omega^2} \qquad \frac{\partial^2 u}{\partial \sigma\, \partial \omega} = \frac{\partial^2 v}{\partial^2 \sigma^2}$$

Since u is differentiable in σ and ω, we can interchange the order of differentiation and thus

$$\frac{\partial^2 v}{\partial \omega^2} = -\frac{\partial^2 v}{\partial \sigma^2}$$

The solution for $\nabla^2 u = 0$ follows similar steps.

Theorem B.4.2 AN ANALYTIC FUNCTION HAS DERIVATIVES OF ALL ORDERS.[8]

Theorem B.4.2 is a remarkable theorem. In the theory of functions of a real variable the existence of a first derivative implies nothing about the existence of a second derivative. In the theory of complex variables the existence of the first derivative implies the existence of the second derivative and of all higher-order derivatives.

B.4.5 The theory of functions of complex variables is remarkable in still another instance. Whereas real variable functions are of greatest use whenever they are continuous and differentiable, complex variables are most useful whenever they fail to be regular.

Definition B.4.5 IF A FUNCTION IS ANALYTIC IN A NEIGHBORHOOD OF s_0 BUT FAILS TO BE ANALYTIC AT s_0, THE FUNCTION IS SAID TO BE SINGULAR AT s_0 OR TO POSSESS A SINGULARITY AT s_0.

This definition leads to the most important theorem for our purposes. The following theorem, together with the Cauchy integral theorem, will give us the most expeditious methods for the inversion of Laplace transforms.

Theorem B.4.3 LET $F(s)$ BE ANALYTIC ON TWO CONCENTRIC CIRCLES C_1, C_2 WITH CENTER AT s_0 AND ANALYTIC IN THE REGION R BETWEEN C_1 AND C_2. THEN $F(s)$ CAN BE EXPRESSED IN AN INFINITE SERIES CONVERGENT IN R.

$$F(s) = \sum_{n=0}^{\infty} a_n (s - s_0)^n + \sum_{n=1}^{\infty} \frac{b_n}{(s - s_0)^n}$$

OR

$$F(s) = \sum_{n=-\infty}^{n=+\infty} c_n (s - s_0)^n$$

The series is called the Laurent series of $F(s)$.[9] In a given region R

there is only one expansion of a function into a Laurent series. The following rules can be used to expand a function.

Rule 1 THE PRODUCT OF TWO POWER SERIES CONVERGES TO THE PRODUCT OF THEIR SUMS.

Rule 2 IN THE REGION OF CONVERGENCE, AS INDICATED BELOW, THE BINOMIAL THEOREM APPLIES IN THE FORM

$$\frac{1}{(a \pm s)^n} = \sum_{m=0}^{\infty} (\mp 1)^m \frac{(n+m-1)!}{m!\,(n-1)!} \frac{s^m}{a^{m+n}} \qquad |s| < a$$

$$\frac{1}{(s \pm a)^n} = \sum_{m=0}^{\infty} (\mp 1)^m \frac{(n+m-1)!}{m!\,(n-1)!} \frac{a^m}{s^{m+n}} \cdot \qquad |s| > a$$

Rule 3 IF A FUNCTION $F(s)$ IS ANALYTIC IN C_2, THEN $F(s)$ CAN BE EXPANDED INTO A TAYLOR SERIES

$$F(s) = F(s_0) + \frac{dF}{ds}\bigg|_{s=0} (s - s_0) + \frac{1}{2!} + \frac{d^2F}{ds^2}\bigg|_{s=0} (s - s_0)^2 + \cdots$$

Rule 4 LET $F(s)$ BE A FUNCTION OF A COMPLEX VARIABLE AND LET $s = f(Z)$ AND $Z = g(s)$. THEN IF THE SUBSTITUTION $s = f(Z)$ GIVES A FUNCTION ANALYTIC IN SOME REGION $R(Z)$ AND IF IT IS POSSIBLE TO EXPAND $G(Z) = F[f(z)]$ INTO AN INFINITE SERIES, THEN THE SUBSTITUTION INTO THE SERIES OF $Z = g(s)$ WILL YIELD A SERIES EXPANSION OF $F(s)$ VALID IN SOME REGION $R(s)$. THE REGION OF CONVERGENCE IS DETERMINED BY THE APPLICATION OF THEOREM B.4.3.

Example B.4.13 Expand into a Laurent series about $s_0 = 0$ and convergent in the ring $1 < |s| < 3$.

$$F(s) = \frac{1}{(s-1)(s-3)}$$

Solution By partial function expansion[10] we find that

$$F(s) = -\frac{1}{2(s-1)} + \frac{1}{2(s-3)}$$

since $|s| > 1$ and $|s| < 3$ we apply Rule 2 to obtain

$$F(s) = -\frac{1}{2} \sum_{m=0}^{\infty} \frac{1}{s^{m+1}} - \frac{1}{6} \sum_{m=0}^{\infty} \left(\frac{s}{3}\right)^m$$

Example B.4.14 Expand the function in Example B.4.13 in the region $3 < |s| < \infty$.

Solution As before

$$F(s) = -\frac{1}{2(s-1)} + \frac{1}{2(s-3)}$$

The expansion of $(s-1)^{-1}$ is still valid since $|s| > 1$ implies $|s| > 3$.

For $|s| > 3$ we can use the second binomial formula on $(s-3)^{-1}$. Thus

$$F(s) = -\frac{1}{2}\sum_{m=0}^{\infty}\frac{1}{s^{m+1}} + \frac{1}{2}\sum_{m=0}^{\infty}\frac{3^m}{s^{m+1}} = \frac{1}{2}\sum_{m=1}^{\infty}\frac{(3^{m-1}-1)}{s^m}$$

Example B.4.15 Expand $\exp(s-1)^{-1}$ into a series valid in $\infty > |s-1| > 0$.

Solution Let $(s-1)^{-1} = Z$ (Rule 4). Then

$$e^Z = 1 + Z + \frac{Z^2}{2!} + \frac{3Z^3}{3!} + \cdots$$

and

$$\frac{1}{e^{s-1}} = \sum_{m=0}^{\infty}\frac{1}{m!\,(s-1)^m}$$

Example B.4.16 Expand the previous function inside the circle $|s| < 1$.

Solution In this region the function is analytic and $s_0 = 0$ is the center. Thus we use the Taylor expansion

$$F(0) = e^{-1} \qquad F'(0) = -e^{-1} \qquad F^N(0) = -e^{-1}$$

and thus

$$e^{1/(s-1)} = e^{-1}\left(1 - s - \frac{s^2}{2!} + \cdots\right)$$

Example B.4.17 Expand $F(s) = [(s-1)(s-3)]^{-1}$ about $s = 1$. What is the region of convergence?

Solution

$$F(s) = -\frac{1}{2(s-1)} + \frac{1}{2(s-3)}$$
$$= -\frac{1}{2(s-1)} + \frac{1}{2}\left[\frac{1}{(s-1)-2}\right]$$

Now let $1 < |s-1| < 2$; then the second term can be expanded by the binomial theorem

$$F(s) = -\frac{1}{2(s-1)} - \sum_{m=0}^{\infty}\frac{(s-1)^m}{2^{m+1}}$$

Example B.4.18 Expand $\sin(1/s)$ about $s = 0$ and $s = 1$.

Solution (a) Let $1/s = Z$; then

$$\sin Z = Z - \frac{Z^3}{3!} + \frac{Z^5}{5!} - \cdots$$

and thus

$$\sin\left(\frac{1}{s}\right) = \sum_{m=0}^{\infty}(-1)^m\frac{s^{-(2m+1)}}{(2m+1)!} \qquad |s| > 0$$

(b)

$$\sin\left[\frac{1}{(s-1)+1}\right] = \sum_{m=0}^{\infty}(-1)^m \frac{[(s-1)+1]^{-(2m+1)}}{(2m+1)!}$$

valid on $|s| > 0$.

Now, if $0 < |s-1| < 1$, then by the binomial theorem

$$\frac{1}{[(s-1)+1]^{2m+1}} = \sum_{p=0}^{\infty}\left[\frac{(-1)^p(2m+p)!}{p!\,(2m)!}\right](s-1)^p$$

and therefore

$$\sin\frac{1}{s} = \sum_{m=0}^{\infty}\sum_{p=0}^{\infty}\left[\frac{(-1)^{m+p}(2m+p)!}{(2m)!\,(2m+1)!\,p!}\right](s-1)^p$$

convergent in $R = [0 < |s-1| < 1]$.

B.4.6 It was previously shown that functions of the form $1/(s-s_0)^n$ $n > 0$ fail to have a derivative at $s = s_0$. Thus it is evident that the complex Laurent series represents a function singular at $s = s_0$ while the Taylor series represents a function which is regular at $s = s_0$. Thus

$$\sin\left(\frac{1}{s}\right) = \frac{1}{s} - \frac{1}{3!\,s^3} + \frac{1}{5!\,s^5} - \cdots$$

is singular at $s = 0$ and

$$F(s) - -\frac{1}{2}\sum_{n=-1}^{\infty}\frac{(s-1)^n}{2^{n-1}}$$

is singular at $s = 1$. If in this last expression we multiply $F(s)$ by $(s-1)$, we obtain a new function

$$G(s) = (s-1)F(s) = -\frac{1}{2}\sum_{n=0}^{\infty}\frac{(s-1)^n}{2^{n-1}}$$

which is everywhere regular in the region of convergence. Note that the function sin $(1/s)$ cannot be made analytic by multiplication by any s^n. $F(s)$ is said to have a *removable singularity* at $s = 1$, and sin $(1/s)$ has an *essential singularity* at $s = 0$.

Definition B.4.6 LET $F(s)$ BE SINGULAR AT $s = s_0$ AND LET THERE BE A NUMBER $N > 0$ SUCH THAT $G(s) = (s - s_0)^N F(s)$ IS ANALYTIC AT s_0 AND SUCH THAT $\lim_{s\to 0} G(s) \neq 0$ EXISTS. THEN $F(s)$ IS SAID TO HAVE A POLE OF ORDER N AT s_0.

Definition B.4.7 LET A FUNCTION $F(s)$ BE SUCH THAT $\lim_{s\to 0}[F(s)] = 0$ AND LET THERE BE AN $N > 0$ SUCH THAT $\lim_{s\to 0}[(s-s_0)^{-N}F(s)] \neq 0$ EXISTS. THEN $F(s)$ IS SAID TO HAVE A ZERO OF ORDER N AT $s = s_0$.

Example B.4.19 Locate the zeros and poles of

$$\left[\frac{s-3}{s+(2-j)}\right]\sin\left(\frac{1}{s}\right)$$

Solution Zero at $s = 3$, pole at $s = -2 + j$ and an essential singularity at $s = 0$. It will be shown later that the zeros and poles of a Laplace transform characterize the inverse completely.

B.4.7 Let $F(s)$ be a single-valued function of a complex variable. It is then "natural" to define an integral as follows:

$$\int_a^b F(s)\, ds = \int_a^b (u + jv)(d\sigma + d\omega)$$
$$= \int_a^b (u\, d\sigma - v\, d\omega) + j\int_a^b (v\, d\sigma + u\, d\omega)$$

It is immediately apparent that such integrals as

$$\int_a^b v(\sigma, \omega)\, d\sigma$$

are not integrable unless a function $\omega = f(\sigma)$ is postulated so as to make v a function of σ or ω alone. Thus a line or curve is specified on the (σ, ω) plane and the integration is said to be performed along a line. The complex integral is a *line integral* and is dependent on the "path of integration."

Example B.4.20 Integrate $1/s$ from the point (1, 0) to the point (−1, 0) along the semicircles C_1 and C_2 shown in Figure B.4.

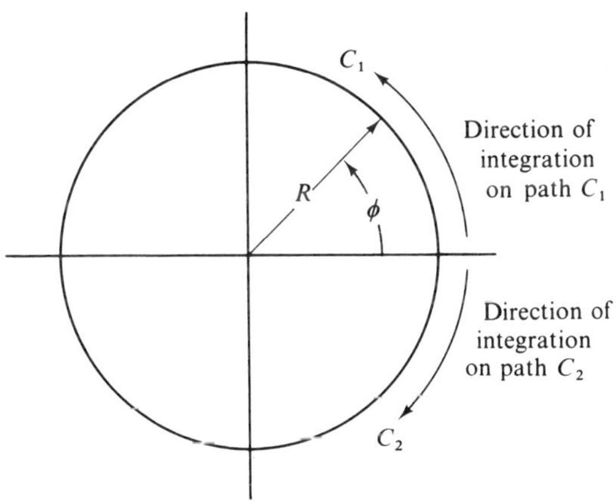

FIGURE B.4 Alternate paths of integration

Solution This integration is more conveniently carried out in polar coordinates. On the circle $|s| = R$ we have

$$s = Re^{j\phi} \qquad ds = jRe^{j\phi}\, d\phi \qquad dR = 0$$

and thus the integral on C_1 becomes

$$j\int_0^{\pi} d\phi = \pi j$$

On C_2 we find that the integral is

$$j\int_0^{-\pi} d\phi = -\pi j$$

The two integrals differ by $2\pi j$.

Theorem B.4.4 LET C BE A CLOSED CURVE AND LET $F(s)$ ANALYTIC ON C AND IN THE REGION R BOUNDED BY C; THEN

$$\oint_C F(s)\,dx = 0$$

The symbol $\oint_C$ indicates that the integration is to be performed on a closed path and that the specified path is C. It must be noted that the path C is not too important. If $F(s)$ is analytic in a region R, and C is any closed curve wholly in R, then the integral along C is always zero.

Definition B.4.8 LET A POINT P MOVE ON A CLOSED CURVE C IN SUCH A FASHION THAT THE INTERIOR REGION R IS ALWAYS TO THE LEFT OF THE VECTOR WHICH REPRESENTS THE MOTION OF P. THEN THE POINT P IS SAID TO BE MOVING IN THE POSITIVE DIRECTION (FIG. B.5).

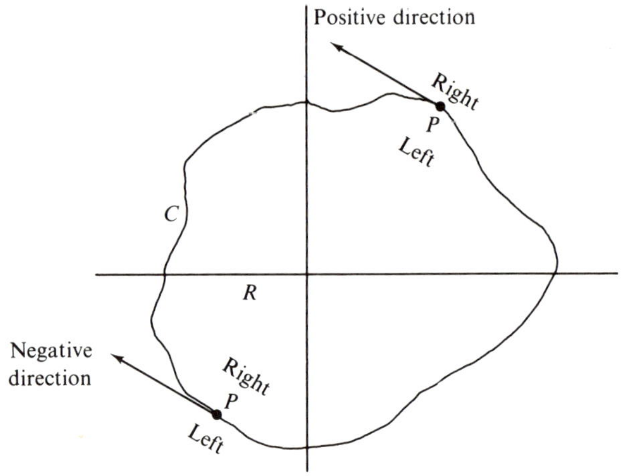

FIGURE B.5 Sense and direction on a closed curve

In Example B.4.20 it was shown that the integral of $1/s$ computed on the upper semicircle was πj while the integral on the lower semicircle was $-\pi j$. It can be seen that the upper path integral is along a positive path while the lower integral is along a negative path. What would be the value of the integral if the integration were carried out completely around the circle in the positive direction? Let us first state a theorem.

Theorem B.4.5 LET $F(s)$ BE ANALYTIC ON A PATH C JOINING TWO POINTS p_1, p_2. THEN

$$\int_{p_1}^{p_2} F(s)\,ds = -\int_{p_2}^{p_1} F(s)\,ds$$

Proof The proof of Theorem B.4.5 can be deduced from Theorem B.4.4. Let the points p_1, p_2 be joined by a closed curve (Fig. B.6). Then if $F(s)$ is analytic on C_1, C_2 and R, it follows that

$$\int_{C_1} F(s)\,ds + \int_{C_2} F(s)\,ds = 0$$

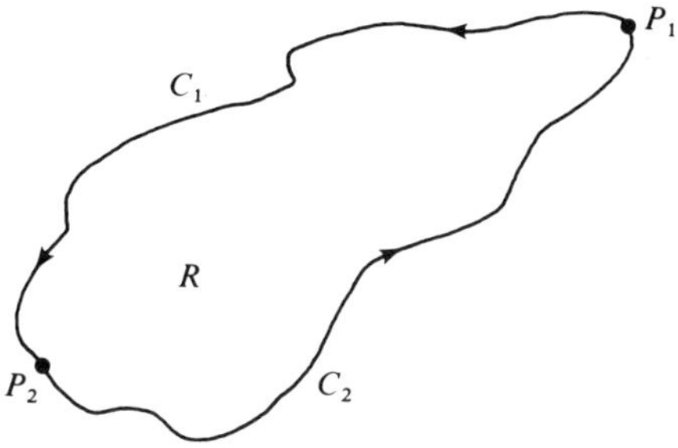

FIGURE B.6 Integration on a closed curve

From this theorem it follows that the integration of $1/s$ along the lower semicircle would be πj and thus that the complete integral around the circle is $2\pi j$. What would be the value of the integral of $1/s$ on some other closed curve surrounding the origin? It can be shown that the result will be $2\pi j$ regardless of the path as long as $s_0 = 0$ is in the interior of the region bounded by the curve.

Theorem B.4.6 LET C_1 AND C_2 BE TWO NONINTERSECTING CURVES AROUND THE POINT s_0 FIG. B.7A) AND LET $F(s)$ BE ANALYTIC ON C_1, C_2 AND IN THE REGION R BETWEEN THEM. THEN

$$\oint_{C_1} F(s)\,ds = \oint_{C_2} F(s)\,ds$$

IN WHICH THE INTEGRALS ARE TAKEN AS POSITIVE WITH RESPECT TO THE REGION CONTAINING s_0. (I.E., R_0 FOR C_2 AND $R + R_0$ FOR C_1).

Proof Let us "cut" the contours as indicated in Figure B.7b so as to make C_1, L_1, C_2, L_2 a complete closed path, with L_1 and L_2 parallel and a distance ϵ apart. Then by Theorem B.4.4

$$\int_{C_1-\epsilon} F(s)\,ds + \int_{L_1} F(s)\,ds + \int_{L_2} F(s)\,ds + \int_{C_2-\epsilon} F(s)\,ds = 0$$

If we make $\epsilon \longrightarrow 0$, the paths L_1 and L_2 approach each other and the integral on L_1 and L_2 become equal in value but opposite in sign (Theorem B.4.5). $C_1 - \epsilon$ approaches C_1 and $C_2 - \epsilon$ approaches C_2 so that

$$\overleftarrow{\oint_{C_1}} F(s)\,ds = -\overrightarrow{\oint_{C_2}} F(s)\,ds$$

But the integral on C_2 is computed on a negative path with respect to

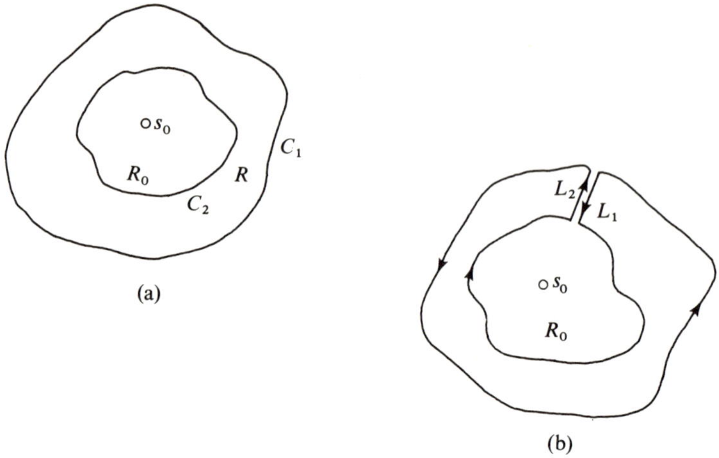

FIGURE B.7 Integrals on nested contours

R_0. Thus

$$\oint_{C_1} F(s)\, ds = \oint_{C_2} F(s)\, ds$$

in which the path of integration is counterclockwise in both instances.

Theorem B.4.7 LET $f(s)$ BE ANALYTIC ON C AND IN R WITH THE EXCEPTION OF A SET OF ISOLATED SINGULAR POINTS s_i WHERE $i = 1, 2, \ldots$; THEN

$$\oint_C F(s)\, ds = \int_{C_1} F(s)\, ds + \int_{C_2} F(s)\, ds + \cdots + \int_{C_n} F(s)\, ds$$

IN WHICH $C_1, C_2, \ldots, C_n$ ARE CONTOURS ABOUT THE SINGULARITIES AND CONTAIN ONLY ONE SINGULARITY EACH.

The proof is analogous to that for Theorem B.4.6. The plane is cut as indicated in Figure B.8 and the previous theorems applied as $\epsilon_1, \epsilon_2, \ldots, \epsilon_n$ are all let go to zero.

Theorem B.4.8 LET $F(s) = (s - s_0)^N$ $N = 0, \pm 1, \pm 2, \ldots$; THEN

$$\oint_C F(s)\, ds = \begin{cases} 2\pi j & N = -1 \\ 0 & N \neq -1 \end{cases}$$

C IS ANY PATH AROUND s_0.

Proof By Theorem B.4.6 the path C can be replaced by a circle of radius ρ with s_0 as center. On this circle, $s - s_0 = \rho \exp j\phi$ and $ds = j\rho \exp j\phi\, d\phi$. Then

$$\oint_C F(s)\, ds = j \int_0^{2\pi} \rho^{1+N} e^{j(1+N)\phi}\, d\phi$$

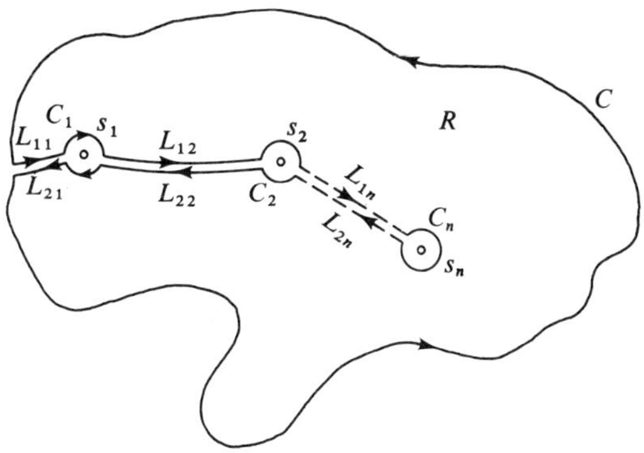

FIGURE B.8 Cut and contours near singular points

For $N = -1$ the integral is obviously $2\pi j$. For $N \neq -1$ we obtain

$$\oint_C F(s)\,ds = \rho^{1+N}\left[\frac{e^{j(1+N)\phi}}{1+N}\right]_0^{2\pi} = 0$$

Theorem B.4.9 LET $F(s)$ HAVE AN ISOLATED SINGULAR POINT AT $s = s_0$ AND LET

$$F(s) = \sum_{M=-\infty}^{m=+\infty} C_m(s - s_0)^m$$

BE VALID IN SOME NEIGHBORHOOD $|s - s_0| < a$ IN WHICH a IS A REAL NUMBER GREATER THAN ZERO. LET C BE SOME CLOSED CONTOUR AROUND s_0 WHICH CONTAINS ONLY THE SINGULARITY AT s_0 IN THE INTERIOR. THEN

$$\oint_C F(s)\,ds = 2\pi j C_{-1}$$

Proof A convergent Laurent series can be integrated term by term and thus the theorem follows from Theorem B.4.7.

Definition B.4.9 THE TERM C_1 OF THE LAURENT SERIES $\sum C_m(s - s_0)^m = F(s)$ IS CALLED THE RESIDUE OF $F(s)$ AT $s = s_0$.

$$\text{Res}\,[F(s);\ s = s_0] = C_{-1}$$

Theorem B.4.10 LET $F(s)$ HAVE A SET OF ISOLATED SINGULAR POINTS s_i $(i = 1, 2, \ldots, n)$ IN A REGION R. LET C BE A CONTOUR ON WHICH $F(s)$ IS ANALYTIC SURROUNDING ALL OF THE SINGULAR POINTS. THEN

$$\oint_C F(s)\,ds = 2\pi j \sum_{i=1}^{n} \text{Res}\,[F(s);\ s = s_i]$$

Proof By Theorem B.4.7

$$\oint_C F(s)\,ds = \oint_{C_1} F(s)\,ds + \oint_{C_2} F(s)\,ds + \cdots + \int_{C_n} F(s)\,ds$$

In each integral on $C_1, C_2, \ldots, C_n$, the function $F(s)$ can be expanded about $s_1, s_2, \ldots, s_n$ respectively into a Laurent series and theorem B.4.10 follows by B.4.9.

It is not necessary to evaluate the Laurent series to find the residue of a function with nonessential singularities. Consider $F(s)$ to have a pole of order M at $s = s_0$. Then we obtain the Laurent expansion

$$F(s) = \sum_{n=-m}^{\infty} C_n(s - s_0)^n$$

If we now multiply $F(s)$ by $(s - s_0)^M$, we obtain

$$K(s) = (s - s_0)^M F(s) = \sum_{n=-m}^{\infty} C_n(s - s_0)^{n+M} = \sum_{n=0}^{\infty} C_{n-M}(s - s_0)^n$$

$K(s)$ is obviously analytic at $s = s_0$ and can be expanded into a Taylor series. Thus

$$K(s) = K(s_0) + \frac{dK}{ds}\bigg|_{s=s_0}(s - s_0) + \frac{1}{2!}\frac{d^2K}{ds^2}\bigg|_{s=s_0}(s - s_0)^2 + \cdots + \frac{1}{(M-1)!}\frac{d^{M-1}K}{ds^{M-1}}\bigg|_{s=s_0}(s - s_0)^{M-1} + \cdots$$

and thus

$$F(s) = \frac{K(s_0)}{(s - s_0)^M} + \left[\frac{dK(s_0)}{ds}\right]\frac{1}{(s - s_0)^{M-1}} + \cdots + \frac{1}{(M-1)!}\left[\frac{d^{M-1}K(s_0)}{ds^{M-1}}\right]\frac{1}{s - s_0} + \cdots$$

Hence we can state the following theorem.

Theorem B.4.11 LET $F(s)$ HAVE A POLE OF ORDER M AT $s = s_0$ AND NO OTHER SINGULARITIES IN A NEIGHBORHOOD OF s_0. THEN

$$\text{Res}\,[F(s); s = s_0] = \lim_{s \to s_0}\left\{\frac{1}{(M-1)!}\frac{d^{M-1}}{ds^{M-1}}[(s - s_0)^M F(s)]\right\}$$

Example B.4.21 Integrate $10/(s + 2)$ on a contour around $s = -2$.

Solution From Theorem B.4.11

$$\text{Res}\left[\frac{10}{s+2}; s = -2\right] = 10$$

and thus

$$\oint \frac{10}{s+2}\,ds = 20\pi j$$

Example B.4.22 Evaluate the integral of $F(s)$ on a contour containing the poles at $s = -1$ and $s = 3$.

$$F(s) = \frac{10s}{(s+1)(s-3)}$$

Solution

$$\text{Res}\,[F(s); s = -1] = \frac{5}{2}$$

$$\text{Res}\,[F(s); s = 3] = \frac{15}{2}$$

Thus

$$\oint F(s)\,ds = 2\pi j\left(\frac{5}{2} + \frac{15}{2}\right) = 20\pi j$$

Example B.4.23 Integrate on a contour around the origin the function $F(s) = (s - 4)/s^2$.

Solution

$$\text{Res}\left[\frac{s-4}{s^2};\ s = 0\right] = \lim_{s\to 0} \frac{d}{ds}[s^2F(s)] = 1$$

$$\oint \frac{s-4}{s^2}\,ds = 2\pi j$$

Example B.4.24 Integrate $F(s) = (\exp st)/(s - 1)^2(s + 2)$ on a contour containing both $s = 1$ and $s = -2$. (t is a parameter.)

Solution

$$\text{Res}\,[F(s);\ s = -2] = \lim_{s\to -2}\left[\frac{e^{st}}{(s-1)^2}\right] = \frac{e^{-2t}}{9}$$

$$\text{Res}\,[F(s);\ s = 1] = \lim_{s\to 1}\frac{d}{ds}\left[\frac{e^{st}}{s+2}\right] = \frac{e^t}{9}(3t - 1)$$

Therefore

$$\oint \frac{e^{st}}{(s-1)^2(s+2)}\,ds = \frac{2\pi j}{9}[e^{-2t} + e^t(3t - 1)]$$

Example B.4.25 Integrate $\sin(1/s - 1)$ on a contour around $s = 1$.

Solution Expand into a Laurent series:

$$\sin\left(\frac{1}{s-1}\right) = \frac{1}{s-1} - \frac{1}{3!\,(s-1)^3} + \cdots$$

Thus

$$\text{Res}\left[\sin\left(\frac{1}{s-1}\right);\ s = 1\right] = 1$$

and

$$\oint_C \sin\left(\frac{1}{s-1}\right) ds = 2\pi j$$

B.5 Some Theorems on Laplace Transforms

The theory of contour integration has a direct bearing on the inversion of Laplace transforms. The initial transformation of a function is simply an exercise in real integration while the inversion is usually an exercise in residue evaluation (as will be shown below). The work is simplified by tables of transform pairs: in practice, the operation $\mathcal{L}f(t) = F(s)$ is accomplished by a table of transforms.[11] An abbreviated table can be found in Appendix C. Inversion, however, is a little more complicated. It is almost impossible to prepare a table with enough entries to meet all the varied transforms one meets in practice. Two methods are met in practice: partial fraction expansion[12] and computation by residues. Most modern textbook writers teach the partial fraction method even though the computations are messy and the contribution of each pole not clear. Recently there has been a swing toward residues, mostly as a result of the popularity of the pole-zero method of circuit design.

B.5.1 The inversion integral can be converted to a contour integral and evaluated by the theory of residues under certain circumstances. A preliminary theorem is needed.

Theorem B.5.1 LET $F(s)$ BE DEFINED IN A REGION R EXCEPT FOR A FINITE NUMBER OF ISOLATED SINGULAR POINTS. LET $s = R \exp(j\phi)$, AND IF

$$\lim_{R\to\infty} |F(Re^{j\phi})| = 0$$

ON THE SEMICIRCLE C_R (FIG. B.9), THEN

$$\lim_{R\to\infty} \int_{C_R} F(s)e^{st}\, ds = 0$$

The proof of this theorem is not important to our development.

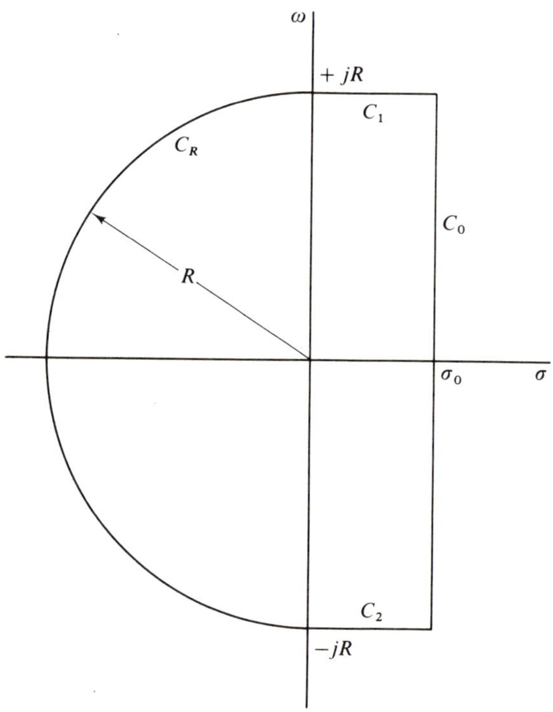

FIGURE B.9 Inversion contour

Corollary to Theorem B.5.1 IF THE CONDITIONS OF THEOREM B.5.1 ARE SATISFIED, THE INTEGRALS ON C_1 AND C_2 DO NOT CONTRIBUTE TO THE INTEGRATION AS $R \longrightarrow \infty$ (FIG. B.9).

Theorem B.5.2 LET $F(s)$ HAVE A FINITE NUMBER OF ISOLATED POLES TO THE LEFT OF $\sigma > 0$ (FIG. B.9) AND LET $\lim_{R\to\infty} |F(Re^{j\phi})| = 0$ ON C_1, C_2 AND C_R; THEN

$$\mathcal{L}^{-1}F(s) = \sum_{i=1}^{n} \text{Res}\,[e^{st}F(s);\ s = s_i]$$

Proof The inversion theorem [Equation (B.2.7)] has the form

$$\mathcal{L}^{-1}F(s) = \frac{1}{2\pi j}\int_{\sigma-j\infty}^{\sigma+j\infty} e^{st}F(s)\,ds$$

Consider now the integral on the path indicated in Figure B.9

$$\begin{aligned}\oint_C e^{st}F(s)\,ds &= \int_{C_R} e^{st}F(s)\,ds + \int_{C_1} e^{st}F(s)\,ds \\ &\quad + \int_{\sigma-jR}^{\sigma+jR} e^{st}F(s)\,ds + \int_{C_2} e^{st}F(s)\,ds \\ &= 2\pi j\sum_{i=1}^{n}\text{Res}\,[e^{st}F(s);\ s = s_i]\end{aligned}$$

In the limit as $R \longrightarrow \infty$ the contributions of C_R, C_1, C_2 vanish and the theorem is established.[13]

Example B.5.1 Invert $F(s) = 3s/(s+1)(s+4)$.

Solution Find the residues of $F(s)e^{st}$

$$\text{Res}\,[F(s)e^{st};\ s = -1] = -e^{-t}$$
$$\text{Res}\,[F(s)e^{st};\ s = -4] = 4e^{-4t}$$

Hence

$$\mathcal{L}^{-1}F(s) = f(t) = 4e^{-4t} - e^{-t}$$

B.5.2 With the basic theory presented in the first part of this Appendix, it is now possible to establish a number of valuable theorems and to make use of Laplace transforms to solve ordinary linear differential equations with constant coefficients.

Theorem B.5.3 LET C_i $(i = 1, 2, 3, \ldots, n)$ BE A SET OF REAL OR COMPLEX CONSTANTS AND LET $f_i(t)$ $(i = 1, 2, \ldots, n)$ BE A SET OF FUNCTIONS IN $\mathcal{L}$. THEN

$$\mathcal{L}\sum_{i=1}^{n} C_i f_i(t) = \sum_{i=1}^{n} C_i \mathcal{L} f_i(t)$$

This theorem and the following are the consequence of elementary integration theory. The proof is obvious.

Theorem B.5.4 LET $f(t)$ BE IN $\mathfrak{L}$ AND LET $f(t) \equiv 0$ FOR ALL VALUES OF t; THEN $\mathfrak{L}f(t) = 0$.

Theorem B.5.5 LET $f(t)$ BE IN $\mathfrak{L}$ AND LET df/dt EXIST ALMOST EVERYWHERE. THEN

$$\frac{df}{dt} = s\mathfrak{L}f - f(0^+)$$

in which $f(0^+)$ is the limit of the value of $f(t)$ as $t \longrightarrow 0$.

Proof We integrate by parts

$$\begin{aligned}\frac{df}{dt} &= \int_0^\infty e^{-st}\frac{df}{dt}\,dt \\ &= s\int_0^\infty f(t)e^{-st}\,dt + [f(t)e^{-st}]_{0^+}^{\infty} \\ &= s\mathfrak{L}f(t) - f(0^+)\end{aligned}$$

Theorem B.5.6 LET $f(t)$ BE IN $\mathfrak{L}$ AND LET $d^n f/dt^n$ EXIST ALMOST EVERYWHERE. THEN

$$\frac{d^n f}{dt^n} = s^n\mathfrak{L}f - \sum_{i=0}^{n-1} s^i\left[\frac{d^{n-i-1}f}{dt^{n-i-1}}\right]_{t=0^+}$$

Proof of B.5.6 is by *induction*. Assume that the theorem is valid for $n = N$. Using integration by parts, show that the theorem is valid for $n = N + 1$. Then since the theorem is valid for $n = 1$ (B.5.5), it must be valid for $n = 2$ and thus for $n = 3$ and hence for all $n \geq 1$.

The details are left for an exercise (see problem B.12).

Definition B.5.1 LET A FUNCTION $u(t - \tau)$ HAVE THE FOLLOWING DEFINITION:

$$u(t-\tau) = \begin{cases} 0 & t \leq \tau \\ 1 & t > \tau \end{cases}$$

THEN $u(t - \tau)$ IS CALLED THE *unit step* FUNCTION.

Definition B.5.2 LET A FUNCTION $\delta(t - \tau)$ HAVE THE FOLLOWING DEFINITION:

$$\delta(t-\tau) = \begin{cases} 0 & \text{if } t \neq \tau \\ \infty & \text{if } t = \tau \end{cases}$$

AND LET

$$\int_0^\infty \delta(t-\tau)\,dt = 1$$

THEN $\delta(t - \tau)$ IS CALLED THE DIRAC DELTA FUNCTIONS.

Theorem B.5.7

$$\delta(t-\tau) = \lim_{h\to 0}\left[\frac{u[(t+h)-\tau] - u(t-\tau)}{h}\right]$$

Proof

(a) For $t > \tau$, $u(t - \tau) = 1$, $u[(t + h) - \tau] = 1$ and thus $\delta(t - \tau) = 0$ for $h > 0$.

(b) For $t \leq \tau$ $u(t - \tau) = 0$, for $t \leq \tau - h$, $u = 0$, and thus $\delta = 0$ for $t \leq \tau - h$.

(c) In the interval $\tau \geq t > \tau - h$, $\delta = 1/n$ and thus as $h \longrightarrow 0$, $\delta \longrightarrow \infty$.

(d)
$$\int_0^\infty \delta(t-\tau)\,dt = \lim_{h\to 0} \int_{\tau-h}^{\tau} \frac{1}{h}\,dt = 1$$

Formally $\delta(t-\tau)$ satisfies a limit equation identical to a derivative. Even though the derivative of $u(t-\tau)$ does not exist in the usual sense,[14] we say that $\delta(t-\tau)$ is the "formal derivative" of $u(t-\tau)$.

Theorem B.5.8

$$\mathcal{L} f(t)\delta(t-\tau) = e^{-s\tau} f(\tau)$$

Proof

$$\int_{0^+}^\infty f(t)\delta(t-\tau)e^{-st}\,dt = f(\tau)e^{-s\tau}\int_{0^+}^\infty \delta(t-\tau)\,dt$$
$$= e^{-s\tau} f(\tau)$$

The second step follows that form the fact that $\delta(t-\tau) = 0$ if $t \neq \tau$.

Theorem B.5.9

$$\mathcal{L} f(t)u(t-\tau) = e^{-s\tau}\mathcal{L} f(t+\tau)$$

Corollary to Theorem B.5.9

$$\mathcal{L} f(t-\tau)u(t-\tau) = e^{-s\tau}\mathcal{L} f(t)$$

Proof

$$f(t)u(t-\tau) = \int_{\tau^+}^\infty e^{-st} f(t)\,dt$$

Let $t = t' + \tau$; then

$$\int_{\tau^+}^\infty e^{-st} f(t)\,dt = \int_{0^+}^\infty e^{-s(t'+\tau)} f(t'+\tau)\,dt'$$
$$= e^{s\tau} f(t+\tau)$$

Theorem B.5.10

LET $f_1(t)$ AND $f_2(t)$ BE IN $\mathcal{L}$; THEN

$$\mathcal{L}^{-1} F_1(s)F_2(s) = \int_{0^+}^t f_1(t-\tau)f_2(\tau)\,d\tau$$

Proof

$$F^* = \int_{0^+}^t f_1(t-\tau)f_2(\tau)\,d\tau = \int_{0^+}^\infty e^{st}\left[\int_{0^+}^\infty f_1(t-\tau)f_2(\tau)u(t-\tau)\,d\tau\right]dt$$

The order of integration can be interchanged[13] and we obtain

$$F^* = \int_{0^+}^\infty f_2(\tau)\left[\int_{0^+}^\infty e^{-st}u(t-\tau)f_1(t-\tau)\,dt\right]d\tau$$

The inner integral can be evaluated by the corollary to Theorem B.5.9, and therefore

$$F^* = \int_{0^+}^\infty f_2(\tau)e^{-s\tau}F_1(s)\,d\tau = F_1(s)\int_{0^+}^\infty f_2(\tau)e^{-s\tau}\,d\tau$$
$$= F_1(s)F_2(s)$$

Theorem B.5.10 is called the superposition integral and is one of the most significant theorems in the theory of linear systems.

B.6 Some Applications of Laplace Transforms

Example B.6.1 Transform the following functions:

(a) $e^{\alpha t}$
(b) $\sinh \alpha t$
(c) $\sin \alpha t$
(d) $k =$ constant
(e) $ku(t - \tau)$ if $k =$ constant

Solution

(a)
$$\mathcal{L}e^{\alpha t} = \int_{0^+}^{\infty} e^{\alpha t} e^{-st}\, dt = \frac{1}{s - \alpha}$$

(b)
$$\mathcal{L} \sinh \alpha t = \mathcal{L}\left(\frac{e^{\alpha t} - e^{-\alpha t}}{2}\right) = \frac{1}{2}\left(\frac{1}{s - \alpha} - \frac{1}{s + \alpha}\right)$$
$$= \frac{\alpha}{s^2 - \alpha^2}$$

(c) It was shown in Section B.4.3 that

$$\sin \alpha t = \frac{1}{j} \sinh (j\alpha t)$$

Replacing $j\alpha$ for α in (b) above gives us

$$\mathcal{L} \sin \alpha t = \frac{\alpha}{s^2 + \alpha^2}$$

(d, e)
$$\mathcal{L}ku(t - \tau) = \mathcal{L}ku(t - \tau) = \frac{ke^{-s\tau}}{s}$$

For part (d), set $\tau = 0$ and then $\mathcal{L}k = k/s$.

Example B.6.2 A system is governed by the differential equation

$$\frac{dx}{dt} + 3x = 2 \qquad x = 1 \quad \text{at } t = 0$$

Find $x(t)$ by Laplace transforms.

Solution By Theorem B.5.4

$$\mathcal{L}\left(\frac{dx}{dt} + 3x - 2\right) = 0$$

By Theorem B.5.3

$$\mathcal{L}\frac{dx}{dt} + 3\mathcal{L}x - \mathcal{L}2 = 0$$

We apply Theorem B.5.5 and the result of Example B.6.1(d)

$$s\mathcal{L}x - 1 + 3\mathcal{L}x - \frac{2}{s} = 0$$

We solve for $\mathcal{L}x$

$$\mathcal{L}x = \frac{s + 2}{s(s + 3)}$$

To invert we use Theorem B.5.2

$$x(t) = \operatorname{Res}\,[e^{st}\mathfrak{L}x;\, s = -3] + \operatorname{Res}\,[e^{st}\mathfrak{L}x;\, s = 0]$$
$$= \frac{2}{3} + \frac{1}{3}e^{-3t}$$

Example B.6.3 Invert by the superposition integral

$$F(s) = \frac{2}{s(s+1)}$$

Solution Let

$$F_1(s) = \frac{2}{s} \qquad F_2(s) = \frac{1}{s+1}$$

Then

$$\mathfrak{L}^{-1}F_1 = 2 \qquad \mathfrak{L}^{-1}F_2 = e^{-t}$$

Hence

$$\mathfrak{L}^{-1}F(s) = \int_0^t 2e^{-(t-\tau)}\, d\tau = 2(1 - e^{-t})$$

Example B.6.4 A system obeys the equation

$$\frac{d^2x}{dt^2} + 5\frac{dx}{dt} + 6x = 3 \qquad x = \dot{x} = 0 \quad \text{at } t = 0$$

Solve for x.

Solution The equation is transformed immediately into

$$\mathfrak{L}x = \frac{3}{s(s+2)(s+3)}$$

To invert we compute the residues at $s = 0$, $s = -2$ and $s = -3$ and add

$$x(t) = \frac{1}{2}(1 - 3e^{-2t} + 2e^{-3t})$$

Example B.6.5 Let $f(t)$ be in $\mathfrak{L}$ and let $f(t) = f(t + n\tau)$ $n = 1, 2, \ldots$, that is, $f(t)$ is periodic in τ. Prove that

$$\mathfrak{L}f(t) = \frac{\mathfrak{L}f_1(t)}{1 - e^{-s\tau}}$$

in which $f_1(t)$ is a function which equals $f(t)$ in $0 < t \leq \tau$ but is zero for $t \leq 0$ and $t > \tau$.

Solution Since $f(t)$ is periodic in τ, we can write

$$f(t) = \sum_{n=0}^{\infty} f_1(t - n\tau)$$

but

$$f_1(t - n\tau) = u(t - n\tau)f_1(t - n\tau)$$

and thus

$$\mathfrak{L}f(t) = \mathfrak{L}\sum_{n=0}^{\infty} u(t - n\tau)f_1(t - n\tau)$$

From the corollary of Theorem B.5.9 it follows that

$$\mathcal{L}f(t) = \mathcal{L}f_1(t) \sum_{n=0}^{\infty} e^{-sn\tau}$$

It is well known that

$$\sum_{n=0}^{\infty} e^{-sn\tau} = (1 - e^{-s\tau})^{-1}$$

and thus

$$\mathcal{L}f(t) = \frac{\mathcal{L}f_1(t)}{1 - e^{-s\tau}}$$

Example B.6.6 Transform the following function (Fig. B.10)

$$f(t) = \begin{cases} k & \text{if } 0 < t \leq a \\ -k & \text{if } a < t \leq 2a \\ k & \text{if } 2a < t \leq 3a \\ \text{etc.} & \cdots \end{cases}$$

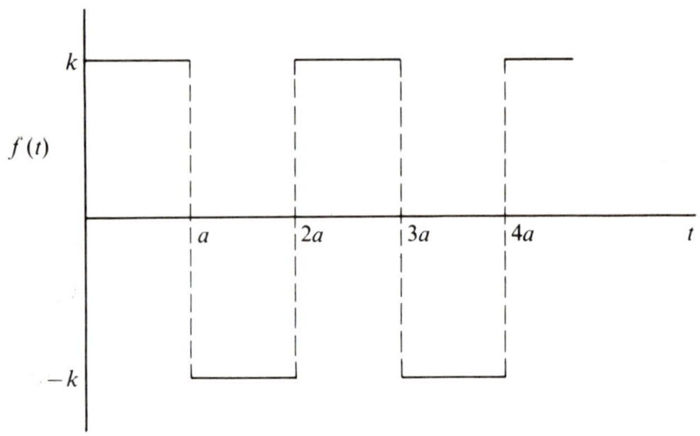

FIGURE B.10 Square wave

Solution This function is periodic in $2a$ and thus it is possible to use the results of the previous example. $f_1(t) = f(t)$ in $0 < t \leq 2a$ but is zero elsewhere. Hence

$$\begin{aligned} f_1(t) &= k[u(t) - u(t-a)] - k[u(t-a) - u(t-2a)] \\ &= ku(t) - 2ku(t-a) + ku(t-2a) \end{aligned}$$

We can use the result of Example B.7.1(d).

$$\mathcal{L}f_1(t) = \frac{k}{s}(1 - 2e^{-as} + e^{-2as}) = \frac{k}{s}(1 - e^{-as})^2$$

Thus

$$\mathcal{L}f(t) = \frac{k(1 - e^{-as})^2}{s(1 - e^{-2as})} = \frac{k}{s}\left(\frac{1 - e^{-as}}{1 + e^{as}}\right)$$

Therefore

$$\mathcal{L}f(t) = \frac{k}{s}\tanh\left(\frac{as}{2}\right)$$

Example B.6.7 A system obeys the equation

$$\frac{dx}{dt} + x = f(t) \qquad x = 0 \quad \text{at } t = 0$$

in which

$$f(t) = \begin{cases} t & \text{if } 0 < t \leq 3 \\ 3 & \text{if } t > 3 \end{cases}$$

Find $x(t)$ by Laplace transforms (Fig. B.11).

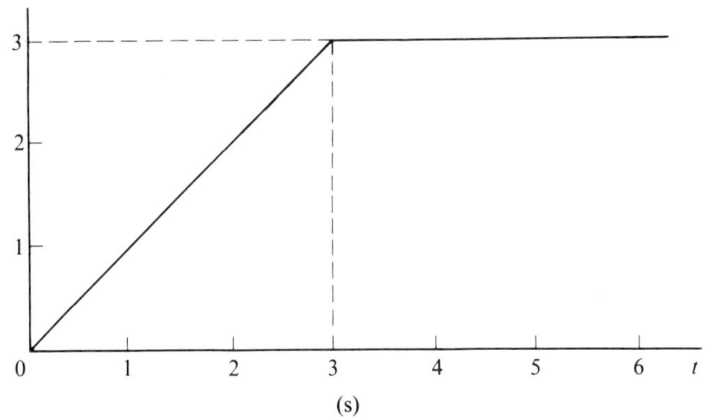

FIGURE B.11

Solution

$$f(t) = t[u(t) - u(t-3)] + 3u(t-3)$$
$$= tu(t) - (t-3)u(t-3)$$
$$f(t) = (1 - e^{-3s})\mathcal{L}t$$

We use the transform pair for $\mathcal{L}t$ in Appendix C to obtain

$$\mathcal{L}f(t) = \frac{1}{s^2}(1 - e^{-3s})$$

whence

$$\mathcal{L}x(t) = \frac{1 - e^{-3s}}{s^2(s+1)}$$

We invert by computing the residue at $s = 0$ and $s = 1$. We note that $1 - e^{-3s}$ simply indicates that there are two additive solutions

$$x(t) = \mathcal{L}^{-1}\frac{1}{s^2(s+1)} - \mathcal{L}^{-1}\frac{e^{-3s}}{s^2(s+1)}$$

The second term will be identical to the first one but shifted 3 seconds. Thus it is necessary only to solve the first part.

$$\mathcal{L}^{-1}\frac{1}{s^2(s+1)} = e^{-t} + (t-1)$$

and thus

$$x(t) = [e^{-t} + (t-1)] - u(t-3)[e^{-(t-3)} + (t-3) - 1]$$

This function has the form

$$x(t) = \begin{cases} e^{-t} + t - 1 & \text{if } 0 < t \leq 3 \\ (1 - e^3)e^{-t} + 3 & \text{if } t > 3 \end{cases}$$

B.7 Some Additional Inversion Theorems

Whenever the poles of a transform are complex, the application of the residue theorem becomes quite messy. It is quite possible to factor the denominator into simple poles of the form $1/(s - a \pm jb)$, but the resulting complex functions involve considerable algebraic manipulation and a chance to make mistakes. It is usually better to factor the pair of complex conjugate poles in the form $1/(s^2 + cs + d)$ and then invert directly by a modified form of the residue theorem.

B.7.1 Before developing the inversion rule let us present the complex translation theorem.

Theorem B.7.1 LET $F(s)$ BE A LAPLACE TRANSFORM; THEN $\mathcal{L}F(s + a) = e^{-at}\mathcal{L}F(s)$

Proof

$$\mathcal{L}^{-1}F(s+a) = \sum_i \text{Res}\,[F(s+a)e^{st};\ s = s_i]$$

$$\text{Res} = \oint F(s+a)e^{st}\,ds = e^{at}\oint F(s)e^{st}\,ds$$

$$\therefore\ \mathcal{L}^{-1}F(s+a) = e^{at}\mathcal{L}^{-1}F(s)$$

Example B.7.1 A system obeys the equation

$$\frac{d^2x}{dt^2} + 4\frac{dx}{dt} + 8x = 2\delta(t) \qquad x = \dot{x} = 0 \quad \text{at } t = 0$$

Find $x(t)$.

Solution

$$(x) = \frac{2}{s^2 + 4s + 8} = \frac{2}{(s+2)^2 + 4}$$

Thus

$$x(t) = e^{-2t}\mathcal{L}^{-1}\frac{2}{s^2 + 4} = e^{-2t}\sin 2t$$

Theorem B.7.2 LET $F(s) = G(s)/[(s+a)^2 + b^2]$ IN WHICH $G(s)$ IS ANY LAPLACE TRANSFORM; THEN

$$\mathcal{L}^{-1}F(s) = \frac{e^{-at}}{b}(P\sin bt + Q\cos bt) + f_0(t)$$

IN WHICH $P + jQ = G(-a + jb)$, THAT IS, $P = ReG(-a + jb)$ $Q = ImG(-a + jb)$, $f_0(t)$ IS A FUNCTION OF TIME COMPUTED FROM

$\sum \text{Res}\,[e^{-st}F(s);\, s_i]$ EXCLUDING THE RESIDUES AT THE POLES OF $1/[(s+a)^2+b^2]$, THE EFFECTS OF WHICH ARE INCLUDED IN THE FIRST TERM $\mathcal{L}^{-1}F(s)$.

Proof

$$F(s) = \frac{G(s)}{(s+a+jb)(s+a-jb)}$$

From the residue theorem we obtain

$$\mathcal{L}^{-1}F(s) = \frac{1}{2jb}[G(-a+jb)e^{(-a+jb)t} + G(-a-jb)e^{-(a+jb)t}] + \sum_{i=3}^{n} \text{Res}\,[F(s)e^{st};\, s = s_i]$$

In the last term the residues at $s_1 = -a + jb$ and $s_2 = -a - jb$ are not included. We now set $G(-a \pm jb) = P \pm jQ$ and remembering that $e^{jbt} = \cos bt + j \sin bt$, we find, after a little reduction, that

$$\mathcal{L}^{-1}F(s) = \frac{e^{-at}}{b}(P \sin bt + Q \cos bt) + f_0(t)$$

Example B.7.2 Solve for x:

$$\frac{dx}{dt} + 4x = \sin 3t$$

Solution

$$\mathcal{L}x = \frac{3}{(s^2+9)(s+4)}$$

The contribution of the imaginary poles is in the form of Theorem B.8.2 with $a = 0$ and $b = 3$. Thus

$$x(t) = \frac{1}{3}(P \sin 35 + Q \cos 3t) + \text{Res}\,[e^{st}x;\, s = -4]$$

$$G = \frac{3}{s+4} \qquad G(3j) = \frac{3}{25}(4 - 3j)$$

Hence

$$x(t) = \frac{4}{25}\sin 3t - \frac{3}{25}\cos 3t + \frac{3}{25}e^{-4t}$$

Example B.7.3 Solve for x:

$$\frac{d^2x}{dt^2} + 4\frac{dx}{dt} + 8x = 2 \sin 3t \quad \text{if } x = \dot{x} = 0 \quad \text{at } t = 0$$

Solution

$$\mathcal{L}x = \frac{6}{(s^2+9)[(s+2)^2+4]}$$

Both factors are composed of a pair of complex poles. Let us first invert the contribution of the $s^2 + 9$ factor.

$$G_1 = \frac{6}{(s+2)^2+4} = \frac{-6}{145}(1 + 12j)$$

Thus $a_1 = 0$, $b_1 = 3$, $P_1 = -(-6/145)$, $Q_1 = -72/145$.

We now invert the contribution of $(s + 2)^2 + 4 = s^2 + 4s + 8$

$$G_2 = \frac{6}{s^2 + 9}\bigg|_{s=-2+2j} = \frac{1}{145}(54 + 48j)$$

Hence $a_2 = 2$, $b_2 = 2$, $P_2 = 54/145$, $Q_2 = 48/145$, from which it follows that

$$x(t) = \frac{2}{145}[e^{-2t}(27 \sin 2t + 12 \cos 2t) - (\sin 3t + 12 \cos 3t)]$$

Theorem B.8.2 is usable whenever the roots of a quadratic factor are complex. If the roots are real, the theorem becomes difficult to apply. Under such circumstances it is better to factor into two real poles and to use the residue theorem directly.

B.7.2 Another difficulty in inversion arises whenever there exist high-order poles in the transform. Evaluation of $(d^n/ds^n)[e^{st}F(s)]$ can be quite messy. The following theorem can be useful in such instances.

Theorem B.7.3

$$\frac{d^N}{ds^N}[e^{st}F(s)] = e^{st}\left(t + \frac{d}{ds^n}\right)^N F(s)$$

WHERE

$$\left(t + \frac{d}{ds}\right)^N = \sum_{r=0}^{N} \binom{N}{r} t^{N-r} \frac{d^r}{ds^r}$$

AND

$$\binom{N}{r} = \frac{N!}{(N-r)!\,r!}$$

Proof By induction.

(a) Let $N = 1$; then by direct differentiation

$$\frac{d}{ds}[e^{st}F(s)] = e^{st}\left(t + \frac{d}{ds}\right)F(s)$$

(b) Assume the theorem is true for $N = M$. Then

$$\frac{d}{ds}\left[e^{st}\left(t + \frac{d}{ds}\right)^M F(s)\right] = e^{st}\left(t + \frac{d}{ds}\right)^{M+1} F(s)$$

Thus if the theorem is true for M, it is true for $M + 1$. But since it is true for $N = 1$, it is true for all N. The expansion of the operator by the binomial theorem is well known. With a table of binomial coefficients (Appendix D) the theorem can be applied very easily. For example, if $N = 3$, the table in Appendix D shows that the binomial coefficients are 1, 3, 3, 1. Thus

$$\left(t + \frac{d}{ds}\right)^3 = t^3 + 3t^2\frac{d}{ds} + 3t\frac{d^2}{ds^2} + \frac{d^3}{ds^3}$$

Example B.7.4 A system obeys the equation

$$\frac{d^2x}{dt^2} + 16x = t^3 \qquad \dot{x} = x = 0 \quad \text{at } t = 0$$

Find $x(t)$.

Solution From Appendix C it is found that

$$x = \frac{3!}{s^4(s^2 + 16)}$$

To invert we compute the residue at $s = 0$

$$\frac{1}{3!}\frac{d^3}{ds^3}\left(\frac{3!\,e^{st}}{s^2 + 12}\right)\bigg|_{s=0} = e^{st}\left[t^3K(s) + 3t^2\frac{dK}{ds} + 3t\frac{d^2K}{ds^2} + \frac{d^3K}{ds^3}\right]\bigg|_{s=0}$$

in which

$$\begin{aligned} K(s) &= (s^2 + 16)^{-1} & K(0) &= 1/16 \\ K'(s) &= -2s(s^2 + 16)^{-2} & K'(0) &= 0 \\ K''(s) &= (6s^2 - 32)(s^2 + 16)^{-3} & K''(0) &= -1/128 \\ K'''(s) &= 16s^2(s - 4)(s^2 + 16)^{-4} & K'''(0) &= 0 \end{aligned}$$

Thus

$$\text{Res}\,[e^{st}\mathfrak{L}x;\ s = 0] = \frac{1}{16}\left(t^3 - \frac{3}{8}t\right)$$

The contribution of $s^2 + 16$ is computed by Theorem B.8.2 with $a = 0$, $b = 4$

$$G(4j) = \frac{3!}{(4j)^4} = \frac{3}{128}$$

and thus $P = 6/256$, $Q = 0$. Finally

$$x(t) = \frac{1}{128}(8t^3 - 3t + 3\sin 4t)$$

B.7.3 There are two Laplace transform theorems which are usually given in the literature as the "initial value" theorem and the "final value" theorem. Neither theorem is general enough to be useful. In certain very limited cases both theorems can be used. They will be included here without proof for the sake of completeness.

Theorem B.7.4 LET $f(t)$ BE IN $\mathfrak{L}$ AND LET $\lim_{t\to 0} f(t)$ EXIST AND BE A CONSTANT. THEN

$$\lim_{s\to 0}\,[s\mathfrak{L}f(t)] = \lim_{t\to\infty} f(t)$$

This theorem applies in very few cases and the ultimate value can be deduced easily without it. In many instances the theorem cannot be applied properly: even though the limit of $s\mathfrak{L}f(t)$ exists, $\lim f(t)$ may not exist. In such very common cases of "instability" it is necessary either to invert $f(t)$ or else to apply some test for stability before accepting the result of Theorem B.7.4. In other words, it is always necessary to prove that $\lim_{t\to 0} f(t) =$ constant.

Theorem B.7.5 LET $f(t)$ BE IN $\mathfrak{L}$ AND LET $\lim_{t\to 0} f(t)$ EXIST AND BE A CONSTANT. THEN

$$\lim_{s\to\infty} s\mathfrak{L}f(t) = \lim_{t\to 0} f(t)$$

In such instances in which the transform of $f(t)$ is known but $f(0^+)$ is not known, $f(0^+)$ can be found from Theorem B.8.5. Such cases are very rare.[16]

B.8 Coupled Linear Systems

There are many systems which are represented by more than one linear ordinary differential equation in several variables. The Laplace transform can be used to reduce the array to one differential equation in one variable. Occasionally the reduced equation can be solved, although a direct solution is usually quite difficult. The characteristic polynomial is usually of a high order so that factoring into poles and second-order factors requires a lengthy approximate numerical operation. Even though there are methods for the discovery of the roots of the characteristic equation,[17] it is usually not worth carrying out a manual solution. If the design methods described in this book fail and it becomes necessary to actually solve the differential equation, it is better to program the equation directly for a digital or an analogue computer.

B.8.1 The most general linear ordinary differential equation array that can be theoretically solved completely consists of n equations in n variables. Let x_i $(i = 1, 2, \ldots, n)$ be a set of n variables and let $m_{\ell k}$ $(\ell = 1, 2, \ldots, n;\ k = 1, 2, \ldots, n)$ be positive integers or zero. Then it is possible to write an array of differential equations as

$$\sum_{i=0}^{M_{11}} a_{11i}\frac{d^i x_1}{dt^i} + \sum_{i=0}^{M_{12}} a_{12i}\frac{d^i x_2}{dt^2} + \cdots + \sum_{i=0}^{M_{1n}} a_{1ni}\frac{d^i x_n}{dt^i} = f_1(t)$$

$$\sum_{i=0}^{M_{21}} a_{21i}\frac{d^i x_1}{dt^i} + \sum_{i=0}^{M_{22}} a_{22i}\frac{d^i x_2}{dt^i} + \cdots + \sum_{i=0}^{M_{2n}} a_{2ni}\frac{d^i x_n}{dt^i} = f_2(t)$$

$$\cdots$$

$$\sum_{i=0}^{M_{n1}} a_{n1i}\frac{d^i x_1}{dt^i} + \sum_{i=0}^{M_{n2}} a_{n2i}\frac{d^i x_3}{dt^i} + \cdots + \sum_{i=0}^{M_{nn}} a_{nni}\frac{d^i x_n}{dt^i} = f_n(t)$$

in which $a_{\ell ki}$ are real or complex constants. A more compact manner of representing the array is

$$\sum_{k=1}^{n}\sum_{i=0}^{M_{\ell k}} a_{\ell ki}\frac{d^i x_k}{dt^i} = f_\ell(t) \qquad \ell = 1, 2, \ldots, n$$

If $f(t)$ and x_k are in $\mathfrak{L}$, it is possible to transform the array. Let $\mathfrak{L}f_\ell(t) = F_\ell$ and $\mathfrak{L}x_k = X_k$. Then remembering that (Theorem B.5.6)

$$\mathfrak{L}\frac{d^i x_k}{dt^i} = s^i X_k - \sum_{p=0}^{i-1} s^p\left[\frac{d^{i-p-1}x_k}{dt^{i-p-1}}\right]_{t=0^+}$$

$$= s^i X_k - \sum_{p=0}^{i-1} s^p[x_k(0^+)]^{i-p-1}$$

The equation array becomes

$$\sum_{k=1}^{N}\sum_{i=0}^{M_{\ell k}} a_{\ell ki}s^i X_k = F_\ell + \sum_{k=1}^{N}\sum_{i=0}^{M_{\ell k}}\sum_{p=0}^{i-1} s^p[x_k(0^+)]^{i-p-1}$$

The triple summation is a set of polynomials in s which depend on the initial conditions of x_k. For algebraic convenience we replace the polynomials by I_ℓ. Similarly $\sum a_{\ell ki} s^i$ is a polynomial in s. We write

$$\sum_{i=0}^{m_{\ell k}} a_{\ell ki} s^i = G_{\ell k}(s)$$

Thus our array has become

$$\sum_{k=1}^{n} G_{\ell k}(s) X_k = F_\ell + I_\ell \qquad \ell = 1, 2, \ldots, n$$

or, in expanded form,

$$\begin{aligned} G_{11}X_1 + G_{12}X_2 + \cdots + G_{1n}X_n &= F_1 + I_1 \\ G_{21}X_1 + G_{22}X_2 + \cdots + G_{2n}X_n &= F_2 + I_2 \\ &\cdots \\ G_{n1}X_1 + G_{n2}X_2 + \cdots + G_{nn}X_n &= F_n + I_n \end{aligned}$$

We can solve this array for a particular X_N by Cramer's rule.[18] If we set

$$\Delta = \begin{vmatrix} G_{11} & G_{12} & \cdots & G_{1n} \\ G_{21} & G_{22} & \cdots & G_{2n} \\ & & \cdots & \\ G_{n1} & G_{n2} & \cdots & G_{nn} \end{vmatrix}$$

and let Δ_N be the determinant Δ with the Nth column $G_{\ell N}$ replaced by $F_\ell + I_\ell$ ($\ell = 1, 2, \ldots, n$). Then

$$X_N = \frac{\Delta_N}{\Delta}$$

or

$$\Delta X_N = \Delta_N$$

But Δ is a function of s, in fact a polynomial, and thus $\Delta = 0$ is the characteristic polynomial and Δ_N is the transform of a driving function. The term Δ_N can be expanded by cofactors into the form

$$\Delta_N = A_{N_1}(F_1 + I_1) + A_{N_2}(F_2 + I_2) + \cdots + A_{N_n}(F_n + I_n)$$

in which the $A_{N\ell}$ are polynomials in s, evaluated from the cofactor of $F_\ell + I_\ell$. From this expansion it follows that

$$X_N = \frac{A_{N_1}}{\Delta}(F_1 + I_1) + \frac{A_{N_2}}{\Delta}(F_2 + I_2) + \cdots + \frac{A_{N_n}}{\Delta}(F_n + I_n)$$

which implies that it is possible to solve for $x_N(t)$ in the form

$$x_N(t) = x_{N_1}(t) + x_{N_2}(t) + \cdots + x_{N_n}(t)$$

in which x_{N_r} is the contribution of $(\Delta_{Nr}/\Delta)(F_r + I_r)$ to the solution. This property of linear equations permits us to determine the effect of each F_ℓ independently of the other inputs.

It should also be noted that every x is solved by inverting a transform with Δ in the denominator and thus that every variable x satisfies a differential equation with the characteristic polynominal $\Delta(s) = 0$.

Conclusions made about x_N by analyzing $\Delta = 0$ are valid for all other x_M, $M \neq N$.

Example B.8.1 A system obeys the equations

$$3\frac{dx_1}{dt} + 2\frac{dx_2}{dt} + 6x_1 + x_2 = 4$$

$$\frac{dx_1}{dt} + \frac{dx_2}{dt} + x_1 + 2x_2 = 0$$

Find the differential equation satisfied by x_1 and find $x_1(t)$. (All initial conditions equal zero.)

Solution The transformed equations are

$$(3s + 6)X_1 + (2s + 1)X_2 = \frac{4}{s}$$

$$(s + 1)X_1 + (s + 2)X_2 = 0$$

By Cramer's rule

$$X_1 = \frac{\begin{vmatrix} 4/s & 2s+1 \\ 0 & s+2 \end{vmatrix}}{\begin{vmatrix} 3s+6 & 2s+1 \\ s+1 & s+2 \end{vmatrix}} = \frac{4(s+2)}{s(s^2 + 9s + 11)}$$

Since $\Delta = s^2 + 9s + 11$, the transformed equation for x_1 is

$$(s^2 + 9s + 11)X_1 = 4 + \frac{8}{s}$$

Inversion gives us the differential equation

$$\frac{d^2x_1}{dt^2} + 9\frac{dx_1}{dt} + 11x_1 = 8$$

$$\left.\frac{dx_1}{dt}\right|_{t=0} = 4$$

We now solve for x_1

$$X_1 = \frac{4(s+2)}{s(s + 7.54)(s + 1.46)}$$

so that

$$x_1(t) = 0.728 - 0.483e^{-7.54t} - 0.245e^{-1.46t}$$

PROBLEMS

B.1 Prove the following trigonometric identities with the complex definitions of sin x and cos x:

(a) $\sin 3x = 3 \sin x - 4 \sin 3x$

(b) $\sin x \pm \sin y = 2 \sin [(x \pm y)/2] \cos [(x \pm y)/2]$
(c) $\tan x \pm \tan y = \sin (x \pm y)/\cos x \cos y$

B.2 Express the following functions in the form $u + jv$:
(a) s^3
(b) $\sin s$
(c) $\sin (e^s)$
(d) $s^{1/3}$

B.3 Show that $\ln (-a) = \ln a + j(2n + 1)\pi$ if a is a real number greater than zero.

B.4 Show that $(\cos \theta + j \sin \theta)^n = \cos n\theta + j \sin n\theta$.

B.5 Noting that $|\cos \theta + j \sin \theta| = 1$, show that $1^{1/3}$ had three values, one real and two complex.

B.6 For what value of s is the following function not differentiable? Are the C-R equations satisfied?

$$F(s) = \frac{1}{s - 3} + s^2$$

B.7 Let $\sqrt{s} = R^{1/2}(\cos \theta/2 + j \sin \theta/2)$. Make $\sqrt{s}$ single valued by requiring $|s| > 0$ and $0 < \theta < \pi$. Show that in this restricted region (Fig. B.12) $\sqrt{s}$ satisfies the Cauchy-Rienman equations.

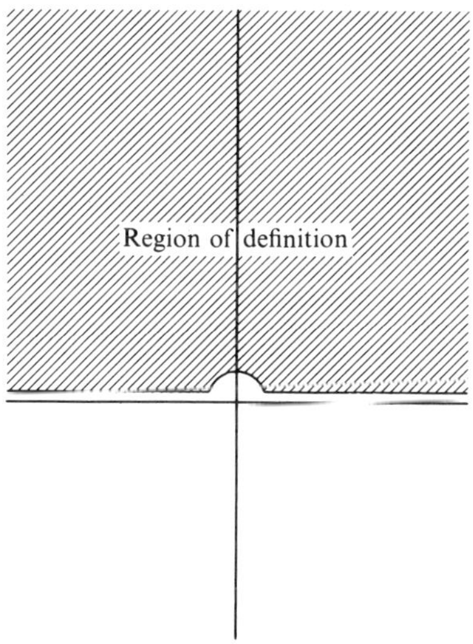

FIGURE B.12 Definition region for $\sqrt{s}$

B.8 Expand the following function into Laurent series in the region $2 > |s| > 0$
(a) $(s - 2)^{-1}$
(b) $\sinh (s^{-1})$
(c) $\exp (s/2 - s)$
(d) $\exp (s^{-1})$

B.9 Expand $(s - 2)^{-1}$ in the region $\infty > |s - 2| > 0$.

B.10 Evaluate the integral of the following functions on a contour surrounding all the singularities:
(a) $[(s - 3 + j)(s + 4)]^{-1}$
(b) $(s - 1)^{-2}$
(c) $\cos (s - 7)^{-1}$
(d) $\exp s^{-3}$
(e) $(\exp st)(s^2 + a^2)^{-1}$ $\quad t > 0, a > 0$ are real

B.11 Invert the following transforms by the residue theorem.
(a) $[(s - 2)(s - 4)]^{-1}$
(b) $[(s + 7)^2(s + 6)]^{-1}$
(c) $(s^2 + 7s + 1)^{-1}$

B.12 Prove Theorem B.5.6.

B.13 Without use of tables, transform the following functions:
(a) $\cos at$
(b) $t \exp at$
(c) $1/a^2(1 - \cos at)$
(d) $(1/b)e^{at} \sin bt$
(e) $u(t - \tau) \cos at$

B.14 Solve by $\mathfrak{L}$ transforms (use tables):
(a) $dx/dt + 7x = 0 \quad x = 2 \text{ at } t = 0^+$
(b) $dx/dt + 7x = \sin t \quad x = 2 \text{ at } t = 0^+$
(c) $dx/dt + 7x = u(t) \quad dx/dt = 1 \text{ at } t = 0^+$
(d) $2(dx/dt) + 5x = t \sin t \quad x = 0 \text{ at } t = 0^+$
(e) $dx/dt + 3x = t^2u(t - 2) \quad x = 0 \text{ at } t = 0^+$

B.15 Solve by $\mathfrak{L}$ transforms (use tables):
(a) $d^2x/dt^2 + 4(dx/dt) + 3x = 1 \quad x = \dot{x} = 0 \text{ at } t = 0^+$
(b) $d^2x/dt^2 + 4(dx/dt) + 3x = 1 \quad x = -1, \dot{x} = 0 \text{ at } t = 0^+$
(c) $d^2x/dt^2 + 5(dx/dt) + 6x = 3u(t - 4) \quad x = \dot{x} = 0 \text{ at } t = 0$
(d) $d^2x/dt^2 + 15(dx/dt) + 50x = \sin 2t \quad x = \dot{x} = 0 \text{ at } t = 0$
(e) $d^2x/dt^2 + 2.7(dx/dt) + 0.83x = \cos 0.2t \quad x = 0, \dot{x} = -0.2$ at $t = 0^+$
(f) $d^2x/dt^2 + 6.3(dx/dt) + 1.11x = e^{0.71t} \quad x = \dot{x} = 0 \text{ at } t = 0^+$

B.16 Solve by $\mathfrak{L}$ transforms (use tables):
(a) $d^2x/dt^2 + 6(dx/dt) + 13x = 0 \quad x = 1, \dot{x} = 0 \text{ at } t = 0^+$
(b) $d^2x/dt^2 + 0.40(dx/dt) + 1.34x = e^{0.2t} \quad x = 0 = \dot{x} \text{ at } t = 0^+$
(c) $d^2x/dt^2 + 2.21(dx/dt) + 5.21x = 3 \sin 1.2t + u(t - 2)$
$x = \dot{x} = 0 \text{ at } t = 0^+$

B.17 Find x:

$$\frac{d^3x}{dt^3} + 3\frac{d^2x}{dt^2} + 3\frac{dx}{dt} + x = 4e^{-0.1t} \qquad x = 1, \dot{x} = 0 \text{ at } t = 0^+$$

B.18 Transform the following function (see Example B.7.5 and Figure B.13):

$$f(t) = \begin{cases} 2t & 0 < t \leq 3 \\ 2(t - 3) & 3 < t \leq 6 \\ 2(t - 3n) & 3n < t \leq 3(n + 1) \quad n = 0, 1, 2, \ldots \end{cases}$$

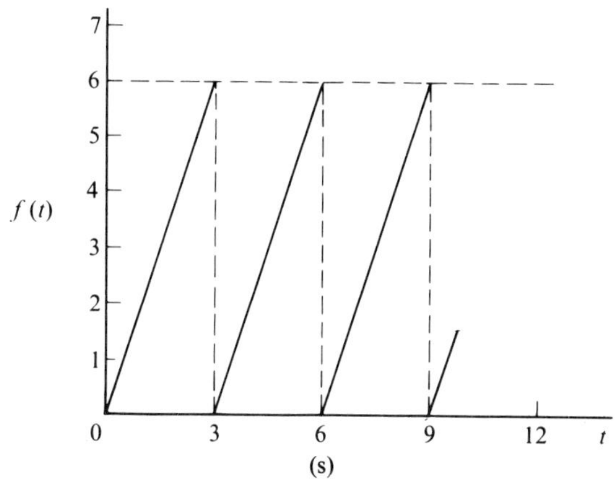

FIGURE B.13 Repetitive ramp functions

B.19 Find x if $x = \dot{x} = 0$ at $t = 0^+$:

$$\frac{d^2x}{dt^2} + 9.12\frac{dx}{dt} + 0.61x = f(t)$$

$$f(t) = \begin{cases} \sin t & 0 < t \leq 4 \\ 0 & t > 4 \end{cases}$$

B.20 Find x if $x = \dot{x} = 0$ at $t = 0^+$:

$$\frac{d^2x}{dt^2} + 6\frac{dx}{dt} + 18x = f(t)$$

$$f(t) = [u(t) - u(t-2)]t + [u(t-2) - u(t-3)](t-2) + u(t-3)$$

B.21 Invert the following transforms:

(a) $\dfrac{3a^2}{(s+b)^3 + a^3}$

(b) $\dfrac{s-b}{(s-b)^4 - a^4}$

B.22 Show by direct inversion that the ultimate state theorem does not apply to the following transform:

$$F(s) = \frac{3}{s(s^2 + 4s - 1)}$$

B.23 Find x_1 and x_2 if all initial conditions are zero:

$$\frac{dx_1}{dt} + 2\frac{dx_2}{dt} + 7x_2 = 3$$

$$\frac{dx_1}{dt} + 6\frac{dx_2}{dt} + 3x_1 + x_2 = 0$$

B.24 Find x_1 if all initial conditions are zero:

$$3\frac{dx_1}{dt} + 6x_1 + x_2 = 2 \sin t$$

$$\frac{dx_2}{dt} + x_1 + x_2 = 1$$

B.25 Find the differential equation obeyed by x_1 assuming all initial conditions to be zero:

$$\ddot{x}_1 + \dot{x}_1 + 1.35\dot{x}_2 + 0.79x_2 = 3.5$$

$$\ddot{x}_1 + 1.11\ddot{x}_2 + 3.71x_1 + x_2 = 0$$

NOTES

[1]As a matter of fact, very little reasearch concerns itself with linear ordinary differential equations. Most linear problems have been examined to exhaustion.

[2]The literature on the theory of Fourier series and Fourier integrals is a formidable subset of mathematical literature. The author follows mainly Ref. 24.

Carslaw, H. S. 1930. *Introduction to the Theory of Fourier Series and Integrals*. 3rd edition. New York, N. Y.: Dover.

The authors of more recent literature use set theory and the Lebsgue integration formalism, both of which make the presentation not only more meaningful but also incomprehensible to most engineers and physicists.

[3]In modern literature the following notation conventions are followed:

$$F(\omega) = \mathfrak{F} f(t)$$

$$f(t) = \mathfrak{F}^{-1} F(\omega)$$

[4]σ is completely arbitrary. If a $\sigma > 0$ can be found, no matter how large, so that the absolute integrability condition is met, then the Laplace transform exists. The engineer does not concern himself with such niceties: if the function can be transformed with the aid of a transform table, he is satisfied. By the nature of physical reality he usually will not be wrong.

[5]The theory and practice of complex variables has been described in hundreds of books. In this work the author has followed the "classical" pattern found in Ref. 25.

Knopp, K. 1945. *Theory of Functions*. New York: Dover.

For the engineer or physicist desirous of a precise but simple rapid presentation, Ref. 27 is recommended:

Churchill, R. V. 1948. *Introduction to Complex Variables and Applications*. New York, N. Y.: McGraw.

A most complete book, full of clever tricks in the "English Tradition" of mathematics and thus useful to applied mathematicians, is Ref. 26: Copson, E. T. 1935. *An Introduction to the Theory of a Complex Variable.* Oxford: Oxford U. Pr.

[6]In recent electrical engineering publications there has appeared the name *phasor* as a replacement for *vector* in the complex plane. The need for this uncomfortable neologism is not apparent to the author.

[7]The discussion preceding Theorem B.4.1 establishes the necessity of the *C-R* conditions. A proof of sufficiency can be found in Ref. 25, p. 30.

[8]Ref. 25, pp. 64–65.

[9]Ref. 25, p. 117ff.

[10]Ref. 30, pp. 291–97.

[11]Ref. 35, p. 307ff.

[12]Ref. 28, p. 111ff; basic details are found in Ref. 30, p. 299ff.

[13]The theorem is valid only if the singularities are poles. It is possible to obtain similar theorems for essential singularities and for multiple-valued functions. Such theorems are not needed here. See Refs. 25, 26, 27, 28 for examples.

[14]The formal derivative (sometimes called the *derivate*) has a genuine mathematical justification in the Lebesgue-Stieltjes theory of integration as applied to function theory. See, for example, Ref. 31, pp. 33–35.

[15]The interchange should not be carried out so blithely. For conditions and cautions, see Ref. 32, p. 518ff.

[16]The author has never used the theorem for anything practical in his whole career.

[17]The oldest known method is Graefe's "Root Squaring" method (Ref. 34, p. 318ff). It is recommended for neither man nor beast. Not only is it very slow but it often yields invalid answers as a result of loss of significance in subtraction. A much easier method, and currently the most popular, is Lim's method (Ref. 29, pp. 519–22). The best modern method, just as simple as Lim's but capable of handling a larger class of polynomials and usually capable of converging much faster than either Graefe's method or Lim's method, is McAuley's backward division procedure (Ref. 33).

[18]Ref. 34, pp. 231–47.

APPENDIX C

TABLE OF LAPLACE TRANSFORMS

	$F(s)$	$f(t)$
1.	$1/s^n \quad (n > 0)$	$t^{n-1}/\Gamma(n)$
2.	$1/(s - a)^n \quad (n > 0)$	$t^{n-1}e^{at}/\Gamma(n)$
3.	$\sin at$	$a/(s^2 + a^2)$
4.	$\cos at$	$s/(s^2 + a^2)$
5.	$e^{bt} \sin at$	$1/[(s - b)^2 + a^2]$
6.	$e^{bt} \cos at$	$(s - b)/[(s - b)^2 + A^2]$
7.	$\sinh at$	$1/(s^2 - a^2)$
8.	$\cosh at$	$s/(s^2 - a^2)$
9.	$e^{bt} - e^{at} \quad (a \neq b)$	$(b - a)/(s - a)(s - b)$
10.	$be^{bt} - ae^{at} \quad (a \neq b)$	$(b - a)s/(s - a)(s - b)$
11.	$\mu(t - a)$	e^{-as}/s

APPENDIX D

BINOMIAL COEFFICIENTS

$$(a+b)^n = \binom{n}{0}a^n + \binom{n}{1}a^{n-1}b + \binom{n}{2}a^{n-2}b^2 + \cdots + \binom{n}{n}b^n = \sum_{p=0}^{n}\binom{n}{p}a^{n-p}b^p$$

where $\binom{n}{p} = \dfrac{n!}{p!\,(n-p)!}$

n	p 0	1	2	3	4	5	6	7	8	9	10
0	1										
1	1	1									
2	1	2	1								
3	1	3	3	1							
4	1	4	6	4	1						
5	1	5	10	10	5	1					
6	1	6	15	20	15	6	1				
7	1	7	21	35	35	21	7	1			
8	1	8	28	56	70	56	28	8	1		
9	1	9	36	84	126	126	84	36	9	1	
10	1	10	45	120	210	252	210	120	45	10	1

INDEX